4/03

Gemini
Steps To The Moon

Springer

London
Berlin
Heidelberg
New York
Barcelona
Hong Kong
Milan
Paris
Santa Clara
Singapore
Tokyo

David J. Shayler

Gemini

Steps To The Moon

Springer

Published in association with
Praxis Publishing
Chichester, UK

PRAXIS

David J. Shayler
Astronautical Historian
Astro Info Service
Halesowen
West Midlands
UK

SPRINGER–PRAXIS BOOKS IN ASTRONOMY AND SPACE SCIENCES
SUBJECT *ADVISORY EDITOR*: John Mason B.Sc., Ph.D.

ISBN 1-85233-405-3 Springer–Verlag Berlin Heidelberg New York

British Library Cataloguing-in-Publication Data
Shayler, David
   Gemini: steps to the moon. – (Springer–Praxis books in
astronomy and space sciences)
   1. Project Gemini (United States) – History 2. Space flight
to the moon
   I. Title
   629.4′54′0973

   ISBN 1-85233-405-3

Library of Congress Cataloging-in-Publication Data
  Shayler, David.
    Gemini: steps to the moon / David J. Shayler.
      p. cm. – (Springer Praxis books in astronomy and space sciences)
    Includes bibliographical references and index.
    ISBN 1-85233-405-3 (alk. paper)
    1. Project Gemini (U.S.) I. Title. II. Series.

  TL789.8.U6 G673 2001
  629.45′4′0973–dc21
                                            2001042668

© Praxis Publishing Ltd, Chichester, UK, 2001
Printed by MPG Books Ltd, Bodmin, Cornwall, UK

Reprinted 2002

Copy editing and graphics processing: R.A. Marriott
Cover design: Jim Wilkie
Typesetting: BookEns Ltd, Royston, Herts., UK

Printed on acid-free paper supplied by Precision Publishing Papers Ltd, UK

Dedicated to the NASA Gemini astronaut team
and to the memory of

Theodore Freeman (1930–1964)
Elliot See (1927–1966)
Charles Bassett (1931–1966)

Table of contents

Foreword

I saw the Gemini programme from beginning to end. At one time I was assigned as Pilot for the first mission, at another time I was scheduled to command the last one, and through a combination of (other people's) ill health and tragedy I flew the fourth (Gemini VI-A, launched 15 December 1965) and the sixth (Gemini IX-A, launched 3 June 1966). I was involved in one of the programme's brightest moments – the first rendezvous in space – and a couple of its dimmer ones, including the Gemini VI-A shutdown at $T = 0$, and Gene Cernan's difficult spacewalk.

First conceived under the name Mercury Mark II in 1961, the Gemini programme was originally expected to be one small step toward the giant leap of the Apollo programme. By the time I arrived at NASA as one of the second group of astronauts in September 1962, Mercury Mark II had become Gemini, a full-fledged programme of its own, designed to answer questions that were vital for any manned flight to the Moon. Could astronauts manoeuvre a spacecraft to a rendezvous and docking? Could they guide the spacecraft to a precise re-entry? Could they function safely and productively for missions lasting a week or even two weeks? Could they leave their spacecraft and work inside bulky pressure suits? These were just the basic human factors. Gemini also faced important technical challenges, such as the development of fuel cell technology, improved tracking and communications, and the first onboard computers. Flight control equipment and methods needed to be invented. Gemini was also designed to allow for a controlled re-entry, greatly reducing the strain on recovery forces.

When it came to programme management, Gemini benefited from the brilliance of two NASA managers, James Chamberlin and Charles Mathews. We had the great good fortune to have continuity in our prime contractor: McDonnell Aircraft had also built Mercury. Lessons that had been learned there, under David Lewis, Walter Burke and John Yardley, were passed on to the next generation. (One of the reasons for the early struggles of the Apollo programme was a shift to a new contractor and NASA management team that sometimes ignored Gemini data as 'old hat'.)

Sixteen astronauts made flights on Gemini, four of them – John Young, Pete Conrad, Jim Lovell and I – flying twice. (Two more gallant men, Elliot See and

Tom Stafford follows Walter Schirra from LC 19 during preparations for the launch of Gemini 6 in 1965. Stafford was selected to NASA in 1962, and was the original Pilot for the first manned flight; but when Al Shepard, his Commander, was medically grounded, he was reassigned with Schirra to serve as back-up on Gemini 3. When Schirra moved to Apollo training after flying Gemini 6, Stafford became back-up Commander of Gemini 9, and flew the mission in 1966 after the deaths of the prime crew during training. Stafford completed two Apollo missions, in 1969 and 1975, and left NASA in 1975.

Charles Bassett, were killed while training for the Gemini 9 mission.) The astronauts took the greatest risks, and enjoyed the greatest acclaim.

But Gemini was the work of thousands of others – engineers and secretaries, accountants and scientists, plumbers and construction workers – employed not only by NASA or McDonnell, but by Martin, by IBM, by David Clark, by GarrettAir, and by hundreds of firms, large and small, scattered across the United States. Gemini would have been impossible without the support of the US Air Force, especially its Space Systems Division in Los Angeles, which managed and launched the Titan II booster, and of the US Navy. These folks made it possible for us to take the ride.

And what a ride it was – from the pad to orbit in 5 minutes 35 seconds. (I shall never forget the illusion of flying through a fireball when our Titan II second stage, as designed, performed its 'fire in the hole', igniting its engines before the first stage had dropped away.) Once in orbit, Gemini was very responsive, and reliable. It was cramped, of course; space vehicles were small in the 1960s. But I found it comfortable enough for up to three days – which was not much of a challenge compared with the flight of Frank Borman and Jim Lovell, who managed to live in Gemini 7 for two weeks.

Thanks to its orbital manoeuvring system, Gemini was capable of changing orbits and, using radar, onboard computer and occasionally the Mark One Eyeball, of manoeuvering to a rendezvous. In that sense, it was the first true spacecraft. And it also brought you home safely.

Today, my two Geminis reside in museums – fond reminders of a fascinating experience thirty-five years past. For those early days of manned spaceflight, as a vital step to Apollo, Gemini was the perfect flying machine, and it is gratifying to see her remembered here in David Shayler's book.

General Thomas P. Stafford
Alexandria, Virginia

Author's preface

On 25 May 1961, President John F. Kennedy gave his historic speech committing NASA and the American people to a manned landing on the Moon by the end of the decade – even though he knew that there was only 15 minutes in the US space log book. On 5 May astronaut Alan B. Shepard had become the first American to fly in space on a short sub-orbital flight onboard Mercury capsule Freedom 7. It was a milestone that came just three weeks after Soviet cosmonaut Yuri Gagarin, onboard Vostok, had become the first man in space, completing one orbit of Earth.

During his speech, Kennedy referred to the monumental task ahead to achieve the lunar goal – in commitment, in funding, and also in technology. The programme charged with achieving the task – Apollo – required the development of new hardware, procedures and, as Kennedy pointed out, 'new materials, some of which have not been invented yet.'

One of the first tasks to be addressed was how exactly to get astronauts from the Earth to the Moon and back again. Over the next few months the problem was resolved with the selection of the most suitable method from three available flight profiles. These were Direct Ascent, using a huge rocket to send the spacecraft straight to the lunar surface and then follow a similar direct flight back to Earth; Earth Orbital Rendezvous (EOR), in which several rockets would launch components into Earth orbit, where they would link up to make the flight to the Moon; or Lunar Orbital Rendezvous (LOR), which was the mode finally selected. The profile of LOR would feature a crew of three, launched into Earth orbit on one vehicle before committing them for the Moon. Once in lunar orbit, a smaller craft would take two of the astronauts to the surface, while the third remained in the parent craft. At the end of the excursion, the two would rejoin the third astronaut for the flight home, lasting at least 7–10 days and possibly as long as two weeks.

To complete this profile, several significant advances in spaceflight experience in Earth orbit would be required which the current one-man Mercury series could not accomplish. Firstly, missions of up to two weeks in duration (simulating the proposed lunar mission duration) with a multi-man crew would be required, to investigate whether humans could not only survive such a duration but also perform useful work and be in a reasonable condition at the end of the flight. Secondly, work outside the protection of the spacecraft would be an advantage in preparing the equipment and techniques that could be used during lunar surface activities. Officially termed Extravehicular Activity (EVA), it was more commonly called 'space-walking'. Thirdly, since two spacecraft were to complete the lunar voyage, the technique to bring them together in space would need to be practised. Bringing vehicles together in space was known as 'rendezvous', keeping them in distances very close to each other was termed 'proximity operations' (prox ops), and linking them together was called 'docking'.

To achieve this, NASA evolved an intermediate programme to bridge the gap between the short Mercury spaceflights and the three-man lunar Apollo missions. Initially called Mercury Mark II, it soon received its own programme identity as Gemini. It featured twelve missions (ten manned) in Earth orbit between April 1964 and November 1966. With the Gemini series, American astronauts finally equalled and surpassed the previous Soviet feats of human space exploration.

Gemini set American duration records of four, eight and finally fourteen days, during which biomedical barriers in human endurance were challenged, met and broken – sufficient, at least, for a flight to the Moon and back. Five astronauts experienced EVA with varying degrees of success, but clearly demonstrated that with adequate training and hardware this technique was a useful skill for the future. Finally, a series of rendezvous and docking experiments conducted, during Gemini, with several target spacecraft simulated the profiles that later Apollo spacecraft would need to achieve around the Moon.

The highly successful Gemini programme also gathered experience of a regular, almost routine spacecraft check-out and launch programme, and the demonstration of pinpoint ocean landings to hasten crew recovery. Using the Mercury experience as a baseline, an expanded astronaut team and training syllabus was matched by enlarging the flight control and mission support network. A broadened programme of flight experiments that not only involved biomedical studies but also featured a range of investigations in space science, technology, and Earth observations, were groundbreaking in their application from manned spacecraft. Gemini also was also to feature prominently in the US Department of Defense space plans of the 1960s, with application for manned space platforms and surveillance missions under the USAF Blue Gemini and Manned Orbiting Laboratory (MOL) programmes.

The American Mercury programme confirmed and supplemented the Soviet Vostok data, which showed that humans could survive in space and return in reasonably good health. Those historic first steps are remembered as pioneering in manned spaceflight, while the dramatic missions of Apollo often place the Gemini programme in its shadow. But Gemini was much more than just a stepping stone to the Moon, as it became the first programme to investigate the techniques and procedures of working in space that would eventually lead to operational spaceflight far beyond that of Apollo in the creation of the first space stations.

The objectives of the Gemini programme can be used to identify the achievements of the programme. Therefore in compiling this book I have briefly reviewed the creation of the programme, the description of the hardware, the selection and training of the astronauts, the flight operations and ground support infrastructure.

These sections are followed by overviews of the major objectives of Gemini – the flight tests, the endurance missions, the rendezvous and docking operations and the EVAs. The final sections review the attempt at land and ocean pinpoint landing, the range of experiments carried on each mission, as well as the DoD plans for Gemini and the proposed applications for the spacecraft that were never fulfilled.

A discussion of the legacy of the Gemini series in space history closes the book, with an Appendix featuring brief biographies of the NASA and MOL astronauts who were selected to train and fly spacecraft of the Gemini design. In compiling this book I have concentrated on the objectives and achievements of Gemini that trailblazed the way to the Moon for Apollo. If the Mercury astronauts began that long journey with the first small pioneering steps that allowed the Apollo astronauts to achieve the giant leap on to the Moon, then surely Gemini was the bridge that carried them there.

David J. Shayler
Spring 2001
West Midlands, England

www.astroinfoservice.co.uk

Acknowledgements

The compilation of this book greatly benefited from the continued support and assistance of a worldwide group of fellow space sleuths, authors, historians, and archivists. Foremost in this group are Rex Hall, Mike Cassutt and Bert Vis, who, over the past twenty years, have been constantly unselfish in sharing information and revelations from their own files and resources that have always added that little extra detail. In addition, my thanks go to Colin Burgess for details of the media coverage of Gemini from an Australian perspective, to balance my own British media files and those of American journalists. Curtis Peebles' earlier research into the USAF MOL programme was of great value in detailing the role of Gemini in that programme, as was Andrew Wilson's research in detailing early US astronaut assignments for a selection of published articles more than twenty years ago.

I also appreciate the comments and suggestions of author David Harland and NASA Chief Historian Roger Launius in directing research and further investigations. The staff at the NASA JSC History Office – originally Janet Kovacevich, Joey Pellerin and Asha Vasha, later David Portree, and more recently Glen Swanson – have, over the past thirteen years, continued to provide access to the many and varied archives of the Gemini programme. At Rice University, Joan Ferry and her staff at the Fondren Library allowed repeated access to the Gemini Collection while stored there (up to 2000), as well as the Faculty's Curt Michel Collection. At the National Archives at Fort Worth, Meg Hacker and her staff were also very helpful in accessing the Gemini Collection when it was moved there in 2000.

I am also indebted to the still photo research staff at KSC (Mary Persinger) and at JSC (Lisa Vasquez, Debbie Dodds, Mary Wilkinson and Jody Russell), and the continued support of Mike Gentry over many years of visits, requests and correspondence. The help of Gwen Pitman at NASA Headquarters photo archive in Washington DC was also essential for tracking down the more elusive Gemini photographs. The extent of my own archive was greatly expanded by the initial assistance of Lee Saegesser, formerly of the NASA History Office at Washington DC, and Iva 'Scotty' Scott, formerly of the PAO at JSC. In addition, Diana Ormsbee and Pete Nubile of the Audio Library at JSC were instrumental in supplying archived audio footage from the Gemini years.

I also appreciate the time and effort given by several former members of the astronaut team in interviews, information and correspondence over a period of several years, on questions pertaining to Gemini and that era of the American space programme. Those interviewed include, among others, Gene Cernan (1988), John Young (1992), Walter Cunningham (1989); and for his splendid Foreword to this book, my special thanks go to Tom Stafford. At JSC, James McBarron provided information on Gemini pressure garments and EVA activities; and at KSC, Dick Button was helpful in detailing aspects of pressure suits and launch-day activities at the Cape. My thanks are also extended to former KSC Historian Ken Nail in providing archive information on early Cape activities, and to John Charles at JSC who, with William Phinney, provided information on the science training of the astronauts during the 1960s.

Special mention should be made of Shirley Jones and the staff (especially Ben Evans), archive and resources of the British Interplanetary Society in London. During the 1960s, the BIS publication *Spaceflight* featured articles and news on the development and flight operations of Gemini. With the generous help of the BIS, this book has been greatly enhanced by original photographs from the Gemini era of which even NASA found it difficult to find the originals. Throughout this book, all photographs are from the Astro Info Service collection, and are reproduced courtesy of NASA unless otherwise credited.

Once again thanks must go to my brother Mike for his initial editing of the manuscript, and for preparing the text for delivery; and to Bob Marriott, who – working on my fourth book in eighteen months – devoted even more countless hours in preparing the illustrations, final editing of the manuscript, proof-reading and indexing. Particular appreciation is extended to Clive Horwood, Chairman of Praxis, for his enthusiasm in supporting my efforts, and to the staff of Springer–Verlag (London), who are always ready to offer post-production support. I would also like to thank Arthur and Tina Foulser and staff at BookEns, for their typesetting skills, Jim Wilkie for his excellent cover design, and MPG Books for such a fine end result.

Clive Horwood's dedicated commitment to provide a broad range of highly detailed space exploration titles for a wider market is both refreshing and welcome. The resulting contributions from a talented group of fellow authors has led to a library of Springer–Praxis space titles that is a credit to all concerned. The success of this series is due to the tireless energy of Clive, and with this project following so closely after two other equally detailed manuscripts, his increased stress level has been balanced only by the depletion of his wine cellar!

List of illustrations and tables

Endurance

Rendezvous and docking

EVA operations

Re-entry and landing

Experiments

Military Gemini

A place in history

Tables

Front cover
Buzz Aldrin performs a stand-up EVA from the open hatch of Gemini 12 during the
final mission of the series in November 1966. During the Gemini programme, EVA
techniques were fully and successfully demonstrated. The photograph also reveals
another element of the Gemini programme important for the Apollo programme to
follow: docking to a target vehicle high above the Earth.

Back cover
Gemini 7 is seen from Gemini 6 during the first rendezvous in space between two
manned spacecraft. Inside Gemini 7 were Frank Borman and Jim Lovell, whose 14-
day mission held the endurance record for five years until it was exceeded by the 18-
day Soyuz 9 mission, and held the American record for more than seven years until
the Skylab missions of 1973.

Acronyms, abbreviations and notes

ABMA	Army Ballistic Missile Agency
AFB	Air Force Base
AFSC	Air Force Systems Command
AMR	Atlantic Missile Range
AMU	Astronaut Manoeuvring Unit
ARPS	Aerospace Research Pilot School
ATDA	Augmented Target Docking Adapter
ATV	Agena Target Vehicle
Ballute	Balloon parachute
CapCom	Capsule Communicator
Cape	Cape Canaveral, Florida, (Cape Kennedy from 29 November 1963).
CB	Astronaut Office
DoD	Department of Defence
ECS	Environmental Control System
ELSS	Extravehicular Life Support System
ETR	Eastern Test Range
EVA	Extravehicular Activity
FO	Flight Operations
FOD	Flight Operations Directorate
G&N	Guidance and Navigation
GATV	Gemini Agena Target Vehicle
G3C	Gemini IVA pressure suit
G4C	Gemini EVA suit
G5C	Gemini long-duration suit
GLV	Gemini Launch Vehicle (Titan II)
GPO	Gemini Programme Office
GSFC	Goddard Space Flight Center
GT	Gemini Titan
GTA	Gemini Titan Agena
HHMU	Hand-Held Manoeuvring Unit
ICBM	Intercontinental Ballistic Missile

LC	Launch Complex (Pad)
LeRC	Lewis Research Center
LaRC	Langley Research Center
LV	Launch Vehicle
M	Orbit in which rendezvous takes place ($=1$, $=2$, and so on)
MCC	Mission Control Center (Houston)
MDS	Malfunction Detection System
MOCR	Mission Operations Control Room
MOL	Manned Orbiting Laboratory
MSC	Manned Spacecraft Center (Houston, Texas)
MSFEB	Manned Spaceflight Experiments Board
OAMS	Orbit Attitude and Manoeuvring System
REP	Rendezvous Evaluation Pod
SLV	Standard (Atlas) Launch Vehicle
SNORT	Supersonic Naval Ordnance Research Track
SOPE	Simulated Off-the-Pad Ejection
SPS	Secondary Propulsion Systems (Agena)
SSD	Space Systems Division (USAF)
STG	Space Task Group (NASA)
SEVA	Stand-up EVA
TPS	Test Pilot School
TDA	Target Docking Adapter
VAT	Vehicle Acceptance Team

When a Gemini mission did not involve the launch of an Agena docking target it was designated GT (Gemini Titan), and those with a dual launch were designated GTA (Gemini Titan Agena): non-Agena missions, GT 1, 2, 3, 4, 5, 7; Agena missions, GTA 6 (originally), 8, 9 ,10, 11, 12. From the fourth mission NASA also adopted a Roman numeral system: Gemini 1, 2, 3, IV, V, VI, VII, VIII, IX, X, XII, XII. When the sixth and the ninth Gemini missions were rescheduled after losing their Agena targets, their flight numbers were suffixed with 'A': Gemini VI-A; Gemini IX-A. The seventh Gemini mission was the sixth launched, and the sixth mission was the seventh launched. This book generally uses the Arabic numeration, although the alternative designations are used as required.

In keeping with the era recalled by this book, the terms 'weightlessness' or 'zero g' are used instead of 'microgravity', and reference is made to 'orbits' rather than 'revolutions'. Time zones are American, and measurements are in imperial units.

Prologue

It is four decades since the first Americans left Earth to enter space, riding on top of rockets converted from ballistic missiles. Now (in 2001) space is usually reached by flying on the Space Shuttle, which 'routinely' leaves the pad. The ascent to orbit – called 'riding the stack' – remains hazardous and exciting, but is somewhat smoother and less demanding than were the launches at the beginning of the space programme. In those fondly remembered days, astronauts rode rockets that, in their earlier unmanned development phase, had a nasty tendency of exploding – which could pose a serious threat to the health of anyone who chose to ride into orbit.

The refinement and improvement of the launch vehicles selected to carry astronauts into space is termed 'man-rating'. For Gemini, the launch vehicle chosen to be man-rated was the USAF Titan II missile; and as with any launch into space, a ride on the Titan was an experience that was never forgotten.

It is generally assumed that a spaceflight begins when the vehicle leaves the pad – or at least, when it enters orbit – but, as many of the first astronauts discovered, a trip into space becomes real when you are sealed inside the pressure suit and are cut off from the rest of the world. With only pumped oxygen for breathing, the sound of a disjointed voice over the radio earpiece and, through the visor of the helmet, a view of technicians mouthing words in a disjointed conversation, you have entered a surreal world. You can feel the suit with your every move, and although aware of your surroundings, you seem to be separated from them. And despite having undergone extensive and repetitive training, today is different. It is launch day.

Holding a portable oxygen and ventilation supply connected to your suit, you waddle out to the transfer van and try to wave at the small, cheering crowd and the cameras. The crowd is enthusiastic but silent, and all you can hear is your own breathing. The short trip to the pad is a time for reflection – with perhaps an occasional joke – and for maintaining the traditions of space explorer folklore. As you step off the bus at the foot of the pad you are greeted by only a few technicians, and you have left the crowd far behind.

As you strain against the suit to stare up at the launcher towering above, it appears to be alive. When you reach the top of the gantry you walk across the crew access arm into the White Room that surrounds the spacecraft entry hatch, and you are ready to be assisted into the cramped confines of your spacecraft.

An artist's impression of the Titan II second-stage powered flight carrying a Gemini spacecraft to orbit. The spent first stage can be seen dropping into the atmosphere.

Grabbing hand-holds, and swinging up your legs to slide into the reclining seat, you lay on your back, looking upwards as a protected boot from one of the launch support team pushes down your shoulder, tightening the seat restraint strap still further. Now securely strapped in your seat, you feel that if you could turn, the spacecraft would turn with you. (The two-man Gemini spacecraft was larger than America's first manned spacecraft, the one-man Mercury, but the increase in size was not very considerable. It was smaller than the later Apollo Command Module, and had nowhere near as much space as would be enjoyed in the Shuttle twenty years later. Those who rode in it likened it to sitting in the front seat of a Volkswagen. It was not a matter of 'getting into' the spacecraft, but rather 'putting it on'.)

With the suit connected to the spacecraft systems, communications links established, and the rocket fuelled, a final handshake is followed by the closing of the hatch, prompting a last look at the departing launch team. Within minutes you and your colleague are left sitting on top of a volatile missile about to be shot off the planet and into space. As the clock ticks towards ignition, the spacecraft controls and switches are set and checked for launch; and you wait. . . and wait. If this is to be a docking mission, then news of a safe launch of the target vehicle will be relayed to you, and you will soon be leaving Earth.

You hear the final ten-second count-down, and feel the Titan engines automatically swivel from side to side far below you as they are prepared for flight. Grasping the ejector seat handle firmly (but not *too* firmly), you instinctively hold your breath. At ignition, you wait for the feel and the sound far below you as the engines ignite. Then, once the thrust is built up and the securing bolts are released, there is movement as the mission event clock begins to count upwards. The thrust builds up, and you realise that you are at last on your way. You can see the blue sky out of your window, but there is no sensation of forward velocity until the cloud layer is approached, reached, and surpassed in the blink of an eye. Now ascending rapidly, the Titan's speed increases, 'hauling the mail' as the g-force pushes you back into your seat.

In little more than a minute you have left the pad, passed 40,000 feet, and increased speed from a standing start to 1,000 mph, pinned down by about 5 g, as the Titan engines gulp the fuel. As the fuel tank empties, it is time for stage separation and ignition of the second stage. Known as the 'great train wreck', it begins with the shut-down of the first-stage engines, and you are thrown forward against the seat harness as the power is cut. Moments later – as the second stage engines ignite and present a brilliant flash which you can see through the window – you are flung back against the seat to begin the next phase of the ascent. As you head out over the Atlantic towards orbit, the sky turns from blue to deepest black.

The Titan trip is quick but rough. You are pushed against your seat, and experience more than 6 g until orbital velocity is reached. At this point the second stage engines shut off, and the spent stage is separated from your spacecraft. Debris floats by the window as you are once again thrown against the harness; but this time you stay there, floating slightly above your seat. You are in orbit, and you, your colleague and everything around you are in the environment of microgravity (more commonly known as weightlessness, or zero g). You are presented with the strange sight of loose screws, nuts, pencils and straps floating in front of your eyes; but through the window there is a spectacular view – and you have a big smile on your face!

If reaching space was an adventure, then returning home is equally as interesting. At the end of the mission the spacecraft is secured for re-entry and landing, and is orientated to travel with you and your colleague's backs facing the direction of flight. The equipment section is jettisoned, exposing the retro-rocket engines that fire to slow the spacecraft to begin the long fall to Earth. With the four engines behind you, each one can be felt as they ignite in turn. Although they incur only ½ g, it seems more like 3 g!

The retro-rocket is jettisoned to reveal the protective heat shield, and slowly the spacecraft arcs down towards ocean splash-down. With gentle firing of the small thrusters to 'steer' the spacecraft and to make small turns to refine the trajectory as the spacecraft rotates, you begin your return to Earth – backwards and upside down. Safe inside your spacecraft and with the g-force increasing, the light show outside begins to attract your attention as the plasma sheath around the spacecraft glows in numerous shades of purple, red, yellow, and white, reaching temperatures well above the melting point of astronauts.

As the descent through the atmosphere slows the craft, the drogue parachute opens at around 50,000 feet, and the spacecraft swings from side to side as the main parachute is deployed at 9,800 feet. At 6,700 feet it drops to a two-point suspension position which, if your harness were to be insufficiently tight, would force you forward towards the control panel. After plummeting like a stone, you are slowed by the parachutes. Clouds flash by horizontally as you approach the ocean, and hopefully you are heading for a soft entry. After splash-down the mission is over, but the adventure continues. Gemini is a great spacecraft, but it is not a boat. As you await recovery by helicopter, the ocean swells around you, and normal gravity takes affect; and as your are still wearing your pressure suit, you begin to perspire, and become hot and sticky. The Gemini adventure will soon be at an end, but the adaptation to life after a spaceflight is just beginning.

Origins

By the middle of the twentieth century it was recognised that the harnessing of rocket technology would be the key to the exploration of space. Research in Europe, Russia and America – both prior to and immediately after the Second World War – identified the strategic advantage of space exploration in addition to the scientific applications made feasible by such exploration. Towards the end of the 1950s, related developments from several fields of aeronautical research (including stratospheric exploration, high-speed atmospheric flight, aerospace medicine, and the creation of pressurised compartments and clothing) culminated in the planning of the first human ventures beyond our atmosphere.

While the military saw the advantages of controlling the 'high ground', scientific bodies promoted the use of artificial satellites for communications, weather forecasting and Earth observations. Several grand ideas for the expansion of spaceflight were reproduced in many scientific journals or in popular science fiction magazines, novels or films of the period. Many of these featured space explorers venturing outside their spaceships and working in the vacuum of space, or building large space stations by joining several smaller spacecraft together, before venturing on even longer journeys to explore the Moon or distant planets.

In order to realise such dreams it was obvious that certain basic skills or techniques would have to be mastered. Initially, the challenge of determining whether living creatures – particularly humans – could actually survive the acceleration of launch, the dangers of spaceflight, and the deceleration of re-entry and landing, would be addressed by the first flights into space. Once it was found that it was possible to survive and operate in space, the next problem was to determine how long a person could safely remain in space. There were other questions, too. Could an astronaut effectively work outside the protection of his spacecraft, as the theorists predicted? Also, just how complex was it to control more than one vehicle in space, or to link them together? These last three main questions would eventually be addressed during the Gemini programme; but first, manned spaceflight itself had to be proven not only possible, but also survivable.

TO THE EDGE OF SPACE

During the early decades of the twentieth century, the primary interest in the exploration of space lay in the realms of science fiction and theory. However, several key developments in military strategy, as well as advances in science and technology, indicated that the exploration of space was closer to reality than most people imagined.

Across the globe, developments in aircraft design and capabilities between the two world wars pointed to the significant military advantages of using the aeroplane in future conflicts. Flying higher, further and faster could secure an early and decisive victory by taking the battle to the enemy's own territory, as well as place the aircraft beyond the range of ground artillery. Observations from such flights would also provide real-time military intelligence, allowing land forces to benefit from the latest information on the territory that they were to enter *before* committing to battle.

In order to sustain such operations, aircraft would require a method of sustaining the crew at great altitudes, where the air is much thinner. The reliability of the aircraft and performance of the engines would need to be improved, so that the operation could be carried out swiftly and safely, and with some degree of comfort. The development of the pressurised cabin and suitable flying garments led to the development, from the 1930s onwards, of stratospheric balloon gondolas, to investigate the environment in which the new air vehicles would operate, and to test pressure suits and life support systems to sustain the crew.

Following World War II, rocket technology developed to the point that its use in ballistic missile and aircraft applications was obvious. Such technology would finally allow an aircraft to not only approach the speed of sound, but surpass it, allowing increased altitudes and greater range than had been envisaged only a few years earlier.

NACA's early X-plane fleet lines up outside a hanger at the High Speed Flight Station at Muroc, in the western Mojave Desert, California. (*Front, from left*) Douglas D-558-2 Skyrocket, Douglas D-558-1 Skystreak, Bell X-5, Convair XF-92A, Northrop X-4.

In the closing years of the 1940s, the famed 'X' series of US rocket planes pushed the barrier of human rocket travel to the very fringes of space, while also developing the technology of pressurised crew compartments and personal survival suits. By the 1950s, further evolution of these aircraft also indicated the need to provide a more suitable crew escape method than jumping out of an aircraft with a parachute. Thus, the ejector seat was developed for use in high performance aircraft.

With the rocket planes, and with the emergence of jet fighters and ever larger missiles to carry nuclear weapons, the military would soon point to space as the next high ground in acquiring security and strategic advantages over other nations. The allied nations of World War II soon became opponents in the Cold War, barely a decade later.

Controlling the high ground
During the decade following the end of World War II, the US armed forces and aircraft industry realised the next goal of aeronautics – to develop 'higher, faster and further' aircraft. The hotbed of military aviation in America during the late 1940s and early 1950s was situated in the deserts of California, at a remote air base called Muroc. Here, a series of experimental aircraft – the X-planes – were routinely tested to their limits and beyond by some of the best pilots in the country. Muroc later became the famous Edwards Air Force Base, home of the rocket pilots and the USAF test pilot school. It was the genesis for the fabled 'right stuff', where pilots constantly pushed the barrier of hypersonic flight to the very edge of the atmosphere.

At the same time, other teams were investigating the potential of ballistic missiles to explore speeds, altitudes and distances beyond any which the rocket aircraft could attain – well beyond the fringes of space and out into orbit. The rivalry between the US Air Force, the Navy and the Army became a race to discover which branch of the armed forces would be the first to break a new barrier or set a new record. In

The NACA High Speed Flight Station at Muroc was completed in 1954, adjacent to the USAF Edwards Air Force Base, home of the USAF Test Pilot and ARPS. Today the HSFS site remains the core facility of the NASA Dryden Flight Research Center.

addition to the armed forces, there was also the primary civilian research organisation, the National Advisory Committee for Aeronautics (NACA), the main post-war objective of which was to channel all this new technology into a more focused national goal.

In the 1950s America possessed a highly varied military aerospace research infrastructure. It included a rocket sled acceleration/deceleration programme, high-altitude balloon and parachute descent research, sounding rocket and hypervelocity research aircraft, emerging ballistic missiles, and nuclear engineering. In addition, there was an infrastructure of government industry and academia, and these, too, were drawn on inter-service rivalry and military objectives, and very little on pure science. The NACA attempted to balance this strategic race for space by assisting the armed forces with a methodical programme of investigation. It would be supported by a network of field research centres and scientific programmes, to promote biological, animal and human medical research, material technology, communications, vehicle control, guidance and navigation, and recovery, and would apply the results to the civilian aviation field for the benefit of the nation.

For the scientists it seemed logical to push the programme that little bit further and attain orbital flight. For some years, discussions had continued concerning the utilisation of the International Geophysical Year (IGY) of 1957–8 to orbit an artificial satellite, to study the Earth at a time of maximum solar activity. The USAF had also long considered space as part of its own domain, and was developing a programme of ballistic missiles to counter any aggression from a foreign power. It had often stated that placing military hardware into space, using these missiles, would be a long-term objective.

For the USAF, it was also imperative to secure its own independent space operations programme in order to keep one step ahead of the Soviets, who were known to be developing the technology to launch objects into orbit, based on their own missile programme. It was obvious that once the first automated objects had been placed in orbit, the next logical goal would be to place a human in space. The first nation to achieve orbital flight would indeed control the 'high ground'.

Three-phase co-operation

Meanwhile, a co-operative programme between the armed forces and NACA was developed to study the results of aircraft travelling at twice and three times the speed of sound at 65,500 feet. From this programme, technology was being developed in the fields of aerodynamics, materials, communications, flight control, and vehicle recovery, to attempt flights that could take manned aircraft into space.

This research featured a three-phased programme. Phase 1 was designed to increase flight speed from Mach 1 towards Mach 6. From there, a new research vehicle would take over during Phase 2. Known as the X-15, this aircraft would fly to almost 50 miles above the Earth, to the fine divide between flying in the atmosphere or in space. Air-launched from under the wing of a converted bomber, the X-15 would climb to altitudes of 275,000 feet, boosted by a powerful rocket engine. At such altitudes, conventional aircraft control surfaces were useless in the thin layers of the atmosphere, and by using small thrusters for attitude control, the aircraft would

Under the wing of a B-52, the X-15 is prepared for a 'drop' to begin the ten-minute flight to the fringe of space before landing on one of the dry lake beds around Edwards AFB.

loiter at this peak altitude for only a few minutes. The pilot would experience a few seconds of weightlessness, and then pitch the nose down to head back into the denser layers of the atmosphere. Here, studies of re-entry heating would be followed by regaining aircraft control from high altitude using conventional aerodynamic surfaces, before completing the ten-minute 'mission' on a desert runway at Edwards AFB.

Phase 3 was called the X-20 (also known as Dyna Soar – DYNAmic SOARing). This was a more substantial vehicle than the X-15 and resembled a delta-winged, flat-bottomed glider aircraft. Launched by ballistic missile, the X-20 would attain orbital flight, and at the end of its orbital mission would re-enter and be flown back to a runway to be recycled for further missions. This programme was intended to be the Air Force's road to space.

It was during this period that the US Navy and US Army were competing to put the first American satellite in orbit by rocket. Within NACA, engineers at Langley Research Center began evaluating the use of blunt re-entry vehicles and ballistic-shaped capsules attached to rockets that were already available, to provide a method of attaining manned orbital flight other than the X-20 aerospace plane concept. Leading this effort at Langley was a small NACA team of engineers headed by Max Faget.

An artist's impression of the X-20 Dyna Soar in Earth orbit. (Courtesy USAF.)

THE DAWN OF THE SPACE AGE

On 4 October 1957, the Soviet Union shook the world by orbiting the first artificial satellite, the 185-lb Sputnik. What was not known at the time was that a much larger satellite was actually planned for orbit; but its development was delayed, and so the smaller, less complex design was selected to ensure that the Soviets were the first to place an object in orbit. Just one month later, on 3 November, a second Sputnik was launched, which this time included the first living creature to orbit the Earth – the dog Laika. This time the payload mass was 1,000 lbs, and remained attached to the launch vehicle's upper stage, demonstrating the capacity to place 7.5 tons in orbit.

US reaction to these launches was summarised by Lt General James M. Gavin, who for a considerable time was associated with the American ICBM programme. He called the event of the first launch 'a technological Pearl Harbour' for America.

On 6 December 1957, a televised launch of America's first test-satellite attempt – the USN Vanguard – ended in failure as the rocket exploded two seconds after launch and just four feet off the launch pad. The next day's headlines highlighted the American 'failure to launch test satellite', while Senator Lyndon B. Johnson echoed the feeling of most of the nation when he described the event as 'most humiliating'. The failure cut deeply into American pride and prestige – so much so that one comment from a Martin Company (launch vehicle contractors) engineer was that the Vanguard project had become a 'whipping boy' for the hurt pride of the American people.

What was not realised at the time was that the Soviets had also encountered setbacks and failures in their launch vehicle programme, but they were not publicly televising or revealing the facts at the time. The opinion of those on the Vanguard programme was that with the amount of criticism hurled at them it was as though they had committed treason. In an attempt to support their efforts, President Dwight D. Eisenhower recalled that even though the Americans were not the first to discover penicillin, they had still made other significant discoveries in similar fields. Vice President Richard M. Nixon wrote to Martin officials, stating that 'at a time when you have been 'catching it' from all sides, I want you to know that I, for one, feel you should have every support. Keep up the good work.'

Success came for the Americans a few weeks later, when the Army Ballistic Missile Agency (ABMA) Redstone Arsenal Jupiter C launched Explorer 1 on 31 January 1958. At last, America was in space; and with this success, they also discovered the radiation belts that surround the Earth, which were named after James Van Allen, the scientist whose instruments detected them.

The American success was short-lived, however, as the third Soviet satellite was orbited in May 1958. This was based on the original design intended for the first launch in 1957. That satellite was lost in a launch failure in April 1958, but its back-up became Sputnik 3. Compared with the first Sputnik and the American Vanguard or Explorer, this third Sputnik was enormous, with a mass of 2,900 lbs. It indicated to the world the ability of Soviet missile technology to lift large payloads into orbit, and by implication, the capability to launch a manned vehicle in the near future. To underline the achievement, Premier Nikita Khruschev pointed out to the Americans that they were now sleeping 'under a Soviet Moon'. For the Americans – especially the military – it was clear that the missile gap between the two countries was becoming wider.

The response from the USAF was a stern warning to military leaders of the might of the Soviet threat and the need to put the USAF in space with a manned vehicle as soon as possible. Four days after the first Sputnik was launched in October 1957, an Ad Hoc Committee of the USAF Scientific Advisory Board suggested the development of second generation ICBMs as spacecraft launch vehicles on missions that could include a manned lunar mission. During 15–21 October, at a NACA conference at Ames Research Center, the three candidate configurations – delta-wing flat bottomed glider, ballistic capsule, and a semi-ballistic blunt end lifting body design – were discussed, to determine which would be the most suitable to place Americans into orbit as soon as possible.

During this same period, as each branch of the service vied for the approval of its programme to put an American (military) man in space, the American Rocket Society in Washington was calling for the creation of a civilian space agency to focus all attention onto putting Americans into orbit. At Langley, at the Pilotless Aircraft Research Division (PARD), engineers Max Faget and Paul Purser were leading a team refining the design for a ballistic manned spacecraft.

The creation of NASA

On 23 January 1958, barely a week before the launch of Explorer 1, Senator (later President) Lyndon B. Johnson presented a summary of the Senate Review of the national space programme. In the seventeen recommendations there was indeed a suggestion for the creation of an independent (from the military) space agency. By the end of that month, eleven leading aircraft and missile companies had outlined their wide-ranging proposals for a manned satellite configuration for both the USAF and NACA.

During April 1958, after the USAF had refused to participate in the Army's plans for an 'interservice' Man Very High space programme, the ABMA instead suggested a co-operative programme with the USN, called Project Adam. In this proposal a modified Redstone launch vehicle, based on the Jupiter design, featured a manned sub-orbital trajectory, with the human subject sealed in a separate 'capsule'.

Instead of co-operating with the USN, however, the Air Force turned to NACA and offered a co-operative two-stage programme that would also adopt a ballistic missile approach to attain early manned spaceflight (the aptly named Man-In-Space-Soonest – MISS). Once this had been demonstrated, the programme would advance to the second phase – the X-20 boost-glide vehicle. The US Navy, however, also proposed its own programme – the (naval) man-in-space project known as Manned Earth Reconnaissance (Project MER). This consisted of a cylindrical spacecraft with spherical ends, which changed shape into an inflatable delta-wing glider when in orbit. Unfortunately, this proposal advanced no further than the feasibility study.

On 14 April 1958, President Eisenhower sent his proposal for a space bill to Congress. During the following month, the details of the MISS design were refined at USAF Headquarters, where both NACA and the USAF agreed to work together on a ballistic manned spacecraft. This resulted in a revised Man-in-Space-Soonest proposal on 16 June. This proposal envisaged the use of the already developed Atlas launch vehicle instead of a new two-stage vehicle. But this project too, proceeded no further than development, with funding for only the life support system being granted before other events changed the course of American space exploration.

On 11 July 1958, the Army's Project Adam was officially rejected, and just five days later, on 16 July 1958, Congress passed the 1958 National Aeronautics and Space Act, creating a new civilian agency to combine both aeronautical and space activities in the civilian field. It was called the National Aeronautics and Space Administration (NASA). On 29 July, Eisenhower signed the Space Act into law.

The following month, the President assigned NASA specific responsibility for the development and operation of the nation's civilian manned spaceflight effort. The MISS concept was cancelled, and its remaining budget was transferred to NASA. However, the military X-20 Dyna Soar programme continued under the auspices of the USAF.

NASA began official operations on 1 October 1958, and for the next six months events moved rapidly. The first NASA Administrator, T. Keith Glennan, approved plans for a manned spacecraft by 7 October, and by 23 October preliminary specifications for such a spacecraft had been passed to industry. On 14 November a second set of specifications was mailed to twenty companies that wished to be

The headquarters of the NASA Space Task Group, at Langley, Virginia, in 1962.

considered for contract bids. On 5 November, NASA formed the Space Task Group (STG) at Langley to manage the manned programme, and on 26 November the project received its official name of Project Mercury. On 9 January 1959, McDonnell Aircraft Corporation was contracted for the development and production of the one-man capsules.

The Mercury Seven
At the same time as development of the Mercury spacecraft was being refined, the programme to select the first American group of astronauts began. The requirements for the selection of candidates to train for the Mercury programme had been established at Langley during November 1958. A joint group, consisting of NASA, and military and DoD representatives, would select a pool of about 150 candidates, and after evaluation a group of 36 would be nominated for physical and psychological testing. From this evaluation, a group of twelve would be nominated for a nine-month astronaut training and evaluation programme, and from those, six would be selected for the Mercury spaceflights.

The problem with the selection was in deciding what previous experience was relevant. Faced with candidates who were qualified to fly aircraft, command a submarine, drive a racing car or climb mountains, in December 1958 President Eisenhower stipulated that the first American astronaut should be selected from the ranks of military jet test pilots. By January 1959, criteria had been established which would have a significant impact on the selection of Americans for spaceflight

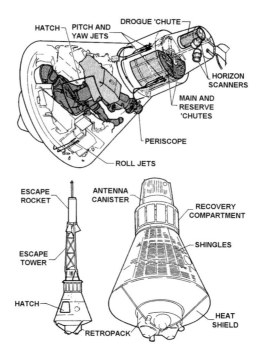

The Mercury spacecraft, showing the external features, the position of the astronaut, and the associated equipment. This was the genesis of Mercury Mark II, which evolved into the Gemini spacecraft.

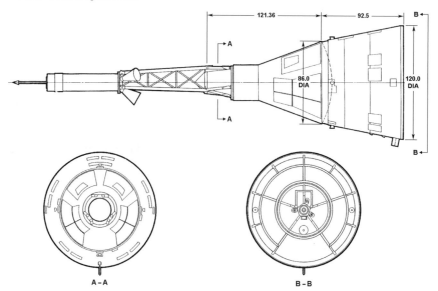

The Mercury Mark II design, revealed in this declassified 1961 McDonnell drawing. (Courtesy McDonnell via NASA.)

training throughout the ensuing decade, until the dawn of the Space Shuttle programme in the mid-1970s.

The resulting group of seven candidates that were selected on 2 April 1959 (and publicly announced on 9 April) became known as the Mercury Seven – or more commonly, the Original Seven. The group included three USAF officers, three USN officers and one USMC officer. There were no representatives from the US Army, no women, and no ethnic minorities. The group's training for Mercury began on 29 April, and would last for two years.

USAF space policy in 1959
Despite having three members of its service selected among the nation's first astronauts, the USAF continued planning for its own independent space programme, and was tasked to provide 'the national forces for offensive and defensive operations, charged with maintaining general air supremacy'. According to the USAF in 1959, this was not confined to any altitude, and 'extends as far from the Earth as the mission demands.' This, of course, meant space.

With a strong history of aerospace technology (especially aerospace medicine), the USAF envisaged their immediate future requirements in a programme that closely paralleled that of NASA, in the development of X-15 technology, applying this to the evolution of operations with X-20 in Earth orbit. There were many objectives offered for putting man into space, and for NASA these included the exploration of one of the last remaining frontiers. There were also political reasons, to enhance national prestige and technological development to benefit the citizens of all nations. Then, there was the scientific knowledge to be gained, which also generated a strong case for sending not just machines, but also man himself into space. None of these were aims of USAF space policy during the late 1950s.

There was only one USAF objective – 'directed toward improving our capability to maintain and safeguard the security and integrity of our nation.' In doing this, the USAF aimed to develop a manned space-vehicle system that both utilised and exploited the unique capabilities of man and thus could be geared towards a greater military effort than could be achieved by unmanned systems. For the USAF, man in space was not to be an 'item of extra payload alone [who] becomes an expensive passenger' (implying that the NASA Mercury astronauts were such!), 'but integrated into the system and enhancing its military usefulness.' Such a programme, the Air Force claimed, would yield 'demonstrable and decisive military advantages over any weapons system in the Air Force armoury.'

To this end the objectives were:

- To establish the technical competence to launch and recover a manned space vehicle.
- To establish the functional usefulness of man in the space weapon system.
- To establish an overall operational reliability of the manned space system, which imposed no greater risks to the crew-member than existed (in 1959) in the operational use of first-line combat aircraft.

Clearly, while taking an active part in the NASA programme, the USAF was pursuing its own long-range space plans for the 1960s. The Soviet threat was being used as a reason for such an increased programme, and the first stage was to develop the X-20 to allow rapid access to space for the military. From here, the USAF was discussing manned space platforms and even journeys to the Moon and Mars, but to place the first USAF officers far above 'the wide blue yonder', the Dyna Soar was, at the time, the programme to achieve it.

MERCURY MARK II

While the USAF looked to its own future manned space programme, NASA was already evaluating what would follow the Mercury programme. As early as May 1959, a NASA committee had begun to consider an advanced Mercury spacecraft that could support two astronauts for up to 72 hours – well beyond the proposed 24-hour capacity of the one-man capsule. By August, the STG New Projects Panel had begun a programme that could lead to a completely new second-generation spacecraft, with significant advances over the Mercury design. At the same time, Mercury's prime contractor McDonnell had also begun studying ways to improve the vehicle in order to expand its capabilities.

In April 1960, STG added a re-entry control navigation system to Mercury's modifications and also indicated a desire for orbital control so that rendezvous techniques would be possible. On 14 April, NASA issued a study contract to McDonnell after a series of joint discussions on improving Mercury. These

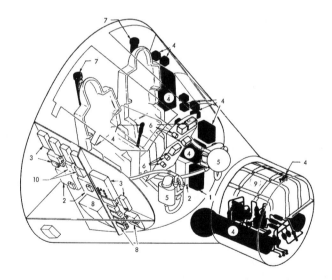

This McDonnell diagram reveals the internal arrangement of equipment in the Mercury Mark II proposal. 1) Sequence and mission profile; 2) electrical and power distribution; 3) communications; 4) stabilisation and control; 5) environmental control system; 6) crew stations; 7) rockets and pyrotechnics; 8) instrumentation; 9) landing; 10) recovery aids.

discussions resulted in the two versions of the envisaged Mercury spacecraft. The first would sustain one man for one orbit to 18 orbits, with only minor changes to the basic Mercury design. The second concept involved much more radical changes, but retained the basic Mercury configuration to support two men. It was this advanced design that became known as Mercury Mark II.

An advanced programme
Concurrent with the studies for Mercury and its follow-on spacecraft in April 1959, a Research Steering Committee on Manned Spaceflight was organised to assist with long-range planning and basic research in the space programme. On 12 August, the STG New Projects Panel, chaired by H. Kurt Strass, met for the first time to discuss future manned programmes, to review the various studies, and to suggest a course of action. The panel recommended that work should immediately begin on a capsule which was even more advanced than proposed in any previous Mercury design. It was suggested that this work should focus on a second-generation capsule, but with three crew-members rather than two, leading to a vehicle capable of achieving near-lunar return velocities. The initial findings of this panel were presented on 28 September 1959.

These recommendations were studied by the Advanced Vehicle Team (headed by Robert O. Piland, and formed by the STG on 25 May 1960), which would conduct preliminary design studies for an advanced vehicle capable of carrying several crew-members. Exactly two months later, on 25 July, this project was named 'Apollo', and included plans for manned lunar landings and a permanent space station. On 28–29 July this announcement was followed by a NASA/Industry Programme Planning conference designed to acquaint the aerospace industry with NASA's plans for advanced spacecraft and circumlunar objectives.

Three ways to go
As the Apollo programme progressed, so talk focused on not just flying past the Moon, but rather on orbiting it and landing on its surface. Engineers had evaluated two main approaches. Direct ascent would require a huge vehicle to take the complete spacecraft from the surface of the Earth to the surface of the Moon without first orbiting. At first, this seemed the most straightforward idea, but it required a massive launch vehicle to lift the spacecraft off the ground, and sufficient fuel both to brake the vehicle for landing on the surface and for the flight back to Earth. The largest vehicle in the planning stages was the Saturn V, but even this could not send a payload to the Moon and back without first entering orbit. The proposed Nova could achieve direct ascent, but was so far in the future that it existed only on paper, and it would probably be ten years before it ever flew.

The second approach was to launch segments of the payload by Saturn V into Earth orbit, join them together, and then take the whole spacecraft to the Moon and adopt the direct ascent return trajectory from the lunar surface. This was called Earth Orbital Rendezvous (EOR) and was well within the range of the Saturn V under development. Since it used several Saturn Vs in its concept, it was the favoured approach of the Marshall Space Flight Center, which was developing the Saturn family of boosters. But there was a third option.

On 10 December 1960, Langley Research Center personnel briefed STG members on an alternative to direct approach and EOR, called Lunar Orbital Rendezvous (LOR). Here, one Saturn vehicle would put the spacecraft into Earth orbit, and then, after a systems check, would boost it on its way to lunar orbit. In this concept, the Apollo spacecraft consisted of two parts – a mother ship, and a separate, smaller lander. Arriving in lunar orbit, two of the astronauts would transfer to the lander and make the descent to the surface, leaving the third in lunar obit. At the end of the surface exploration, the lander would launch and rendezvous with the mother craft, which contained the main engine for the burn for home. In this profile, considerable weight was saved in not taking the return hardware and consumables to the surface and in not bringing back the landing hardware to re-enter the atmosphere, thus reducing the weight supported by the recovery parachutes.

The disadvantage of this approach was the risk of being stranded on the surface, or in lunar orbit. But this could have happened with any method chosen if the hardware failed, as there was no viable system of rescue available in any of the options. However, there were several key stages where a Go/No Go decision could be made with LOR, and certain elements could be provided with a back-up or redundancy capability. But the overriding factor in favour of LOR was that rendezvous would have to be conducted in lunar orbit. Rendezvous and docking was proposed for Mercury Mark II, but it would have to be an operational requirement if either EOR or LOR was adopted for Apollo. It was during this period that the Grumman Aircraft Engineering Corporation in Bethpage, New York, began in-house studies of the LOR concept and the requirements for the proposed landing vehicle.

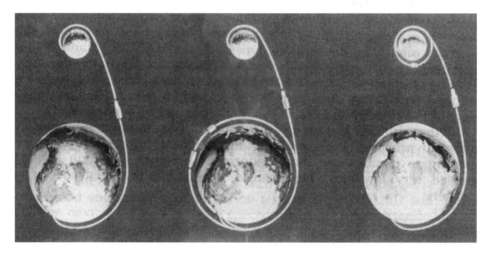

The three competitive modes under consideration for Apollo lunar landing missions: (*Left–right*) Direct Ascent, favoured by NASA HQ and the Space Task Group; Earth Orbit Rendezvous, favoured by Marshall Space Flight Center; and Lunar Orbit Rendezvous, supported by Langley Research Center. In July 1962, LOR was selected to take Americans to the Moon by 1969.

On 19 January 1961, NASA awarded study contracts to Douglas and Chance Vought, to evaluate the concept of EOR in future manned lunar and interplanetary missions. The following month, on 7 February, a manned lunar landing task group suggested that a manned landing on the Moon was indeed possible 'in the 1960s' using EOR or direct ascent, but with no mention of the third, LOR option.

Kazakhstan, Florida, Washington and the Moon
In the mid-1950s, in the barren deserts of Kazakhstan in Soviet Central Asia, a former copper mining area had been converted to launch the Soviet nuclear carrier missile, the R-7. Close to a small railway stop named Tyuratam, a vast complex of launch pads, servicing and preparation buildings, a tracking and control network, machine shops, a railway network, and even a small town grew over the years to become known across the world as the Baikonur (Tyuratam) Cosmodrome. From these launch pads the first Sputniks were launched in the late 1950s, and by early 1961 a much more significant advance was about to be despatched from the site.

On the eastern coastline of central Florida in the United States, at the Cape Canaveral facilities of the Eastern Test Range (more commonly known as 'Missile Row'), the Mercury astronauts were preparing for a series of sub-orbital flights on the Redstone rocket onboard the Mercury spacecraft. These were short hops down the Atlantic Missile Range, lasting about 15 minutes and reaching approximately 115 miles altitude. A series of unmanned and primate flights preceded the three Redstone manned flights planned, before the programme would move on to orbital flights using the larger Atlas launchers (again proceeded by unmanned and primate flights).

It had been hoped that the first manned Redstone flight could be attempted in late March 1961, as indications from the CIA had suggested that the Soviets were about to launch a man into space. A successful, but delayed, Mercury Redstone 1A unmanned flight on 19 December 1960 seemed to indicate that the opportunity to launch the first American into space was indeed very close. However, on 31 January 1961, during the sub-orbital flight of MR-2 carrying the chimpanzee Ham, the thrust of the propulsion system was considerably higher than expected, and the spacecraft separated from the Redstone a few seconds earlier than planned. This resulted in a higher altitude and longer range than planned. Although Ham was recovered successfully, it was clear that some work still needed to be completed on the Redstone before placing a man on board. Then, on 21 February, Mercury Atlas 2 was launched unmanned to evaluate the maximum heating and its effects for a worst-case survivable re-entry from a sub-orbital profile. An 18-minute flight that attained 114 miles and travelled 1,431 miles down range indicated that all test objectives had been met and that the heat shield had performed well and appeared to be in excellent condition following recovery.

Despite the astronauts' desire to fly a late March mission, the Marshall Space Flight Center Redstone team, led by the former German V2 Programme Director Wernher von Braun, was not so confident. They decided that one more test – originally termed MR-2A Booster Development Flight – would be needed to provide the data to ensure that a manned flight on the smaller Redstone was possible. NASA indicated that data from the two earlier flights showed that several problems needed to be resolved before committing the vehicle to manned flight. The manned flight was rescheduled for 25 April at the earliest (but was subsequently delayed until early May) – a choice that was to determine who became the first to put a man into space. What should have been the day that the first American and first man flew in space – 24 March 1961 – became the launch date for the unmanned MR Booster Development Flight (MR-BD) to an altitude of 113 miles and 307 miles down range. It was highly successful in clearing the Redstone for manned flight.

Twenty-four fours later, on 25 March, the Soviets launched their fifth Korabl Sputnik, with a dog on board, in the final test of their manned spacecraft. Then, at 09.07 am Moscow Time on 12 April 1961, the Soviets launched Vostok, carrying Yuri Alexeyevich Gagarin into Earth orbit on a mission that lasted 108 minutes and heralded the dawn of a new mode of transport – human spaceflight. Although the Soviets were known to be developing the capability for manned spaceflight, the launch of Gagarin, like the Sputniks more than three years earlier, came as a shock to the American people, an embarrassment to the American military, and a frustrating disappointment for the American astronauts.

If this was not a big enough disappointment to the Americans, the launch of Mercury Atlas 3 on 25 April with a 'mechanical astronaut' on board was an even more bitter pill to swallow. The intention was to orbit a Mercury spacecraft, but the Atlas failed to roll and pitch to the correct heading. The abort sensing system activated the escape tower prior to destruction of the launch vehicle, some 40 seconds into the flight at 16,400 feet. The spacecraft was recovered intact from the Atlantic Ocean, but instead of flying around the Earth, the mission ended 2,000 yards north of the Cape, just seven minutes after leaving the pad.

Despite this, just ten days later NASA was confident enough to put a man on the next Redstone. 'Ahh, roger; lift-off and the clock is started,' exclaimed Al Shepard as he finally left the pad on his way to becoming the first American in space, on 5 May 1961. He would complete a 15-minute 22-second flight in which he was weightless for five minutes and travelled 302 miles down range. The launch was carried live on TV with approximately 45 million Americans watching the beginning of their manned spaceflight programme. At last America could celebrate a man in space, if not in orbit.

What was also noted was a considerable change in the attitude of the American public to their space programme. In light of Gagarin's success and the disappointments with Mercury launches before Shepard flew, and following the fiasco at the Bay of Pigs in Cuba, the Kennedy administration looked to capitalise on the 'space fever' that seemed to ignite the nation. Since early April 1961, the US Congress had been looking at spaceflight plans and proposals not only for what would follow Mercury at NASA (Mercury Mark II and Apollo), but also at the USAF ideas (mainly X-20) in providing an effective strategic capability from orbit.

After reviewing the options and consequences, President Kennedy made his famous speech to Congress just twenty days after Shepard flew. On 25 May 1961,

America was challenged to land a man on the Moon and return him safely to the Earth, and do it by the end of 1969. It was a bold and vast undertaking that would take almost all of the timescale the President had proclaimed. On the day of the speech, the total of manned spaceflight experience in America was 15 minutes, and the Russians had only 108 minutes. Over the next four years, spacecraft of the Mercury and Vostok types would provide valuable knowledge and experience in human spaceflight on short missions, each lasting just a few days.

Manned spaceflight achievements, April 1961–March 1965

Date	Spacecraft	Crew	Country	Achievement
1961				
April 12	Vostok	Gagarin	USSR	First manned spaceflight (108 minutes) 1 orbit
May 5	Mercury Redstone 3	Shepard	USA	First US manned spaceflight (15 minutes) (sub-orbital)
July 21	MR-4	Grissom	USA	Second US sub-orbital spaceflight (15 minutes)
August 6–7	Vostok 2	Titov	USSR	First man to spend a day in space (25 hours) 18 orbits
1962				
February 20	Mercury Atlas 6	Glenn	USA	First American to orbit the Earth (5 hours) 3 orbits
May 24	MA-7	Carpenter	USA	Second American to orbit the Earth (5 hours) 3 orbits
August 11–15	Vostok 3	Nikolayev	USSR	First extended spaceflight (4 days–94 hours) 64 orbits
August 12–15	Vostok 4	Popovich	USSR	Joined Vostok 3 in orbit (3 days–71 hours) 48 orbits
October 3	MA-8	Schirra	USA	Extended US experience (9 hours) 6 orbits
1963				
May 15–16	MA-9	Cooper	USA	First US day-long flight (34 hours) 22 orbits
June 14–19	Vostok 5	Bykovsky	USSR	Set solo endurance record (119 hours) 81 orbits
June 16–19	Vostok 6	Tereshkova	USSR	First female in space (71 hours) 48 orbits
1964				
October 12–13	Voskhod	Komarov Feoktistov Yegorov	USSR	First crew flight (24 hours) 16 orbits
1965				
March 18–19	Voskhod 2	Leonov Belyayev	USSR	First spacewalk (10 minutes) (26 hours) 18 orbits

MERCURY MARK II DEVELOPMENTS

While NASA progressed with the Mercury programme, developments in the upgraded design of the Mercury capsule also continued. From the earliest discussions, the ability to manoeuvre a vehicle in orbit was seen to allow the techniques of rendezvous and docking in space to be explored. This was an advantage which the military realised was useful from a strategic point of view, aside from the engineering advantage for going to the Moon. In addition there would be advantages in allowing the spacecraft to be recovered on land, in the continental US instead of in the ocean, saving the cost of dispatching a huge fleet of naval vessels to cover the Pacific and Atlantic.

In September 1959, a 300-page report indicated six 'experiments' that a follow-on to Mercury should investigate. These were astronaut-controlled recovery and land landing, manoeuvring in orbit, mission duration of up to 14 days (to simulate the calculated duration of a flight to the Moon and back), manned reconnaissance of objects in orbit, and re-entry from lunar orbital speeds. All were no more than technically supported suggestions but, in the words of the report, 'could be conducted with practicable modifications to the Mercury capsule.'

Discussions at NASA over the next 18 months refined these 'experiments' to incorporate them into an advanced version of the Mercury spacecraft. In January 1961, an STG Capsule Review Board considered the type of missions for such a follow-on programme. Primary suggestions included rendezvous, long duration, artificial gravity, and the ground landing capability. All of these would in some form be incorporated into the Mercury II design that became Gemini (each of their developments is discussed in more detail in the following chapters) and the meeting began a detailed and on-going phase of what evolved into the 'objectives of Gemini'. In March 1961 the question was raised that if a two-man design was being developed, why not provide for one of the crew to open a hatch and step outside? EVA was thus also integrated into the plans for the Mercury follow-on design.

Mercury had been a 'first step' programme designed to place a man in space for up to 24 hours, support him while there, and then bring him home again. What was now being suggested for the advanced version was far more than an 'experimental' spacecraft. It was a move towards an operational system, and so discussions were held into changing equipment locations to improve pre-launch preparations and to relocate hardware not required for re-entry and landing to a position outside of the pressurised vessel.

On 1 May 1961, a working group located at NASA HQ in Washington produced a staff paper costing a programme of integrated research and applied orbital operations, and a co-ordinated programme with the Department of Defense that included 'orbital inspection, and ferry missions'. Such a diverse and wide-ranging programme needed to be managed as a programme separate from Mercury or Apollo, with its own Project Office.

Three days after Shepard flew, on 8 May 1961 Martin staff briefed NASA officials on the systems of their Titan ballistic missile used in the USAF programme, and in its possible application to manned flight by adapting the Mercury II vehicle to fly on it. The idea of using the Titan II as a manned launcher was first proposed for the lunar programme. It was also the launch vehicle selected by the USAF for Dyna Soar, and therefore man-rating was already in progress. Its additional thrust in two stages also offered far more lift capability than the Atlas being developed for launching Mercury into orbit. During the summer of 1961, it was agreed that Titan II would be a suitable launcher, and that LC 19 at the Cape, formerly used for Titan I launches, should be converted for space applications.

On 17 May 1961, the Space Task Group issued a design study for a controlled ground-landing concept for an advanced version of Mercury. A paraglider landing system was envisaged, and three design studies by the Goodyear Aircraft, North American Aviation, and Ryan Aeronautical, were initiated to constitute Phase One of the paraglider development for the advanced Mercury. Development of the follow-on spacecraft was rapidly proceeding in several areas: the design of the capsules; its landing capabilities; its launch vehicles; the objectives; and in ground support facilities (including the development of a new home for the Space Task Group).

Way out west

When the manned spaceflight programme was authorised in October 1958, the location of the 35-strong Space Task Group heading the project in the Langley Research Center in Hampton, Virginia, was only a temporary measure. It had been planned to move the ever-expanding team to a new facility, constructed on the site of an agricultural experimental farm in Beltsville, Maryland, to become the new centre for NASA manned spaceflight programmes. By February 1959, this site had become known as the Beltville Space Center (though without any construction being started), and it formed the nucleus of what from May became known as the Goddard Space Flight Center. As the Mercury and then the Apollo programmes grew during the ensuing two years, and with the prospectus for a programme emerging as Mercury Mark II, it was the Presidential goal of committing to a manned landing on the Moon that necessitated the search for a new, separate facility to achieve those aims.

In August 1961, a shortlist of twenty cities was evaluated and judged using ten criteria. The Manned Spacecraft Center Site Selection Committee passed its recommendations to James Webb, who in February of that year had become the second NASA Administrator. On 19 September 1961 Webb announced that a site near Houston, Texas, had been selected. With Project Mercury in flight operations and soon to support orbital manned spaceflights, a commitment for Apollo to reach for the Moon, and a near decision on a third programme (Mercury Mark II) on 1 November 1961, the Space Task Group was renamed the Manned Spacecraft Center, with STG's Robert Gilruth retained as its first Director.

Design work began the following month, and focused on a 1,620-acre site (1,020 acres donated by Rice University and 600 acres purchased by NASA) in Clear Lake vicinity, about 20 miles south-east of Houston. For the city of Houston the dollar signs began appearing almost as soon as the official announcement was made. But for 750 NASA employees and their families who would have to move 1,500 miles from the pleasant climate of Virginia to the humid heat of the Gulf Coast, the idea was not so appealing, especially as the area had just been hit by Hurricane Carla.

The 1,000-acre site of the proposed Manned Spacecraft Center, south-east of Houston, near Clear Lake, Harris County.

Choosing the way to go to the Moon

When Kennedy committed the Americans to go to the Moon there were still uncertainties about how the flight should proceed. The Lundin Committee, formed on 25 May (the day of Kennedy's speech to Congress), reported on 10 June in favour of EOR, stating that the use of two or three Saturn V launchers would make a manned lunar landing feasible by 1970. By August, an Ad Hoc Task Group for Studying Manned Lunar Landings by using rendezvous techniques (as proposed for Mercury Mark II) concluded that EOR offered the easiest option and the best chance of landing, over the more technically demanding Direct Ascent.

John Houbolt, of Langley Research Center, was not convinced, however, and thought that LOR offered a much simpler and more achievable alternative to EOR. He pushed his case forward by writing directly to Associate Administrator Robert C. Seamans in Washington, on 15 November 1961, pointing out the advantages of LOR over the EOR method.

In the midst of these discussions, the case for the Direct Ascent approach continued. On 20 November 1961, the Director of Launch Vehicles stated that the use of the Nova in a Direct Ascent trajectory still offered the most promising method to reach the Moon. A week later, on 28 November, North American Aviation was selected to construct the Apollo parent spacecraft which would be used to carry three astronauts to the vicinity of the Moon and back, in whichever mode was finally selected.

After the seasonal holidays, in January 1962, Grumman began a six-month in-house study of LOR, as a possible application for a separate landing craft which LOR would require if selected by NASA. On 6 February, Houbolt made his presentation on LOR to the Manned Spaceflight Management Council. This was followed on 2–3 March by a meeting at NASA Headquarters, reviewing LOR as a possible option for the Apollo lunar landing mission profile, using a single Saturn V instead of at least two in the EOR proposal. Then, on 7 June, in a presentation made at MSFC, von Braun became a strong convert and recommended that LOR be adopted for the Apollo profile, as the best chance of success (and funding) within the Presidential time-frame. On 22 June, the NASA Manned Spaceflight Management Council announced that it favoured LOR for Apollo, which was the final hurdle to securing the method of reaching the Moon.

Direct Ascent with Nova was recognised as a highly difficult and long-term project that would not see a landing by 1970. Earth Orbital Rendezvous was an option, but required the use of two Saturn V launches for each mission, and carried excessive weight (and cost) penalties at the lunar distance with the consumables required for the return. LOR offered a staged approach that took only the supplies needed for the landing on the surface, while the return spacecraft remained in lunar orbit. On 11 July 1961, NASA announced that this mode had been selected for Apollo, and four months later Grumman's early work on LOR and a separate lander paid off with the award of the contract for the Lunar Excursion Module (later the Lunar Module) on 7 November 1962. With the method selected to take Apollo to the Moon, the need to develop the experience and techniques required to achieve LOR became a priority for the follow-on programme to Mercury, which had also grown to a full programme in the previous year.

The heavenly twins

On 7 December 1961, NASA Associate Administrator Robert C. Seamans approved the Mercury Mark II development plan. This called for a two-man version of Mercury on longer missions, launched by a modified Titan II. It would conduct orbital rendezvous and docking with an Agena target (launched by Atlas) and would include a controlled ground landing. A programme of unmanned ballistic flights would be included, alongside several manned orbital flights, in a programme consisting of twelve production capsules.

The objectives, announced on 11 November, were:

- To perform Earth orbital flights lasting up to 14 days, in order to train pilots for flights on long-duration circumlunar or lunar landing missions with Apollo.
- To determine man's ability to function in a space environment during extended missions.
- To demonstrate rendezvous and docking (with an Agena-stage target vehicle launched into Earth orbit) as an 'operational technique'.
- To develop simplified countdown procedures and techniques for the rendezvous missions, compatible with spacecraft launch vehicle and target performance.
- To make a controlled landing in the primary recovery mode.

It was also stated that although the two-man spacecraft would retain the general aerodynamic shape and basic concept of Mercury, it would feature several important changes: an increase of internal volume to accommodate two spacesuited astronauts; the use of ejector seats instead of an escape tower; an Adapter Module that contained the equipment not required for the re-entry and landing (saving recovery weight, as it would be left in orbit). Most of the systems hardware would be housed

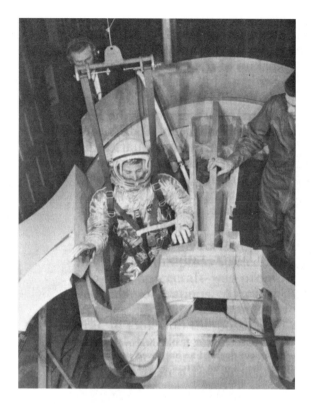

The spacesuited test subject in this 1961 simulation of proposed EVA from Mercury Mark II is attached to a harness to simulate weightlessness in orbital flight.

outside the pressurised compartment, and it would be modular in design instead of integrated, for ease of pre-launch access. It would also include spacecraft rendezvous and docking systems and an orbital manoeuvring capability, as well as a controlled ground landing facility. The programme was to be completed by October 1965.

On 15 December 1961, McDonnell was awarded the contract to develop twelve Mark II spacecraft. In the final contract, signed on 2 April 1962, a total of thirteen spacecraft was agreed, with one to be used in ground testing. Meanwhile, on 26 December, the Air Force Space Systems Division passed on a late Christmas present from NASA to notify Martin to begin work on modifying the Titan II for Mercury Mark II missions. The contract letter was issued on 19 January 1962. Four days earlier, on 15 January, the Project Office was established at MSC, and at end of the month MSC notified Marshall that it should procure, through the USAF, eleven Agena target vehicles and Atlas boosters for rendezvous and docking missions.

What's in a name?
While the contracts were being settled, the Mercury Mark II designation had been replaced with a new identification. On 3 January 1962 the official designation for the new programme was announced: Gemini.

The Mercury and the Gemini spacecraft.

The Gemini programme emblem.

When the programme was officially endorsed there was still uncertainty about exactly what it should be called. When the 'Advanced Mercury' designation was changed to Mercury Mark II upon the suggestion of Glen Bailey of STG and John Brown of McDonnell, it simply became known as Mark II. In order to decide on a more offical name an *ad hoc* programme-naming committee was established to select a suitable designation from suggestions made by NASA Headquarters personnel. The person whose suggestion was finally accepted would receive a token reward (apparently a bottle of Scotch).

The name suggested by Alex Nagy was Gemini – Latin for Twins – which seemed to reflect a two-man crew, a rendezvous between two vehicles, the use of the Titan II, and also the original Mercury II symbol. The name Gemini was selected from several suggestions that included Diana, Valiant and Orpheus.

Nagy did not claim to link astronomy or astrology to the suggestion, but the connections were evident. The zodiacal constellation Gemini consists of the twins Castor and Pollux, who were considered to be the patron gods of voyagers. It also seemed fitting that (in astrology) Gemini is controlled by Mercury. Ironically, its spheres of influence include adaptability and mobility, which for the NASA programme were very apt, as well as zodiacal links to communications and transportation. The correct pronunciation, however, has continued to be debated in NASA to this day. Is it Gemin*ee* or Gemin*eye*? Many suggest the former, while there are others who disagree and are convinced that it is the latter!

Co-operation and division
Throughout the following months, discussions between the USAF and NASA examined how the two could co-operate in the Gemini programme. The USAF would handle launch preparations with Titan and Atlas, and would provide experiments to fly on Gemini, as well as tracking and recovery support. While NASA looked at Gemini as a stepping stone to Apollo and the Moon, the USAF was also beginning to evaluate the spacecraft for its own plans, in conjunction with the X-20, which could have led to a series of USAF Gemini missions in conjunction with, or possibly separate from, NASA's Gemini missions.

The interest in rendezvous and docking and proximity operations, suggested Earth observations from orbit, conducting small scientific experiments, and the prospect of a spacewalk capability, all interested the USAF for strategic objectives aside from the announced NASA objective for Apollo. While the X-20 continued for the time being, the use of Gemini in USAF plans would feature even more strongly as the programme developed (see p. 361), and in Gemini, NASA had also found a spacecraft that offered much more than the stated objectives from 1961 (see p. 385).

Conflict and frustration
With official approval, the Gemini programme could now progress to the construction of the hardware, the creation of the infrastructure to support the missions, and the training of astronauts for flight operations. However, this would not always be a straightforward issue. With Gemini targeted to achieve several objectives linked to the Apollo programme, there was a tight schedule in which to

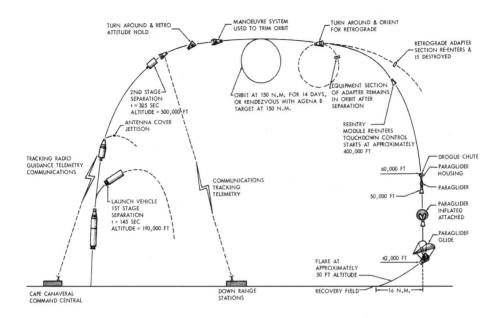

TURN AROUND & RETRO
ATTITUDE HOLD

MANOEUVRE SYSTEM
USED TO TRIM ORBIT

TURN AROUND & ORIENT
FOR RETROGRADE

RETROGRADE ADAPTER
SECTION RE-ENTERS &
IS DESTROYED

2ND STAGE
SEPARATION
t = 325 SEC
ALTITUDE = 500,000 FT

ORBIT AT 150 N.M. FOR 14 DAYS,
OR RENDEZVOUS WITH AGENA B
TARGET AT 150 N.M.

EQUIPMENT SECTION
OF ADAPTER REMAINS
IN ORBIT AFTER
SEPARATION

ANTENNA COVER
JETTISON

REENTRY
MODULE RE-ENTERS
TOUCHDOWN CONTROL
STARTS AT APPROXIMATELY
400,000 FT

TRACKING RADIO
GUIDANCE TELEMETRY
COMMUNICATIONS

DROGUE CHUTE

PARAGLIDER
HOUSING

60,000 FT

PARAGLIDER

COMMUNICATIONS
TRACKING
TELEMETRY

50,000 FT

PARAGLIDER
INFLATED
ATTACHED

LAUNCH VEHICLE
1ST STAGE
SEPARATION
t = 145 SEC
ALTITUDE = 190,000 FT

PARAGLIDER
GLIDE

FLARE AT
APPROXIMATELY
50 FT ALTITUDE

42,000 FT

CAPE CANAVERAL
COMMAND CENTRAL

DOWN RANGE
STATIONS

RECOVERY FIELD   16 N.M.

A 1962 diagram of the Gemini mission sequence, showing the primary objectives of 14-day missions, rendezvous and docking with an Agena, and the land-landing objective.

conceive, deliver, test and fly the twelve missions and to deliver what was promised. Several key hurdles were to feature throughout the programme into the flight phase – some of which almost threatened the whole programme. Most of these were successfully overcome, while a few were not.

Each of the more important items (discussed in more detail later) included:

- Keeping the development and operations within a tight budget.
- Qualification of the Titan II missile and conversion for manned flight.
- Development, testing and suitability of the Agena as a target vehicle.
- Integration of the fuel cell over batteries for spacecraft power supply.
- Development of the crew rescue and recovery system – the paraglider concept.
- The desire of the USAF to become more involved in Gemini to a point of a proposed programme takeover from NASA and the space agency's reluctance and resistance to such a move.

The clock is running

When Gemini was officially approved in December 1961, there remained eight years to the end of the decade – eight short years in which to design, build, test and fly not only the hardware but create the infrastructure for Apollo *and* complete the Mercury, Gemini and unmanned pathfinder missions – Ranger, Lunar Orbiter and Surveyor.

The lift-off of a USAF Titan II ICBM – the launch vehicle proposed for the Gemini missions.

This race against time (and, as America believed, against the Soviet Union) was complicated by the fact that the opportunity to launch a payload to the Moon occurs for only a few days every four weeks, when the Earth and its natural satellite are in a particular alignment. This time period (launch window) also ensured adequate lighting conditions at a lunar landing site and for recovery at the end of the mission. This meant that there were only 104 opportunities to reach the Moon before the end of 1969, and every month that figure would all to quickly decrease.

With the final goal shining full and bright in the night sky each month, it would be Gemini's role to provide the experience that would blaze the trail for Apollo to secure the prize.

Hardware

By early 1962, the Gemini series of manned missions had become the next official NASA manned spaceflight programme after Mercury. With the first unmanned launch planned for the second half of 1963 (subsequently delayed into the spring of 1964), there remained much to be accomplished and only a short time in which to complete it.

Over the next two years, the design of the vehicle and the mission profile it was to complete were redefined, as the first elements of hardware were prepared for testing and shipment. There were several areas in this development that presented hurdles along the way – in particular, the use of the Titan II, the method of astronaut escape, the development of the ground landing capability, the choice of target vehicle, and the role of the Department of Defense in the project.

In addition, construction of the new Manned Spacecraft Center in Houston, conversion of the launch facilities at the Cape, and the organisation of the management infrastructure, all needed to be completed before anything could fly. After evolving the advanced Mercury over the previous three years, there remained less than two years to put all the elements in place and start flying, while still restricting the escalating costs. As a 'follow-on' from Mercury, the 'new' programme also did not have to go before Congress for sanctioning before it was authorised, as it could be funded from money already allocated to NASA's budget. The programme would go before Congress in formal budget requests for Fiscal Year 1963, beginning in 1962. The costing for the project had been only a preliminary estimate, but when more defined fees started accruing, and problems began occurring in several areas of development, costs escalated as the programme developed.

THE GEMINI BUDGET, 1962–1967

Defining a complete costing for one specific space programme is difficult due to the support of other elements funded by separate budget allocations. The following table is based on official NASA figures, and reveals how the Office of Manned Space Flight (OMSF) Gemini budget, or most of it, was spent on such items as spacecraft, some launch vehicles, and supporting development. NASA has estimated the total cost of Gemini – including facilities, salaries, research and development, operations and hardware – at $1.283 billion.

The Gemini budget

Total funding ($million)

Year	Request	Authorisation	Programmed
1962	–	–	54.959
1963	203.2	–	288.09
1964	306.3	306.3	418.9
1965	308.4	308.4	308.4
1966	242.1	242.1	–
1967	40.6	40.6	–

The 1963 request was taken from the advanced manned spaceflight budget. Gemini was not listed as an item in the FY 1968 budget estimate, but it was estimated in the FY 1967 budget that $226,611 million would be programmed in FY 1966. There was no FY 1969 budget estimate for Gemini, and after FY 1967 no funds were programmed.

Spacecraft funding

Year	Request	Authorisation	Programmed
1962	–	–	30.329
1963	131.35	–	205.045
1964	196.206	–	280.52
1965	168.9	168.9	165.3
1966	122.7	122.7	–
1967	19.1	19.1	–

The 1963 request was from the advanced manned spaceflight budget. In the FY 1967 budget estimate it was expected that $107,211 million would be programmed in FY 1966 for the spacecraft. No funds were programmed after FY 1967.

Operations and support funding

Year	Request	Authorisation	Programmed
1962	–	–	239
1963	–	–	3.936
1964	15.3	–	15.68
1965	28.2	28.2	27.7
1966	30.8	30.8	–
1967	13	13	–

The 1962 programmed figure was for supporting development. The 1964 request includes $700,000 for supporting development. In the FY 1967 budget estimate, $30,800 million would be programmed in FY 1966 for Gemini support. No funds were programmed after FY 1967.

Launch vehicle funding

Year	Request	Authorisation	Programmed
1962	–	–	24.391
1963	71.85	–	79.109
1964	94.8	–	122.7
1965	111.3	111.3	115.4

Launch vehicle funding (contd.)

Year	Request	Authorisation	Programmed
1966	88.6	88.6	–
1967	8.5	8.5	–

The 1963 request is from the advanced manned spaceflight budget. It was estimated that $88,600 million would be programmed for launch vehicles in FY 1966. No funds were programmed after FY 1967.

Gemini Program Office total accrued costs ($million)

	FY 1962	1963	1964	1965	1966	1967	Total
Spacecraft	20.0	178.6	223.5	177.0	77.5	19.5	696.1
Paraglider	–	9.1	15.8	2.3	–	–	27.4
Atlas (SLV)	–	–	4.7	10.1	11.4	4.9	31.4
Agena (target)	0.9	14.5	26.1	21.7	31.1	5.8	100.1
GLV (Titan II)	12.8	71.2	77.5	60.9	39.6	21.3	283.3
Support	–	0.6	0.9	2.3	4.6	0.9	9.3
Totals	33.7	274.0	348.5	274.5	164.2	52.4	1,147.3
Cumulative	33.7	307.7	656.2	930.7	1,094.9	1,147.3	

Adapted from *On the Shoulders of Titans*, NASA SP4203, 1977 p. 582; and from *NASA Historical Data Book Volume II, Programs and Projects, 1958–1968*, NASA SP4102, 1988, pp. 126–127.

GEMINI SPACECRAFT FEATURES

The Gemini spacecraft was constructed in three sections. The pressurised crew compartment was located in the forward *Re-entry Module*, while in the rear was the unpressurised *Adapter Module*, which itself was divided into two parts – the *Equipment* and *Retrograde* sections.

The two-crew positions were forward facing in the 'centre' of the Re-entry Module. This also featured the 90-inch-diameter heat shield at the rear, and recovery parachutes and rendezvous equipment in the front. Behind the Re-entry Module was the Adapter Module, with the forward element of this, the Retrograde section, housing the retro-rockets for the end-of-mission return-to-Earth burn. At the rear, the Equipment Module carried the onboard supplies for power, manoeuvring and other equipment not requiring pressurisation or inclusion in the Re-entry Module. The whole Adapter Module was separated prior to the re-entry burn. Overall, the Gemini spacecraft measured 18 feet 5 inches in length, with a 10-foot maximum diameter at the widest part of the Adapter Module, narrowing to just 39 inches at the nose of the spacecraft. The combined Equipment and Retrograde sections were 7 feet 6 inches in length, and the Re-entry Module measured just less than 11 feet in length. Weight varied across the twelve spacecraft, but ranged from 7,100 lbs to 8,350 lbs.

Although larger than Mercury, with about a 50% increase in internal volume in the crew compartment, Gemini was still a compact vehicle – and the crew had to

The primary components of the Gemini spacecraft.

spend up to two weeks in it, in weightlessness! Gemini was twice as heavy as its predecessor, and was 20% larger.

Construction of the Re-entry Module
As McDonnell (the prime contractor for the Mercury spacecraft) had completed much of the preliminary design work, and since Gemini retained the general aerodynamic shape and basic systems concepts, there was no need for NASA to seek any competitive bids, and on 22 December 1961, Company President James S. McDonnell Jnr signed the contract that assigned the McDonnell Aircraft Corporation the task of delivering the Gemini capsule to NASA. In turn, McDonnell sub-contracted elements of the design to other vendors.

Although outwardly the shape resembled the Mercury capsule, the design of the Gemini capsule was a departure from its predecessor in several ways. Using the experience of the one-man spacecraft in design, construction, testing, launch preparations, and flight operations, a number of improvements were incorporated into the larger vehicle.

Only systems that were required to support the crew were located within the pressurised compartment. These were the instrument panels, the flight controls, the environmental control systems, the ejection seats, the food and waste systems, communications, and all guidance controls.

The internal atmospheric environment was the same as for Mercury: 100% pure oxygen at a pressure of 5–5.3 psi. An unpressurised volume surrounded an inner pressurised crew compartment featuring the two crew stations. This outer area

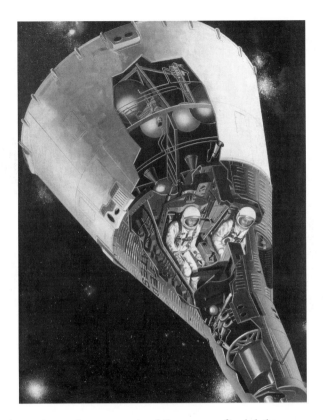

The primary internal components of the spacecraft, and the crew positions.

contained equipment and systems that did not require pressure to function. These hardware units were constructed in the form of modular packages that, should they need replacing, could be exchanged from the outside via easy-to-remove ground access panels, without the need to disturb the crew compartment or affect systems already checked. This saved construction and launch preparation time, which was crucial in assuring that delivery and launch dates were met. By adopting the modular equipment system, the two-month pre-flight check-out was a significant advance over the 6–7 months required during Mercury.

In fabricating the spacecraft there was a desire to obtain the best strength-to-weight ratio, and as a result McDonnell chose to use titanium and magnesium as the principal construction materials. Almost all the axial loads were carried by ring-frame stabilised stringers made of these two metals. The pressure shell was constructed from a fusion-welded titanium frame attached to side panels and fore and aft bulkheads constructed from double-thickness, thin sheeted titanium (0.010 inches). To ensure maximum stiffness of such a thin material, the outer sheet was beaded.

It was important that construction remained lightweight and structurally strong to allow the vehicle to be lifted into orbit, and to sustain a pressurised environment

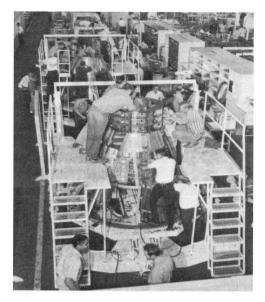

The assembly of the spacecraft at McDonnell.

This view of the Gemini spacecraft without the outer panels reveals the modular components surrounding the pressure vessel. One of the open hatch doors (at left) has a protective cover plate on the window. (Courtesy McDonnell.)

for the crew inside. The vehicle also had to withstand aerodynamic forces during ascent and entry, and be heat-resistant behind the heat shield during re-entry. To solve this, Gemini was fitted with overlapping beaded shingles of René 41 (0.016-inch) alloy that covered the outer surface of the Re-entry Module, as with the Mercury spacecraft. These shingles were thermally isolated from the stringers, and as an added protection between the inner pressure compartment and the outer skin of the spacecraft, a layer of insulation was fitted. On the front of the spacecraft, where the rendezvous equipment was located, the nose was fitted with unbeaded beryllium shingles and a laminated nose cap fairing made from reinforced plastic and glass fibre. This was jettisoned upon entry into orbit, revealing the rendezvous equipment.

The re-entry heat shield
The heat shield was located at the wide, blunt base of the Re-entry Module. As the spacecraft descended into the atmosphere, the heat shield absorbed heat (at around 3,000°) by ablation and increased loads. It therefore had to both dissipate the heat and at the same time retain structural strength. The structure of the heat shield was based on a sandwich of glass fibre, formed from two five-ply face-plates of a resin-impregnated glass fibre cloth, separated by a glass fibre honeycomb core (0.65 inches thick). On the outer, convex side of this structure was bonded an additional glass fibre honeycomb.

Engineers monitor the pouring of the Dow-Corning DC-325 organic compound mixture onto the glass fibre honeycomb forming the outermost layer of the ablative shield. (Courtesy McDonnell.)

A coating of Dow-Corning DC-325 organic compound was applied as a paste, and was allowed to harden in normal temperatures to form the ablative surface to absorb and then disperse the re-entry heat. A Fibrerite ring formed a seal from the edge of the heat shield to the side structure of the spacecraft.

Escape tower or ejection seats?

Gemini (unlike Mercury and Apollo) did not feature an escape tower to pull the astronauts clear of any potential explosion of the launch vehicle on the pad or during ascent. Instead, the spacecraft adopted the technique used in military jet aircraft: the individual ejection seat. This type of escape system was similar to the system which the Vostok cosmonauts had available. As with Vostok, it allowed crew escape either during ascent or during the latter stages of entry, whereas an escape tower is only a viable means of escape during the first few minutes of the ascent.

Initially, the design of Mercury II featured both an escape tower and a deployable landing bag attached to the heat shield to cushion water recovery, as no-one was really sure that ejecting from a spacecraft was survivable, or that a landing with the paraglider was soft enough to prevent injury. With the experience gained using the

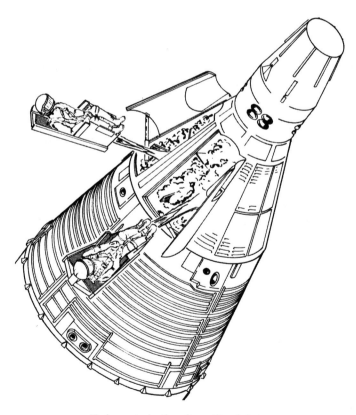

Twin-seat ejection from Gemini.

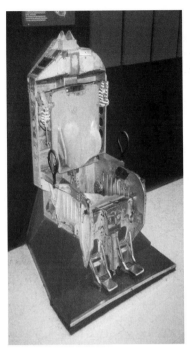

The wear on this Gemini ejector seat was not caused by astronauts, but by thousands of visitors to the Johnson Spaceflight Center (formerly MSC), Houston. (Astro Info Service collection.)

escape tower and the landing bag on Mercury, these were now flight-proven and reliable designs. Indeed, the development of both the ejector seat and the paraglider would be troublesome. However, the inclusion of the escape tower and landing bag on Gemini was short-lived. Analysis of the launch profile of the Titan indicated that ejection seats would provide 'adequate crew escape [using] their quick reflexes, because Titan's non-explosive propellants merely burn and allow time for human reactions.'

The use of an ejector hatch, and the need to provide EVA access, meant that the spacecraft hatches of Gemini had to be capable of being opened during flight (at up to 60,000 feet during ascent and below during entry), either under maximum dynamic conditions for ejection, or in the vacuum of space for EVA. An additional requirement was that, in the event of ejection, the hatch must stay locked open to allow a clear passage for the escaping astronaut, whereas on EVA the astronaut needed to easily open and close the hatch on his own, as access from the other crew position was difficult.

A window on the world
As Gemini would feature rendezvous, proximity operations, docking, and crew observations, the windows had to be of a suitable optical clarity. There were two

windows in the crew compartment – one in each hatch – supplied by the Corning Glass Company. Both featured a three-pane construction with a sealed air space between each, but each consisted of different panes. The Command Pilot window had outer and middle glass panes of 96% silica glass (each 0.22-inch thick), with the innermost pane made of a temper-toughened alumino-silicate glass. This was suitable for most operations, including rendezvous and docking. The Pilot window was constructed for extra optical clarity and for scientific observation experiments and instruments. The outer and central panes remained the same, but the inner pane was replaced with 96% silica glass (increased to 0.38-inch thickness), which allowed 99% optical clarity for observation and photography through all three panes.

On the early missions it was discovered that staging during ascent caused a residue of smoke and soot to settle on the outer panes, so that when the craft finally entered orbit the windows appeared smeared and dirty. This problem was solved by adding an outermost transparent pane, which would be jettisoned by the turn of a thumbscrew upon reaching orbit.

Re-entry Module power supply and control

Power for the spacecraft during the re-entry phase came from four 45 amp/hour silver–zinc batteries in the equipment bay of the crew compartment. In addition there were three 15 amp/hour squib batteries used to trigger small pyrotechnics. The set of sixteen 25-lb thrusters, in two rings of eight in the Re-entry Module also gave attitude control during the return to Earth, and were located in the nose of the spacecraft in front of the pressurised crew module and behind the recovery and rendezvous section.

Features of the Adapter Module

The two separate elements of this section of the spacecraft were both jettisoned prior to re-entry at the end of the mission. The larger equipment section at the rear supported the whole spacecraft on top of the Titan on the pad, and during boosted ascent it housed the main spacecraft supplies. On the early spacecraft, three batteries provided 400 amp/hour units for primary power, and on the duration and later rendezvous missions, the power supply consisted of two hydrogen–electric fuel cells, supplied by General Electric. The attitude control system was also housed in this compartment and was fuelled by monomethyl hydrogen and nitrogen tetroxide. There were eight engines providing thrusts of 25 lbs, two of 85 lbs, and six of 100 lbs.

The Adapter Module engines included a set of eight 25-lb thrusters arranged in four pairs around the base of the Equipment Module, aligned to provide attitude control in the three axes. Four 100-lb thrusters were located 90º apart on the exterior of the Adapter Module pointing directly outwards, and allowing movement left, right, up or down. Two 100-lb thrusters, 180º apart, were located in the aft of the Adapter Module for forward movement, while a second pair of 85-lb thrusters pointed forward and pushed the spacecraft directly backwards when fired.

Inside the retrograde section, on a cruciform beam assembly, were four solid-propellant retro-rockets used for de-orbiting at the end of the mission. Once the equipment section had been jettisoned, these four rockets were fired sequentially,

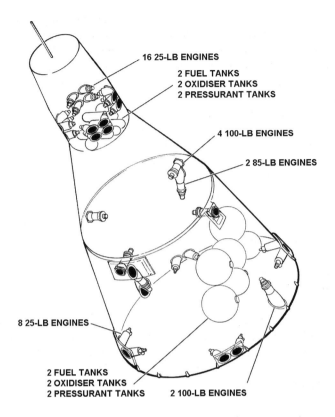

16 25-LB ENGINES

2 FUEL TANKS
2 OXIDISER TANKS
2 PRESSURANT TANKS

4 100-LB ENGINES

2 85-LB ENGINES

8 25-LB ENGINES

2 FUEL TANKS
2 OXIDISER TANKS
2 PRESSURANT TANKS

2 100-LB ENGINES

The OAMS and RCS thruster arrangement.

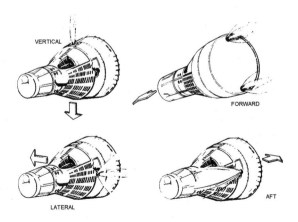

VERTICAL

FORWARD

LATERAL

AFT

The manoeuvring of the spacecraft with combinations of thrusters.

each providing 2,490 lbs thrust, and like the OAMS were capable of firing automatically or manually.

Additional thermal protection across the spacecraft came from heat-protective paint coatings, giving the Re-entry Module a dark grey appearance. The Adapter Module was finished in white to allow maximum reflective use of heat brought to the surface by coolant channels beneath the skin of the module.

Construction of the Equipment Module featured circular aluminium alloy frames, with stringers from extruded magnesium alloy and a skin of magnesium. Each of the stringers were T-shaped, and had a hollow bulbous section which allowed liquid coolant to pass through and transfer heat generated inside the spacecraft to dissipate to the skin of the module and radiate into space. Several experiments were located in this section, and the Astronaut Manoeuvring Unit was housed centrally at the rear of the Equipment Module. The open end of the Equipment Module had a cloth cover and insulation blankets which protected internal equipment from solar radiation after separation from the Titan. This was constructed of Vitron-impregnated glass cloth, and was treated with vapour-deposited gold.

Spacecraft sub-systems

Computer Described as a 'shoebox'-sized computer, it weighed 57.6 lbs and occupied one cubic foot of cabin space. It was small by the standards of the time, and although antique by today's standards it was an example of high technology in the early

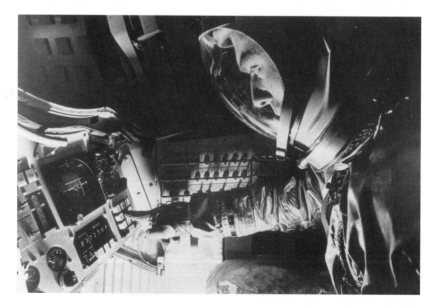

The onboard computer used for navigation during rendezvous and docking and re-entry manoeuvres. The astronauts pressed keys on the numbered keyboard, and the computer instantly displayed navigational information on the dial to the left of the keyboard. (Courtesy British Interplanetary Society/NASA.)

1960s. The system allowed the astronauts to compute the intricate programme of manoeuvres for rendezvous and docking with the Agena. It was a digital design, was capable of storing 159,744 bits in 4,096 computer 'words', and had the capacity to add, subtract or conduct a transfer operation in 140 microseconds. It could also multiply in only 420 microseconds, divide in 840 microseconds, and could conduct three operations at the same time.

The computer could store, implement or display information to the crew, and allowed an astronaut to select from seven control modes: rate command; direct; pulse; re-entry rate command; platform; horizon scan; and re-entry. These could be selected through the fully automatic control system, by full manual control, or by combinations of both. The use of the computer certainly alleviated a significant amount of mundane and time-consuming operations at a time when the crew's attention was diverted elsewhere. But there still remained the need to punch a large amount of data into the computer by means of a push-button keypad. The computer was used in conjunction with the inertial measurement unit and tape memory system.

Communications Onboard the spacecraft were three main communication systems: voice (built by Collins Radio, and consisting of one HF and two UHF transmitter/ receivers), which also included an intercom between both astronauts; a receiver for command signals and updated orbital information from MCC; and data collection tapes and their relay transmitters for automatic transmission (dumping) of reports to Earth. Two C-band radar beacons (one in the Re-entry Module and the other in the Adapter Module) provided ground signal response for orbital flight tracking and recovery tracking.

Ejection seats Based upon a jet aircraft ejection system, the Gemini seats were of a lightweight construction, and were fabricated by Weber Aircraft. Evaluation of launch stresses revealed that a maximum of 7 g would be encountered by the pilots during launch, and a maximum of 4 g during descent. This meant that the form-fitting contoured couches designed for each Mercury astronaut would not be required on Gemini, allowing the use of a standard seat to support the astronaut. The system had to include the capacity to eject the astronauts clear of the Titan from the crew module at a height of about 150 feet off the ground. The propulsion had to produce both a clear trajectory for clearance of an exploding booster, *and* a lifting trajectory high enough for the parachutes to open and not drive the seat and its occupant into the ground!

There was no such seat in production in 1962, and NASA therefore selected Rocket Power Inc to provide a rocket catapult (rocat) to fit onto the Weber seat structure. This was tested at the Naval Ordnance Test Station at China Lake, in the Mojave Desert in California. The test tower was a 150-foot structure originally built for tests of the Sidewinder missile. The tests began in July 1962 and were identified by the term SOPE (Simulated Off-the-Pad Ejection – see Chapter 3).

Electrical power This was originally to have been provided by chemical batteries, but as missions increased in length, these could not provide sufficient power within the weight restriction. Therefore, in January 1962 it was decided to use a fuel cell,

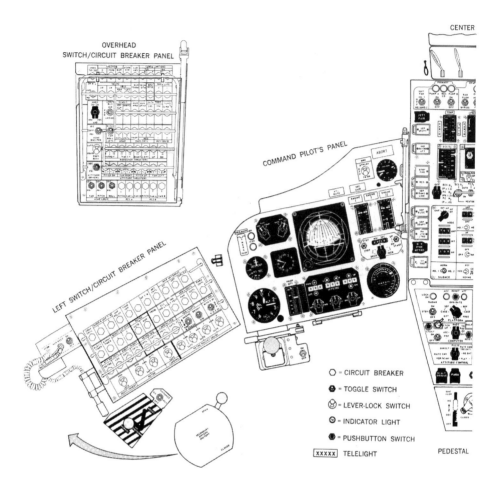

The Gemini instrument panel, showing the layout of crew displays, controls and instruments (Courtesy British Interplanetary Society/NASA.)

developed by the General Electric Co, on longer missions. Light, simple in design and well suited to the requirements of Gemini, the fuel cell used hydrogen and oxygen in a reaction that produced both water and heat. The cell design used a solid ion-exchanging membrane, which chemically bonded electrolyte and water instead of diffusing gases into liquid electrolyte, as featured in other designs of cells. Using a separate stream of coolant, the water produced at the cell was condensed and removed through a series of wicks, enabling the reaction to continue at a constant rate. This in turn used very little of the cell's own power, rendering it more efficient and simpler in design. Two groups of fuel cells could each provide 1,000 W, which was sufficient for the spacecraft's total electrical needs. As a by-product, water was produced for drinking and spacecraft cooling. Using a fuel cell as the primary power source was also planned for the Apollo missions, and although testing the system

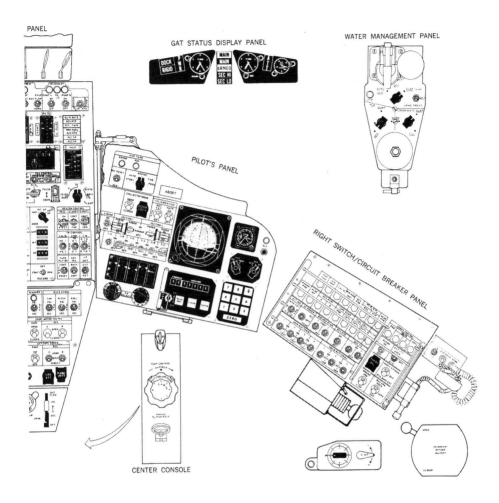

PANEL

GAT STATUS DISPLAY PANEL

WATER MANAGEMENT PANEL

PILOT'S PANEL

RIGHT SWITCH/CIRCUIT BREAKER PANEL

CENTER CONSOLE

on Gemini would be a challenge in terms of the tight schedule, it would also offer valuable experience for the lunar missions.

Environmental control system Provided by AiResearch Manufacturing Company, the ECS provided the astronauts with oxygen to breathe, and eliminated exhaled carbon dioxide by using lithium hydroxide. A carbon cartridge supplemented the removal of the fouled air, and cooled the spacecraft, its equipment and the spacesuits, as well as purifying drinking water. Although similar to the Mercury system, significant engineering changes were required to allow its inclusion on Gemini. These included an improved method of dissipating unwanted heat into space; storage of oxygen (Mercury's stored bottled gas was replaced by a liquid oxygen supply which required less storage of the maximum 14-day supply); and a system of redundancies allowing each astronaut to be supplied by two parallel suit

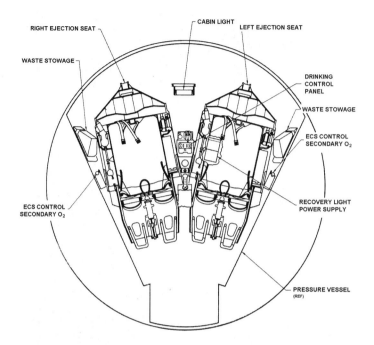

The two-astronaut ejection seat, located in the crew module. (Courtesy British Interplanetary Society/NASA.)

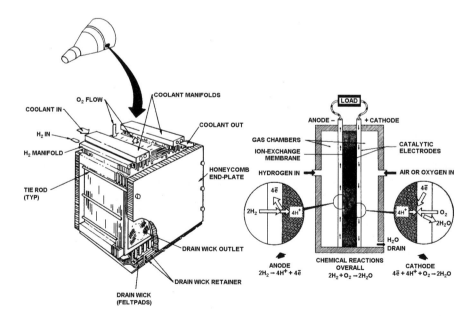

Gemini's ion-exchange membrane fuel cell.

This view of the Pilot station reveals the confined conditions to which a crew was subjected for up to fourteen days.

circuits for oxygen. They also had the option of using the spacecraft's cabin system while partially unsuited.

Food and water A basic diet of 2,550 calories per man per flight day consisted of a selection of freeze-dried foods, including meats, soups, desserts and fruit. The addition of water restored the food to an edible form. On 6 June 1962 the contract to supply both the food and the waste management system was awarded to Whirlpool Corporation Research Laboratories. Requirements included the water dispenser, food storage and waste storage systems. The US Army Quartermaster Food and Container Institute provided the food and feeding devices, under the direction of MSC Life Sciences Systems Division. Water came from a two-tank reservoir – a 42-lb capacity tank in the Adapter Module and a 14.5-lb tank in the Re-entry Module. Initially supplied by Canada Dry, these would be replenished by the water supplied as a by-product of the fuel cells during the mission.

Guidance and navigation The internal guidance system, supplied by IBM, recorded every motion of the spacecraft from launch to orbital insertion. Once there, the system was linked to the onboard computer, radar, electronic controls, attitude thrusters and propulsion systems, enabling the astronauts to use the combined systems to achieve rendezvous with the target vehicle. The inertial guidance system was controlled by the inertial measurement unit and by three gyros and three

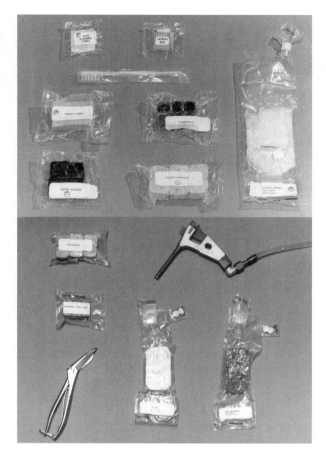

Selection of bite-sized and rehydratable foods consumed during Gemini missions. The scissors were used to open the packages, and the water gun was used to supply water for reconstituting the food. The toothbrush was packaged with the first meal eaten in space, and the chewing gum was also used for dental hygiene. (Courtesy NASA via British Interplanetary Society.)

accelerometers. The accelerometers recorded the spacecraft's movement about the three primary axes (X, Y, Z) and at what rate this motion was being completed. The three gyros, using a pre-set attitude reference, indicated which way the spacecraft was pointing. All this information was then passed to the onboard computer and the ten-mode tape memory library (pre-launch, ascent, catch-up, rendezvous, relative motion, orbit navigation, orbit prediction, orbit determination, touch-down prediction, and re-entry).

Pressure suits The contract for producing the Gemini spacesuits was awarded to the David Clark Company on 13 June 1963 (after an earlier design by BF Goodrich was rejected) and resulted in the production of 75 assemblies. The suits were designed for each astronaut, using a soft, body-forming design which permitted long-wear

A technician demonstrates how astronauts handled the problem of eating during their missions. (Courtesy British Interplanetary Society/NASA.)

A technician works on an IMU which, during flights, provided information on the spacecraft's position relative to the Earth's surface. (Courtesy British Interplanetary Society/NASA.)

John Young models an early version of the Gemini pressure suit (1963). Behind him are the form-fitting couch liners for each of the Mercury astronauts.

John Young – wearing a Gemini training suit – takes a break from water egress training exercises in 1965.

comfort and an increased mobility range to meet increased mission tasks over the Mercury missions. There were designs for both internal and external activity, and all were to be fully pressurised to work in a vacuum for several hours without jeopardising the efficiency of the crewman. There were five basic designs configured for Gemini: G1C and G2C were early versions of the G3C suit, and were used for suit qualification and astronaut training; G3C (see below) was an IVA suit that afforded four hours of continuous operation in an emergency decompression of the cabin; G4C was modified from the G3C, with added protective layers, and was used for EVA (and on later missions without the protective layer for IVA – see Chapter 8); and the G5C suit (see Chapter 6) was an IVA suit designed for maximum comfort and mobility for the very long fourteen-day missions, as well as maximum support in the event of decompression.

The G3C suit weighed 25 lbs maximum, and was custom built for each astronaut (helmets were one size with custom liners). Each was pressurised to 3.7 psia for up to four hours in flight, or 7 psia for fifteen minutes under test conditions. The suit also incorporated a 4.6 $\pm$ 0.3 psia relief valve. Donning the helmet and gloves would take three minutes unassisted, and the suit was worn for the duration of the mission. Each was supplied with 100% oxygen; extracted carbon dioxide at 3.8 to 7.6 Hg partial pressure; and H_2O ventilated at 4.75 in. maximum. They maintained an internal temperature of 50º–80º F.

From the inside outwards, an astronauts suit consisted of the *underwear garment assembly* – a collarless, one-piece lightweight unit with integrated foot 'socks' designed for comfort and for absorbing perspiration. Pockets were included for dosimeters and other bioinstrumentation, and with access for using the waste management devices. The *torso assembly* was the basic pressure garment covering the body except for the head and the hands. It consisted of four layers (assemblies):

1) The *inner liner assembly* was an Oxford nylon garment attached to the gas container assembly by snaps and Velcro. It was designed to provide a smooth inner surface to help in putting on the suit (donning) and removing it (doffing) to minimise pressure points and snagging.
2) The *gas container assembly* was made of neoprene ripstop-coated nylon, with flocked rubber boots integrated at the lower leg. The main entry slide fastener extended from the upper back down through the crotch to the abdominal area, with a dual redundant pressure closing device that maintained the inner integrity of the gas container while the outer seal maintained the pressure seal.
3) The *restraint assembly*, formed from link-net fabric, restrained the gas container to the astronaut's body contours when pressurised, and provided structural support.
4) A single layer of Nomex, the main function of which was to protect the gas container and restraint layers from snagging and abrasion, and to insulate the suit from IVA heat and cold.

The suit assembly also included the *helmet*, attached to the torso by rotating it on a neck bearing/disconnect ring. The helmet included a foamed pre-polymer resin

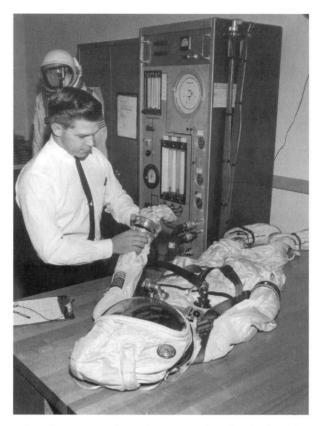

A G7C long-duration spacesuit undergoes a bench check. (Courtesy British Interplanetary Society/NASA.)

anti-buffet liner with a leather layer that interfaced with the astronaut's head. A drink port was also provided for use with a sealed visor. The helmet featured a compact helmet shell to provide more headroom in the capsule and a visor assembly that could pivot and be opened in a pressurised crew compartment, but remained sealed for unpressurised use. The helmet also included communication connections for voice communication when sealed inside the garment.

Completing the pressure garment integrated with the torso assembly and the helmet were the *glove and boot assemblies*. The gloves had a gas-containing inner glove made of elastomer, and an outer glove for protection and restraint. They were individually formed to fit the astronaut's hands, and were attached at the wrist by a 360° rotation bearing. Later models included fingertip lights and an articulating palm restraint. The boot assembly fitted over the flocked rubber boot, and was laced to the leg of the torso assembly. Constructed of a four-layer assembly, it consisted of (inside to out) Oxford nylon, cotton-covered sponge, high temperature fabric, and a laminated protective cover. It had a side fastener for donning and doffing, and laces and straps for adjustment and personal comfort. Protective rubber was provided for

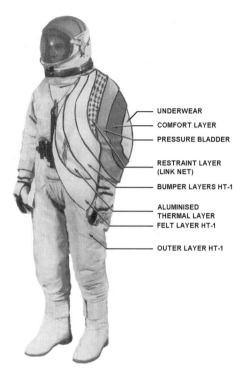

UNDERWEAR
COMFORT LAYER
PRESSURE BLADDER

RESTRAINT LAYER
(LINK NET)
BUMPER LAYERS HT-1

ALUMINISED
THERMAL LAYER
FELT LAYER HT-1

OUTER LAYER HT-1

The G4C suit was worn over underwear, and consisted of five basic layers plus two layers of micrometeroid protective cover layer.

protection and interface with the spacecraft and ejection seat foot bar, and a strengthened sole and heel offered support for walking in 1 g.

The IVA assembly was completed with accessories for *blood pressure cuffs* for the arms and *venous cuffs* for the legs; a *neck dam assembly* to keep water out of the torso when not wearing the helmet in the water after splashdown; and a *flotation device* – an inflatable life jacket to assist in the flotation of the astronaut in water immersion.

Other systems included the *ventilation system*, including ducts from inlet fittings in the lower abdomen area of the suit to the head, hands and feet. The gas was then passed back over the body and exhausted through a second fitting adjacent to the inlet. Anti-block devices prevented closing the paths between the underwear liner and supply channels between the inner and gas container mated at the connections with the helmet and glove ventilation channels. The *biomedical/communication system* was a dual independent earphone and microphone system connected by a helmet side connector to an internal harness leading to an external connector in the chest area. This also fed the biomedical sensors to the same external connector. There were also batteries for the fingertip lights, a suit pressure gauge on the arm, and a pressure relief valve on the thigh.

The suit was designed for up to fourteen days' continuous wear (but mostly

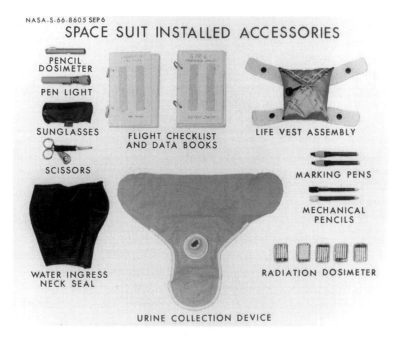

NASA-S-66-8605 SEP 6
SPACE SUIT INSTALLED ACCESSORIES

Some accessories worn inside the suit or installed in the pockets. (Courtesy British Interplanetary Society/NASA.)

without the helmet and gloves) in an environmental parameter of 0°–160° F (or 250° F for five minutes during ejection), operating at an ambient pressure of 5 psia normally for a maximum of four hours. The wearer could also sustain acceleration at up to a maximum of 15 g 'eyeballs in' (forces from the back) for 30 seconds during re-entry, or 40 g 'eyeballs out' (forces from the front) for 1 second during ejection, as well as a shock level of 15 g maximum in each axis at 36.09 feet per second during landing.

Pressure suit assignments

Flight	Command Pilot	Suit number	Pilot	Suit number
GT-3	Grissom	G-3C-1	Young	G-3C-4
GT-4	McDivitt	G-4C-3	White	G-4C-8
GT-5	Cooper	G-4C-10	Conrad	G-4C-15
GT-6	Schirra	G-3C-3	Stafford	G-4C-21
GT-7	Borman	G-5C-5	Lovell	G-5C-6
GT-8	Armstrong	G-4C-24	Scott	G-4C-27
GT-9	Stafford	G-4C-17	Cernan	G-4C-32
GT-10	Young	G-4C-19	Collins	G-4C-36
GT-11	Conrad	G-4C-39	Gordon	G-4C-40
GT-12	Lovell	G-4C-41	Aldrin	G-4C-42

Propulsion The sixteen thrusters in the Adapter Module provided for all attitude and orbital manoeuvring operations. The Orbital Attitude and Manoeuvring System (OAMS) used the hypergolic rocket motors installed in and around the Adapter Module, with fuel, oxidisers and gas pressurisation tanks linked to the thruster motors. The astronauts used a central hand-controller located between the seat stations, which could control either the RCS in the nose of the spacecraft, or the OAMS system in the Adapter Module. Translation control could be accessed via the hand-controller from either crew station. When Walter Schirra visited the McDonnell plant in St Louis in July 1961 he sat in the early plastic mock-up of the two-man spacecraft. He stated that the designers had finally found a place for a left-handed astronaut (on Mercury the hand-controller was on the right of the pilot seat) although when he flew on Gemini 6 he was in the Command seat on the left and had to use his right hand again!

Recovery Although the paraglider technique was intended for use on Gemini, and was tested (see Chapter 9), all Gemini spacecraft that were intended for recovery were recovered by parachuting into the ocean. Had any parachute recovery failed, there was still available the option of ejector seat evacuation. A 10.7-foot drogue parachute was deployed to control descent from 24 miles altitude down to two miles, and then, once stabilised, an 18.3-foot pilot canopy was deployed, which also pulled away the 40-inch long rendezvous and recovery section from the nose of the Re-entry Module. At 1.8 miles, an 84-foot diameter ring-sail main parachute was deployed, and then, supported by a two-leg support structure between the two hatches, the parachute would be reconfigured to a two-point suspension from the nose and just forward of the heat shield, pitching the spacecraft down 35 of horizontal for a more comfortable sitting position entry into the ocean at about 20.4 mph. After splash-down, the parachutes would be jettisoned and recovery aids deployed, including a flashing recovery light, dye markers, life rafts if required, HF and UHF rescue communication and recovery systems, and splash curtains that could be pulled up over the open hatch to prevent flooding of the crew compartment. Emergency electrical supplies were also part of the recovery kit, along with a hoist loop and flotation material to surround the spacecraft.

Rendezvous radar Located in the nose of the spacecraft, the high-definition radar presented the crew with range (distance), bearing (direction and angle of approach) and rate (closing speed) of the spacecraft and target from a distance of 250 miles apart. At a maximum range of 50 miles, the high-intensity flashes of the light beacon on Agena became visible to the crew. By using a combination of radar tracking and visual clues, the crew followed a manually guided approach to arrive at the Agena target. The radar equipment weighed less than 70 lbs, and required less than 80 W of power.

Survival kit In addition to the personal parachute each astronaut had as part of the ejection system, and the spacecraft parachutes (all of which could be used for shelter and signalling in the event of an emergency ground landing), each crewmen had individual life rafts with a CO_2 bottle for inflation, with a sea anchor and dye

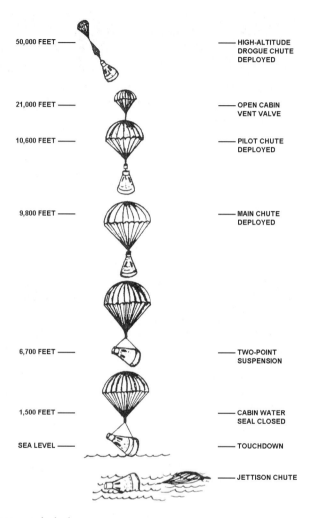

50,000 FEET ——	—— HIGH-ALTITUDE DROGUE CHUTE DEPLOYED
21,000 FEET ——	—— OPEN CABIN VENT VALVE
10,600 FEET ——	—— PILOT CHUTE DEPLOYED
9,800 FEET ——	—— MAIN CHUTE DEPLOYED
6,700 FEET ——	—— TWO-POINT SUSPENSION
1,500 FEET ——	—— CABIN WATER SEAL CLOSED
SEA LEVEL ——	—— TOUCHDOWN
	—— JETTISON CHUTE

The parachute ocean splash-down sequence.

markers. Each 23-lb survival assembly also included 3.5 lbs of drinking water, a Sun bonnet, a small fishing kit with hooks and 14 feet of nylon line, a compass, a sewing kit, a strobe and a flash light, a signalling mirror, a whistle and a lighter. There was also a medical kit, a salt water desalination kit to accommodate eight pints of water, a survival knife, a machete, emergency food rations, and a radio with homing beacon and voice reception.

Waste management One of the constant challenges of any spacecraft design supporting a human crew was to provide a suitable system to collect and store waste products. For the extended duration of Gemini, altering the diet to reduce solid waste, or providing a simple urine collection device would both be impractical, and a new system was therefore devised for Gemini missions. A Crew System Design

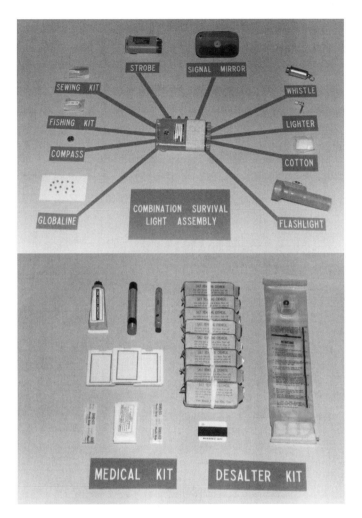

A selection of survival items issued to each astronaut.

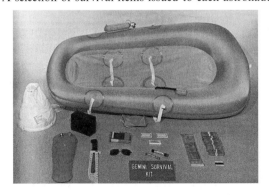

The one-man dinghy and support equipment.

Review was held on 9–10 April 1964, at which prototypes of the systems were reviewed. By December, refinements to these systems led to a four-day comfort test evaluated by a human test-subject using the food system, water collection and bio-instrumentation. Qualification of the combined food, water and waste systems for flight was completed by 29 January 1965.

The Gemini system needed to be reliable, sanitary, and psychologically acceptable to the crew. There were three waste management systems used on Gemini – two for urine and one for solid waste. The first urine collection device was used for pre-launch and launch phases, and was a Y-shaped neoprene-coated nylon bag which fitted inside the suit around the pelvic region. The first of two openings was the inflow receptacle with fitted rubber sleeves – one inside the bag and one outside – which ensured that the urine flowed only one way, into the bag.

Upon reaching orbit, the zipper that extended from the lower front of the suit between the legs to the back was unzipped, and the bag was removed. The rubber sleeve was clamped off with a hose clamp, completely sealing off the device. A second opening was located above the inflow receptacle but on the opposite side. To this, the astronaut connected a hose which was connected to the overboard dump system. Once emptied, the bag was folded and stored.

The pre-launch and launch device could not be worn for long periods in orbital flight, because of discomfort, and because of the need (and inconvenience) to empty it. For orbital periods, the astronauts used a second device – the Inflight Urine Transport System, which was designed to be fitted immediately prior to use and then afterwards removed. The urine was dumped overboard, and the system was then stored until needed again. The Transport System consisted of a urine receiver with a one-way valve and a small wire mesh to prevent any solid particles from flowing into the system. A manually operated shut-off valve also served as a preventive against urine back-flow. The rectangular urine collection bag was made of neoprene–nylon with a 37.18-fluid ounce urine sampler valve, allowing a 2.5-fluid ounce sample to be drawn off via a tube, for medical purposes. There was also a quick-disconnect fitting.

An aluminium tee connected to the bag, one end of which was connected to the roll-on rubber receiver and shut-off valve, while the sampler valve (also used to empty the device) and quick-disconnect fitting was located on the other end. The system could also feature the Chemical Urine Volume Measurement System (CUVMS) as an integral element, to measure the volume of urine and in sampling for post-flight analysis, by inserting a known amount of tracer chemical into the urine. The quick-disconnect fitting was attached to a hose for overboard dumping controlled by manual valves. When emptied, the system was disconnected from the overboard dump, and was stored until required.

Faecal waste management was more of a problem on a flight, as it could not be flushed overboard. The solid waste had to remain on board for the duration of the mission, which created a storage as well as a sanitation problem. Collection in microgravity is also a problem, especially if the waste is of a diluted consistency. Gemini astronauts used a lightweight nylon–polyethylene laminated plastic collection bag, measuring 7 inches wide at the base and 12 inches in length, with a 4-inch circular opening at the top of the bag. This opening included a $1\frac{3}{4}$ -inch flange of wax paper-covered adhesive surgical tape that attached to the astronaut's body. Each crew-member was provided with one bag per flight day, plus 20% excess for contingency.

Use of the bag began with the insertion of a germicidal pouch inside the collection bag. The protective wax paper was then removed, and the collection bag was firmly sealed against the buttocks. A finger thimble was built into the side of the bag near the opening for correct positioning of the bag. After use, the bag was removed from the body. Used toilet tissues were also inserted into the bag, and the top opening was sealed by folding the seal lip in half against itself. The germicide pouch was then ruptured by external pressure, and its 0.35-fluid ounce liquid content mixed with the faecal material to prevent build-up of gas during storage. The used and sealed bag was then stored in a waste product storage container.

LAUNCH VEHICLES

For Gemini, two launch vehicles were employed: Titan II for the manned launches, and the Atlas booster as the target vehicle.

Titan II

The two-stage launch vehicle selected for Gemini was chosen for its simplified operation, greater lifting capability and availability at the time of selection. Standing 90 feet high, with a Gemini capsule on top it reached 110 feet. The widest diameter was 10 feet, and it had a launch weight of 340,000 lbs.

Stage one measured 63 feet in length and 10 feet in diameter, and was powered by two Aerojet-General LR87AJ7 engines. These were fuelled by nitrogen tetroxide (N_2O_4) and Aerozine 50, with a half-and-half combination of hydrazine and unsymmetrical dimethyl hydrazine (UDMH). The total thrust was 430,000 lbs.

Stage two was 27 feet long and also 10 feet in diameter, with a single LR91AJ7 engine using propellant of nitrogen tetroxide (N_2O_4) and Aerozine 50, producing a thrust of 100,000 lbs.

The use of storable hypergolic (self-igniting) propellants allowed a much shorter count-down period than experienced with Mercury Atlas, which was crucial in achieving the tight launch schedule in the time period allowed. The major changes from the Titan I ICBM were in the use of the hypergolic propellants on Titan II, instead of the liquid oxygen and kerosene (RP-1) used on Titan I.

Man-rating Titan II Before allowing two astronauts to sit on top of a Titan, it had to be made safe enough (man-rated) to ensure its suitability to carry a human cargo. Man-rating encompassed a broad and rigid series of disciplines including aspects of both technical and non-technical issues from construction to launch. Martin Marietta defined Gemini Titan man-rating as 'The philosophy and plan for marshalling the disciplines necessary to achieve a satisfactory probability of crew

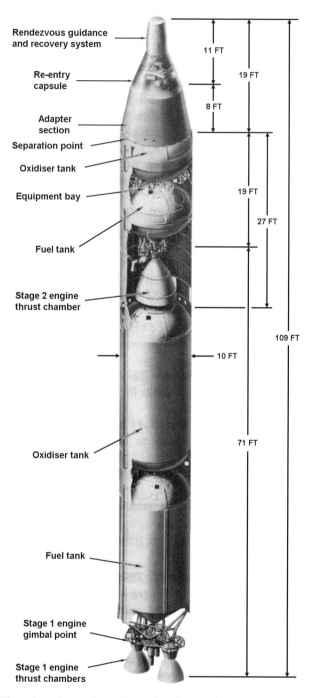

Rendezvous guidance and recovery system

Re-entry capsule

Adapter section

Separation point

Oxidiser tank

Equipment bay

Fuel tank

Stage 2 engine thrust chamber

Oxidiser tank

Fuel tank

Stage 1 engine gimbal point

Stage 1 engine thrust chambers

11 FT

19 FT

8 FT

19 FT

27 FT

109 FT

10 FT

71 FT

The Gemini Titan launch configuration, showing major components and vehicle dimensions.

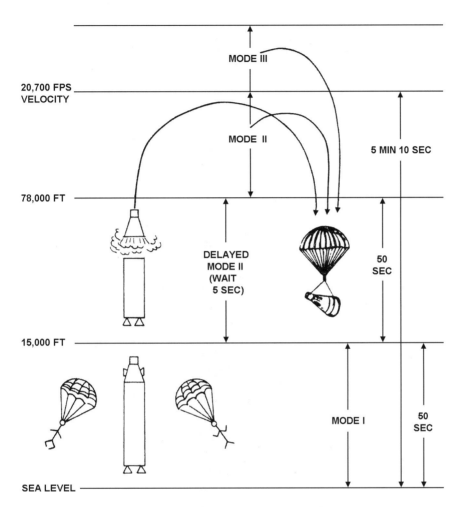

The three Gemini Titan II abort modes.

safety and survival [which] may be expressed in terms of reliability of the launch vehicle and the reliability of the Malfunction, Detection and Escape System.' That is, during the boost phase of a Gemini mission, astronaut decision-making and control was limited to abort-type problems, and not in thrust control, propellant management, or guidance control effecting a chance of survival.

In order to achieve man-rating, consideration was given to component and system redundancy and analysis of the Titan failure modes, followed by the design of a reliable Malfunction Detection System (including the development of the astronaut's personal escape system). Further consideration was given to the use of the crew as part of the MDS, and to improvements on launch vehicle check-out (to reduce the probability of launching 'a bad vehicle'), trading off a complex system against reliability and improved test, count-down, and launch procedures to maximise the chance of launching 'a good vehicle'.

Titan II failure mode analysis

System	Relative failure probability (%)	Possible application of redundancy
Engines	33.7	No
Hydraulics	21.2	Yes
Flight controls	18.5	Yes
Propellant and pressurisation	13.7	No
Guidance	4.8	Yes
Electrical	4.3	Yes
Ordnance	2.8	Already redundant
Structure	1.0	No

Martin conducted a comprehensive analysis of the Titan to determine exactly what modifications would be required to allow astronauts to ride a reliable Titan II with a given degree of safety. After reviewing the probability of failure of each system, and by applying current redundancy, investigation of past failures and investigations of the vehicle, the comparison was made between an automatic versus a manual malfunction detection and abort system, the concept of an escape system, what role the astronaut should play, and what decisions he should make as an integral part of this system. (The results of the analysis appear in the table above.)

A failure in one or more of the first six systems would result in loss of thrust or thrust vector control. This could be caused (amongst other things) by single and both engines hardover (gimballed), drifting engines, or loss of attitude signals. In analogue tests, 150 simulated launches of a Titan II were programmed, flying normal ascents through stage one (aside from a variable wind profile and inserted malfunctions). The results of these ascents, and the data from varying malfunctions along the trajectory, yielded further data on the ascent time history and conditions where structural and crew safety could be exceeded. It was determined that the crew could experience a physiological hazard due to excessive turning rates, or the break-up of the vehicle due to structurally excessive loading caused by high angle of attack, diverging from the prescribed profile. During stage two boosted flights it was determined that crew safety was at risk from excessive pitch and yaw.

The study also revealed that in the event of an ascent engine hardover, the malfunction would place physical limits on the crew within just 1 second. If a problem occurred during high dynamic pressure (75 seconds after lift-off), the structural limits of the vehicle would be exceeded in just 2 seconds. The warning time that the crew would have to initiate ejection would therefore be shorter than this. Medically, the tests also evaluated that a maximum of 6 g 'eyeballs out' or 11 g 'eyeballs in' could be endured before serious medical damage occurred. Finally, in addition to studying the flight profile, engine performance, and astronaut rescue, the tests also included the levels of pressurisation from the supply tanks prior to structural failure. The data were fed into the onboard Malfunction Detection System.

Malfunction Detection System (MDS) Perhaps the most important system for astronaut safety on Gemini, from lift-off to orbit insertion and spacecraft separation, was the MDS, which constantly monitored the vehicle's performance during operation. Warning signals on the crew display panel (in the form of flashing lights and meters) enabled the crew to react and use the escape system. Having experience of the automated system for abort sensing and implementation on Mercury, and using the results of failure analysis on Titan flight histories and computer studies, the system used on Gemini was manual. Selected performance parameters were presented to the flight crew on the spacecraft display console for their decision and action. The parameters used were:

- Engine chamber under pressure (low thrust) of stage one.
- Engine chamber under pressure (low thrust) of stage two.
- Oxidiser tank pressure, stage one and stage two.
- Fuel tank pressure, stage one and stage two.
- Over-rates in turning in pitch, yaw and roll.
- Staging command signal received (and actual separation occurrence).
- Switch-over to redundant flight control system.

The system was active from lift-off through to the separation of the spacecraft, and included back-up circuitry for redundancy (which also provided a second system and prevented false malfunction signalling). The Titan therefore featured two MDS systems. To test the systems, several of the Mercury astronauts participated in simulations of possible failure modes, testing their response to the malfunction and the speed of their decision to determine whether or not to activate the escape system. It was soon found that a one-second warning – for MDS detection, instrument panel display, observation, reaction, escape activate, and escape implementation – was obviously insufficient. A redundant guidance and flight control system was incorporated into the vehicle, and Gemini therefore flew with two complete guidance and flight control systems. This meant that if a failure was expected in the primary system, switch-over was achieved automatically in 15 milliseconds, notifying the crew of a pending problem and, if required, initiating the ejection sequence. The system could also be switched manually, allowing more time for an accurate crew decision if the data indicated an erroneous signal.

Atlas–Agena D

This was a two-stage vehicle, with the second stage also serving as the docking target. Including the launch shroud, the vehicle stood just over 113 feet tall. The first stage was the Atlas Standard Launch Vehicle (SLV), derived from the USAF Atlas ICBM, and similar to that used to orbit the Mercury astronauts in 1962 and 1963.

Stage one (Atlas) was 77 feet long and 10 feet in diameter and was powered by two Rocketdyne LR89NA7 and one Rocketdyne LR105NA7 engines, offering a combined thrust of 389,000 lbs (2 x 160,000-lb 'boost' engines and a single sustainer engine of 57,000 lbs, plus two smaller vernier engines for trajectory and final velocity control). The propellant was liquid oxygen and kerosene (RP-1), and the total weight of the stage was 260,000 lbs.

The launch of the Atlas–Agena from LC 14 at the Cape on 17 May 1966. The target vehicle was intended for Gemini 9, but was lost in the launch failure which occurred only a few minutes after this photograph was taken.

Stage two was 27 feet 7 inches long and 5 feet in diameter (and with the launch shroud was increased to 36 feet 3 inches long). A single Bell 8096 engine with propellant of inhibited red fuming nitric acid (IRFNA) and UDMH provided a maximum thrust of 16,000 lbs. The secondary propulsion system featured 2 x 200-lb (Unit II) and 2 x 16-lb (Unit I) thrust engines, with a propellant of mixed oxides of nitrogen (MON)/UDMH. The total fuelled weight in orbit was 7,000 lbs.

The Agena was chosen due to its versatility and its proven use in many previous missions. It had been developed from a family of launch vehicles that had propelled Ranger spacecraft to the Moon and early Mariner spacecraft to Venus and Mars.

TARGET VEHICLES

Agena D
At the forward end of the propellant stage was the docking target, covered by the launch shroud. By using the onboard propulsion system, Agena could be boosted to

higher-inclination orbits, as well as being used as an experiment carrier and second vehicle for tethered operations. Modifications made to accept it as a target vehicle for Gemini included adding the docking adapter and equipment to permit a mechanical connection with the spacecraft; a radar transponder compatible to that on Gemini; a display and instrumentation panel; acquisition lights; a secondary propulsion system for small orbital changes; an auxiliary equipment rack; command and control equipment to allow either the astronauts on the Gemini or the ground controllers to control the vehicle; multi-start engines to provide in-orbit manoeuvring capability; and three protruding copper 'fingers' installed on the Agena docking cone to serve as static charge conductors upon first contact with the spacecraft. Any charge would be carried to a ground inside the Agena and then dissipated at a controlled rate. In the docking adapter an electrostatic charge-monitoring device was installed to measure the potential of difference in the charge between the Agena and the Gemini.

The Augmented Target Docking Adapter (ATDA)
When development problems plagued the Agena target, a back-up target was constructed in the event of a primary Atlas–Agena vehicle loss. The ATDA was the same docking target, but without independent propulsion, and so no manoeuvring or high-altitude missions could be attempted while the Gemini was docked. At the rear was located a Gemini rendezvous and recovery section which produced an overall length of 10 feet 11 inches and a diameter of 5 feet. Launched on an Atlas D, the overall height was 95 feet 6 inches. The weight at launch was 2,400 lbs, and in orbit was 1,700 (without the shroud).

GROUND SUPPORT

In addition to the development of the spacecraft, targets, and its launch vehicles, Gemini required an infrastructure of management, mission control, and launch and recovery operations:

Gemini mission management
Gemini was managed by the Manned Space Flight Center in Houston, Texas, under the direction of the Office of Manned Spaceflight, at NASA Headquarters in Washington DC. For most of the duration of the programme the Associate Administrator for Manned Spaceflight, George Mueller, also served as Gemini Program Director, assisted by Samuel Hubbard and by various departmental directors for programme control, system engineering, testing, flight operations, and reliability and quality.

The Gemini Program Office opened on 3 January 1962. The first Program Manager was James Chamberlin, who had worked on the early designs of Mercury Mark II at McDonnell. On 19 March 1963, when several hardware problems pushed the programme over budget and behind schedule, Chamberlin was relieved as Program Manager by MSC Director Robert Gilruth, who replaced him with Charles

Mathews, initially as Acting Manager but from 22 December 1963 as the new Program Manager. Mathews set about reorganising the programme office and he and his deputy, Kenneth Kleinknecht, were supported by a team of managers for program control, spacecraft management, vehicle and missions. In addition, personnel from MSC monitored the progress at the major contractors: at McDonnell in St Louis (Gemini spacecraft), at Martin Company in Baltimore (Titan II), and at Lockheed in California (Agena target vehicle), as well as launch operations at the Cape in Florida.

The GPO supported a mid-programme conference at MSC during 23–24 February 1965 to discuss the progress to date and prospects for the future. The Gemini Program Office was officially abolished on 1 February 1967, after five years of operations; and on the same day, a two-day summary conference (1–2 February) was held at MSC to discuss the results of the programme following the end of flight operations. (Gemini management is listed in the table on p. 63)

The Manned Spacecraft Center, Houston, Texas
In September 1961, the site for the Manned Spacecraft Center was selected: near Clear Lake, about 30 miles south-east of Houston. This location has become known as the major NASA facility for the selection, training and preparation of flight crews from Gemini to the current International Space Station. The site was renamed Johnson Spaceflight Center after the death of President Lyndon B. Johnson in 1973, but until then it was known around the world by its call-sign 'Houston'. The design of the campus-like centre began in December 1961, and construction began in April 1962. By September 1963, the first permanent facilities were completed, and the initial personnel moved in during the following month. The majority of the major complex was finished by April 1964. There were 1,620 acres of land available at MSC, of which 85% was allocated for future facility development.

Initially, during the Gemini years of 1963–1966, MSC featured 6,900 feet of concrete tunnels containing steam, chilled water and compressed air piping. A central heating and cooling plant provided 12,000 tons of air conditioning and 240,000 lbs/hour of steam as required. There was also a sewage treatment plant with a capacity to treat 140,000 gallons per day, a water treatment facility for a potable water supply, a fire station, a medical facility, a thermo-chemical test area for developing and testing small scale propellants and explosive devices, a garage for ground vehicle maintenance and storage, logistics support buildings, a support shop and warehouse facility, numerous car parks, and several staff and visitor canteens. Several of the main buildings have the capability to withstand hurricane-force winds Key buildings related to Gemini were:

Building 2 Project Management Building, providing office space (194,191 square feet on nine floors) and facilities for the Center Director, Deputy Director, Assistant Directors for Administration, flight operations, flight crew operations, engineering and development). The Gemini programme office was located on the fifth floor. This is currently known as Building 1.

Gemini management

NASA Headquarters Washington DC

Administrator	James E. Webb
Deputy Administrator	Robert C. Seamans Jr
Associated Administrator for Manned Space Flight	Dr George E. Mueller (serving as Acting Director of the Gemini programme)
Deputy Director of the Office of Manned Space Flight for Mission Operations	William C. Schneider (served as Mission Director for all Gemini flights, beginning with Gemini 5)

Manned Spacecraft Center, Houston, Texas

Director	Dr Robert R. Gilruth
Program Manager	Charles W. Mathews
Flight Director	Christopher C. Kraft
Director of Flight Crew Operations	Deke Slayton
Chief Astronaut	Alan B. Shepard
CB Gemini Branch Chief	Gus Grissom (1963 to September 1965)
	Gordon Cooper (September 1965 to February 1966)
	Thomas Stafford (from February 1966)

Kennedy Space Center, Cape Canaveral, Florida

Director	Dr Kurt Debus
Deputy Mission Director for Launch Operations	G Merritt Preston

Department of Defense

Lt Gen. Leighton I. Davis, USAF National Range Division	Commander and Manager of Manned Spaceflight Support Operations
Maj Gen V.G. Huston, USAF	Deputy Manager
Col Richard C. Dineen USAF Systems Command	Director, Directorate Gemini Launch Vehicles, Space Systems Division,
Lt Col Jon G. Albert	Chief, Gemini Launch Division 6555th Aerospace Test Wing, Air Force Missile Test Center, Cape Kennedy, Florida
Rear Admiral B.W. Sarver, USN	Commander, Task Force (Recovery) 140

Building 1 Auditorium Building, adjacent to the Project Management Building. This was divided into two parts. The front portion was an 800-seat auditorium and lobby for public presentations and displays. This was the primary visitor centre and display area until the mid-1990s, when Space Center Houston was opened on an adjacent lot at the edge of the site on NASA Road 1. At the rear was the one-storey building for Public and Media services, including the press conference room where most of the Gemini press conferences were held. This is currently known as Building 2.

Building 4 was a three-storey office and laboratory facility that housed the crew training and support divisions, and also the location of the Astronaut Office. It was

The Manned Spacecraft Center, Houston, Texas. This 1,600-acre site is located near Clear Lake (top right).

Computers play a vital role in the activities of MSC. The Computation and Analysis Division used these machines to support the Gemini missions and many other activities.

not just one 'office', but a series of smaller offices where three or four astronauts would work between crew assignments. When assigned to a crew, the members would team up in one office to prepare for their mission. Currently (2001) there are two buildings – 4 and 4 South – as the astronaut group has expanded to more than 100 active members, including representatives from other nations, besides staff.

Building 5 housed the Gemini mission simulators used for astronaut training. *Building 7* was the Life Systems Laboratory, where the Crew Systems Division tested and developed spacesuits and life support systems. *Building 10* was the Technical Services Shop that provided craftsmen and facilities to construct wood, plastic and metal spacecraft mock-ups and three-dimensional prototypes for experiments, locations of crew compartment, guidance equipment, and other flight equipment for simulation, training and implementation. *Building 12* was the location of the Central Data Office, and *Building 45* housed the project engineering facilities.

To the north-west of the MSC site, on Highway 3, was the Ellington Air Force Base used for storage and maintenance of a fleet of T-38 jets. Used by the astronauts to fly around the country to conduct public appearances, and to visit contractors and other NASA sites, these aircraft also maintained their required monthly pilot proficiency levels. Ellington was also the location of a large water tank, used for egress training and housed in *Building 260A*.

Technical Services Shop, Building 10, MSC. This facility produces wood, plastic and metal spacecraft mock-ups for the evaluation of equipment location, experiment integration, crew training and future hardware development. Note the Gemini mock-up in the aisle (centre right).

Building 30: Mission Control Center, Houston (MCC-H)
Perhaps the most recognisable (at least in name) building on the MSC site was, and still is, Mission Control, located in Building 30. In 1964 this was described as 'a focal point to all Americans as well as the rest of the world in years to come' – and this proved true. Most of the world is aware of Mission Control Houston, which took command of its first mission – Gemini 4 – in 1965, and has assumed the control of every subsequent American domestic manned spaceflight up to the current ISS operations. The original control building operated through the Apollo, Skylab and ASTP programmes and on to the Space Shuttle until the mid-1990s, when a new mission control was opened for the Shuttle. This was later supplemented by a Space Station mission control, all in the adjacent and expanded building.

Mission Operations and Control Room (MOCR)
A team of flight controllers manned rows of consoles during training, simulation and actual spaceflights, supported by teams of specialists in adjacent rooms, and a duplicate control room. They provided support in areas of recovery, control and communications, meteorology, trajectory, network support, simulation, flight planning, monitoring of life support, crew and vehicle systems, and a vast computer and communication systems.

Gemini Mission Operations Control Room (MOCR)
This was located on the second and third floor of Building 30 (see the illustrations below and opposite).

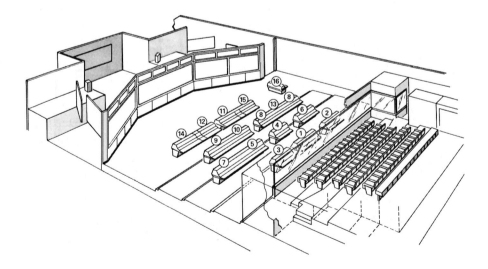

Mission Operations Control Room, Building 30, MSC. Mission Control Houston during the Gemini programme.

A familiar view of Mission Control, during what appears to be a change of shift for the flight controllers.

Mission Control, showing the controllers manning their consoles during a mission. This facility was first used operationally for Gemini 4 in June 1965, and continued to be used into the 1990s.

1   *Mission Director* held overall responsibility for test operations. He also was responsible for mission scrubs and rescheduling flights, as well as making real-time decisions on alternative flight plans when problems arose.

2   *Department of Defense Representative* (DoD) was the focal link between NASA MCC and the worldwide DoD support team. This included the deployment of recovery forces, the operation of the recovery communications network, and the search, location and recovery of the flight crew.

3   *Public Affairs Officer* (PAO) was the 'voice' of Mission Control, providing status information to the public over the open vocal communications loop during aspects of missions operations, from just prior to launch, to crew recovery at the end of the mission.

4   *Flight Director* (FD) was the 'Boss' of Mission Control – called 'Flight'. He was responsible for the operational control of the mission from lift-off to splash-down, with ultimate responsibility for mission success and crew safety. He was also *locum* for the Mission Director when necessary. For Gemini, the first Flight Directors were assigned in August 1964 to join Chris Kraft, who had shared flight control responsibility on Mercury missions with the now retired Walt Williams. With flights lasting several days, rotational teams of controllers would be required, and colours were adopted to identify each team. The FD took a colour for his entire career in MCC, and it was then 'retired' when the FD retired. The 'original' Flight Directors were Chris Kraft (Red), Gene Kranz (White), John Hodge (Blue), and Glynn Lunney (Black). Kraft moved over to Apollo after Gemini 7, while Hodge transferred to Apollo flight control preparations after Gemini 8 and was replaced by Cliff Charlesworth (Green). Kranz transferred after Gemini 9, and all three worked on Apollo 1. Lunney and Charlesworth handled the last three Gemini flights as Flight Director, and supported by other FDs as and when necessary.

5   *Assistant Flight Director* assisted the Flight Director during the mission, and acted as FD in his absence.

The first four Gemini Flight Directors: (*front left*) Gene Kranz (White Team), (*front right*) Chris Kraft (Red Team), (*rear left*) Glynn Lunney (Black Team), and (*rear right*) John Hodge (Blue Team).

6 *Network Controller* (Network) was responsible for detailed operational control of the Ground Operational Support System (GOSS) network.

7 *Operation and Procedures Officer* (Ops) was the link that handled detailed operational control and implementation of MCC and GOSS procedures.

8 *Vehicle Systems Engineer* (Systems) was the flight controller responsible for the monitoring and performance evaluation of all electrical, mechanical and life support equipment onboard the Gemini, and also, during the rendezvous and docking missions, onboard the Agena.

9 *Flight Surgeon (Surgeon)* was responsible for the direction of all medical activities, and for monitoring the crew's health during the flight.

10 *Spacecraft Communicator (Capsule Communicator, or CapCom)* was the person who conducted most of the direct voice communication with the crew during the mission. Acting as a point of contact between the spacecraft and MCC, this position was held, by tradition, by a fellow astronaut. The word CapCom derived from the days of Mercury, during which astronauts flew space *CAP*sules (although the astronauts disliked that term, and insisted that they be referred to as manned space*craft*), and those who talked to them were termed *COM*municators. (For Gemini CapCom assignments, see p. 124)

11 *Flight Dynamics Officer (Fido)* monitored and evaluated the flight parameters required to achieve a successful orbital mission, and gave the relevant Go/NoGo recommendations to Flight as the mission progressed.

12 *Retro-fire Office (Retro)* constantly monitored and updated retro-fire times, managing impact prediction displays and updating the options for retro-fire during the mission for both contingency and nominal end of mission operations.

13 *Guidance Officer (Guido).* During the ascent of the Titan, this controller monitored Stage 1 and Stage 2 booster deviations, as well as programmed operations. During the missions he verified nominal performance of the Gemini Internal Guidance System, and recommended appropriate actions to the FD.

14 *Booster Systems Engineer (Booster).* During ascent, this flight controller monitored the propellant tank pressurisation systems, advising both the flight crew and the Flight Director of system anomalies in order to prepare for abort if required.

15 *Experiments Officer (Experiments)* replaced the Booster controller after separation of the spacecraft, was responsible for monitoring and updating experiments prior to re-entry, and also acted as a link to the experiment support teams and DoD PIs.

16 *Assistant Flight Dynamics Officer* monitored and evaluated Gemini launch vehicle systems and reported any abnormalities to Flight.

17 *Maintenance and Operations Supervisor* was responsible for monitoring MCC equipment and the ability of the system to support the mission in progress, and was to ensure that if any problem surfaced, it was solved as soon as possible.

Staff Support Rooms were adjacent to both the MOCR facilities on the second and third floors of Building 30. In these rooms, technicians, specialists and other flight controllers would analyse data and project the long-term performance of systems, and would compare them with baseline data to reflect systems projections constantly throughout the mission. This information was relayed to MOCR controllers on each console in six main areas:

Flight Dynamics studied aspects of the boosted phase of the mission from launch to orbital insertion that affected crew safety. They also recommended refinements to trajectories to meet mission objectives, and constantly reviewed future manoeuvring requirements for both a nominal and contingency situation.

Vehicle Systems evaluated the trends recorded on the various systems onboard the spacecraft in an effort to avoid corrections or to work around potential systems failure.

Life Systems support team used telemetry from the spacecraft to monitor and evaluate the physiological and environmental aspects of the crew and the crew compartment.

Flight Crew co-ordinated non-medical aspects of flight crew activities, including the control of the spacecraft and the operation of assigned experiments.

Network operated a programme of ground station scheduling, monitoring, and direction of all tracking stations around the globe. This included readiness checks prior to the spacecraft entering the range of a specific station (by verifying remote site pre-pass equipment checks), and directing the handover from one station to another as the spacecraft progressed around the globe.

Operation and Procedures Technical and administration. This group provided support for all mission plans and procedures and MCC communication plans and procedures, documenting changes, and notifying all network stations and each flight controller in order to keep the crew up to date as conditions changed in training, simulation and real time.

Cape Kennedy (Canaveral), Florida

Cape Canaveral (the Cape), near the Banana River, was a barren promontory on the Atlantic coast of Florida when it became the site of a missile launching range in 1949. During the 1950s the USAF established launch facilities there, and from 1951 this became the Air Force Missile Test Center (AFMTC), and the test area stretching out across the Atlantic Ocean became known as the Atlantic Missile Range (AMR) . After NASA was formed in 1958 it established the Launch Operations Directorate and began to construct its own facilities on a place called Merritt Island, and liaised with the AFMTC in using DoD facilities for its own launches. This co-operation extended to the use of LC 19 for Titan, and LC 14 for Atlas Agena launches from the AMR within the Gemini programme, as well as other facilities.

On 28 November 1963, less than a week after the assassination of President Kennedy, President Johnson announced that the new facilities at Merritt Island would be known as the JFK Space Center, and that Cape Canaveral would be renamed Cape Kennedy. By May 1964 the AFMTC became the Air Force Eastern Test Range, and the Atlantic Missile Range the Eastern Test Range. These names

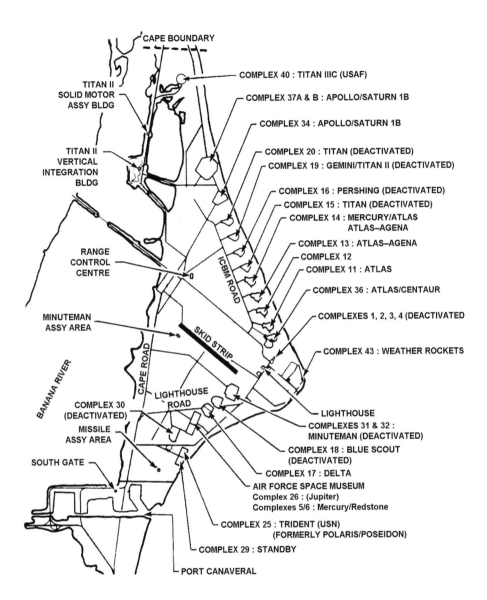

CAPE BOUNDARY

TITAN II
SOLID MOTOR
ASSY BLDG

TITAN II
VERTICAL
INTEGRATION
BLDG

RANGE
CONTROL
CENTRE

ICBM ROAD

MINUTEMAN
ASSY AREA

BANANA RIVER

CAPE ROAD

SKID STRIP

LIGHTHOUSE
ROAD

COMPLEX 30
(DEACTIVATED)

MISSILE
ASSY AREA

SOUTH GATE

COMPLEX 40 : TITAN IIIC (USAF)

COMPLEX 37A & B : APOLLO/SATURN 1B

COMPLEX 34 : APOLLO/SATURN 1B

COMPLEX 20 : TITAN (DEACTIVATED)
COMPLEX 19 : GEMINI/TITAN II (DEACTIVATED)

COMPLEX 16 : PERSHING (DEACTIVATED)
COMPLEX 15 : TITAN (DEACTIVATED)
COMPLEX 14 : MERCURY/ATLAS
ATLAS–AGENA
COMPLEX 13 : ATLAS–AGENA
COMPLEX 12
COMPLEX 11 : ATLAS

COMPLEX 36 : ATLAS/CENTAUR

COMPLEXES 1, 2, 3, 4 (DEACTIVATED

COMPLEX 43 : WEATHER ROCKETS

LIGHTHOUSE
COMPLEXES 31 & 32 :
MINUTEMAN (DEACTIVATED)
COMPLEX 18 : BLUE SCOUT
(DEACTIVATED)
COMPLEX 17 : DELTA
AIR FORCE SPACE MUSEUM
Complex 26 : (Jupiter)
Complexes 5/6 : Mercury/Redstone

COMPLEX 25 : TRIDENT (USN)
(FORMERLY POLARIS/POSEIDON)

COMPLEX 29 : STANDBY

PORT CANAVERAL

A 1972 map of the Cape Canaveral Air Force Station, showing the major launch complexes used during the previous decade at the dawn of the American space programme. LC 19 for Gemini Titan, and LC 14 for Atlas Agena are shown. (LC 39, developed for Apollo, is to the north, and is not on this plan.)

remained for the duration of the Gemini programme. In 1973 the name of Cape Canaveral was reinstated, but the NASA facilities on Merritt Island retained the name JFK Space Center.

Inside the launch control blockhouse at LC 19 at the Cape, contractor and NASA technicians monitor the final stage of the launch as the count-down moves towards zero and lift-off.

Blockhouse This was the Launch Control Center at the Cape for the preparation, count-down and launch of the Agena and the Titan. During preparations for each mission, a team of controllers and contractors monitored the progress towards launch. Several astronauts also participated in this process as monitors or a CapCom ('Stoney') with the crew before launch, until Houston MCC took over control of the flight from the moment the vehicle cleared the tower.

The original blockhouse was a uniform igloo-shaped building located about 750 feet from the pad. It housed communications, control consoles and various instruments as well as providing protection for the launch crew at launch. Excavated out of the ground, the hole was lined with sand, and supported reinforced concrete flooring and walls. These were poured in two layers with a cushioning of sand in between to absorb the shock of launch or an exploding vehicle. As with LC 14 for the Atlas launch, the inside walls of the twelve-sided building were 10.5 feet thick at the base, with 40 feet of sand around them and a retaining wall to hold the sand in place. At the apex of the dome was an inner wall, 5.5 feet thick, with a 7-foot barrier of sand between the two walls. The internal 60-foot diameter floor space was filled with the launch consoles, displays, relays, cables and other elements required to lift the vehicle off the pad.

Launch Complex 19 When the Titan II was assigned to Gemini, it was required that the conversion of LC 19 to accept a Titan II would be completed by August 1963. Initially it seemed fairly straightforward to change facilities from one version of the missile to another, but it was to become one of the most ingenious face-lifting projects of the early space programme.

The initial task was to remove the fuelling system for the Titan I (liquid oxygen and kerosene) and replace it with the hypergolic fuel used by the Titan II. The next problem was that the launch tower used for the Titan I was too short (with the missile standing 98 feet tall, compared with the Titan II /Gemini at 108 feet). At the top of the new structure, a 'white room' also had to be installed for access to the upper stage and the spacecraft. This involved the addition of 28 feet to the original height but (and here was the challenge) the extended tower could weigh no more than the original tower, as it had to be swung from a horizontal to a vertical position, and a top-heavy tower could not be raised.

The US Army Corps of Engineers, which carried out the conversion, had to cut off the top 19 feet of the erector and replace it with both a four-storey white room structure and an elevator and crane at the top, without increasing weight or losing structural strength. The key to achieving this was in the extensive use of aluminium, which was critical in weight-saving measures. Even so, every piece of the new structure, including every rivet, was pre-weighed before assembly.

The construction method was similar to that used for aircraft, and featured a stressed-skin principle, which took some of the stress of the whole structure and become an integral part of the central framework. The 5-ton lift that hoisted each Gemini to the top of the Titan was made of aluminium, and ran along aluminium rails. Aluminium was also used in the rolling access door enclosing the white room, and in most of the metalwork on the work access platforms to the Titan II. The elevator had to be made of steel, but to reduce weight it ran on a single rail instead of the normal two rails.

At the completion of the conversion programme, the erector at LC 19 had

LC 19 at the Cape. (Courtesy NASA/British Interplanetary Society.)

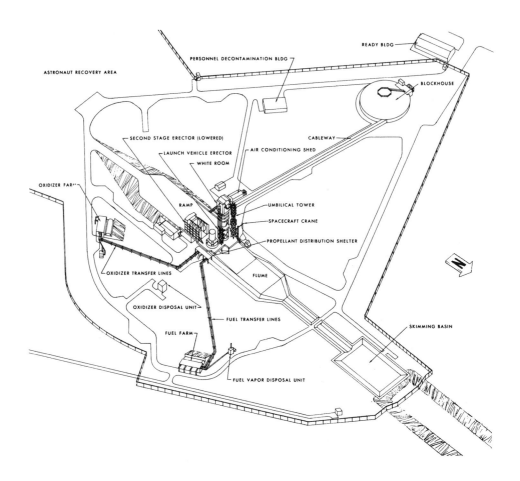

The key facilities at LC 19 (see the previous photograph). (Courtesy British Interplanetary Society/NASA.)

increased by 110 feet to 138 feet, including the 25-foot square, 47-foot high White Room. Combined with the initial cost of LC 19, the modifications exceeded $4 million (1963). The pad was demolished in the 1970s.

LC 14 Atlas Agena

As with all early missile facilities, a 'launch complex' derived from test facilities was used in the missile programme. Each complex consisted of a missile check-out and assembly area, access to the launch pad, the pad itself (designed to meet the thrust and flame characteristics from the missile that it was intended to launch), a gantry service tower, a blockhouse for on-site launch command and control, and a network of power, fuel and communication feeds, and as fire and security devices and facilities.

LC 14 was constructed for the testing of Atlas missiles, and was used during project Mercury for the launch of the four Mercury orbital missions (Glenn, Carpenter, Schirra and Cooper). It was then converted to support the launch of the Atlas Agena for the Gemini missions. In 1972, ceremonies were held to mark the site as the point from where the first American in orbit left Earth. Four years later, in 1976, it had to be blown up when salt wind corrosion was found to have rendered the site dangerous.

Tracking and communications
The network of Gemini tracking stations expanded from the Mercury tracking network. Two types of stations were available. Primary stations were those which could give direct commands to spacecraft systems, while Secondary stations were those used mostly for radar tracking and for receiving telemetry information. The overall management of the network during Gemini simulations and missions was the responsibility of the Manned Spacecraft Center in Houston. The Goddard Space Flight Center handled the planning and implementation, as well as the technical operation of the tracking and data acquisition in response to mission requirements. Astronauts would be located at some of these remote stations to provide support during early missions (up to Gemini 4) as part of their CB technical assignments and mission support duties.

Under a co-operative agreement, the Australian Weapons Research Establishment, Department of Supply, handled the maintenance and operation of the three sites on that continent, while the US DoD provided maintenance and operational control of American facilities required to support Gemini operations.

For the Gemini programme, tracking stations were located as follows:

Primary Bermuda (call-sign BDA) British Island in North Atlantic; Cape Kennedy (CAPE), Florida; Carnarvon (CRO), north-western Australia; *Coastal Century Quebec* (CSQ), a tracking vessel; Corpus Christi (TEX), at Rodd Field, Texas; Grand Canary (CYI), a Spanish-owned island off the Atlantic coast of Africa; Guaymas (GYM), Mexico; Kauai (HAW), Hawaii; Mission Control Center (HOU), at MSC, Houston, Texas; Point Arguello (CAL), Santa Barbara, California, USN operated; and *Rose Knot Victor* (RNV), operated by the AF Eastern Test Range.

Secondary Antigua (call-sign ANT), British West Indies; Ascension (ASC), a British island in the South Atlantic; Canton Island, US/British co-dominion in the Pacific Islands; Eglin (EGL), Air Force Gulf Test Range, Florida; Goddard (GSFC), NASA, Greenbelt, Maryland; Grand Bahamas (GBI), British Bahamas Islands; Grand Turk (GTK), British West Indies; Kano (KNO) northern Nigeria, Africa; Perth (MUC), Western Australia, operated by WRE personnel (it originally used Muchea station, operated during Mercury when equipment arrived late; when activated, it kept the original call-sign); Pretoria (PRE), South Africa; *Range Tracker* (RTK), located in the Pacific Ocean, west of Midway, operated by AF Western Test Range; Tananarive (TAN), Malagasy Republic; Wallops Island (WLP), off the coast of Virginia; White Sands (WHS), north of El Paso, Texas, on the US Army Missile Range; Woomera (WOM), South Australia, operated by WRE personnel on the rocket test facility.

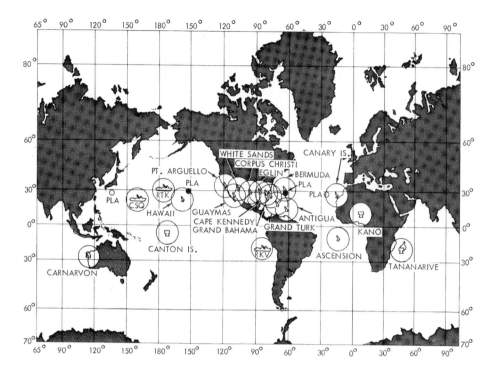

The Gemini tracking station network.

Recovery support

Recovery support was primarily provided by the US Navy in a flotilla of ships and command aircraft. Each mission was planned with two types of recovery area: planned and contingency.

Planned areas were those in which ships and aircraft were pre-positioned to recover both the spacecraft and flight crew within a short period of time and featured two sites each in the western or eastern Atlantic and the western or mid-Pacific. These included a primary landing area in either the Pacific Ocean or Atlantic Ocean, where the primary recovery vessel (an aircraft carrier) was pre-positioned. A secondary area covered another three locations, where other surface ships were deployed.

For launch aborts, two areas were covered. Firstly, for a pad abort or an abort encountered early in flight, up to 41 miles down-range from the Cape and three miles from LC 19 towards Banana River, Florida. In the event of a launch abort later in the ascent, this extended from 41 miles from the Cape to the west coast of Africa.

Contingency landing areas were located on the spacecraft ground track, other than planned landing areas and required aircraft and pararescue support for recovery within 18 hours from splash-down (or touch-down in the event of descent onto land).

All recovery forces were provided by the US military services, and for the duration of the mission were under the control of the DoD Manager for Manned Spaceflight Support Operations.

A typical deployment of recovery vessels included eleven surface ships (up to orbital insertion), with eight remaining on station through to mission completion. Approximately one dozen amphibious and shallow draft vessels and helicopters supported launch operations in the event of a pad or early ascent abort, initially inside a 27-mile footprint from the Cape. In the event of a parachute failure, ejection seats would be used, as in the case of a potential ground landing.

Spacecraft primary contractors

Spacecraft	
Prime contractor	McDonnell Aircraft Corp., St Louis, Missouri
Battery	The Eagle Pitcher Co., Joplin, Missouri
ECS	AiResearch Manufacturing Co., Los Angeles, California
Computer and guidance	IBM Corporation, New York
Parachutes	Northrop Corp. Newbury Park, California
OAMS, RCS	Rocketdyne, Canoga Park, California
Retro-rockets	Thiokol Chemical Corp., Elkton, Maryland
Ejection seats	Weber Aircraft Corp., Burbank. California
Rendezvous radar	Westinghouse Electric Corp.
Crew systems	
Spacesuits	David Clark Co., Worcester, Massachussetts
Principal food contractors	Swift and Co., Chicago; Pillsbury Co., Minneapolis, Minnesota
Food procurement, processing and packaging	Whirlpool Corp., St Joseph, Michigan
Atlas launch vehicle	
Airframe and systems integration	General Dynamics, Convair Division, San Diego, California
Propulsion systems	Rocketdyne Division, North American Aviation Inc., Canoga Park, California
Guidance	General Electric Co., Syracuse, New York
Titan II	
Airframe and systems integration	Martin Co., Baltimore Division, Baltimore, Maryland
Propulsion systems	Aerojet-General Corp., Sacramento, California
Radio command guidance system	Burghs Corp., Paoli, Pennsylvania
Systems engineering and technical direction	Aerospace Corp., El Segundo, California
Agena D	
Airframe and systems integration	Lockheed Missiles and Space Co., Sunnyvale, California
Propulsion systems	Bell Aerosystems Co., Niagara Falls, New York
Target docking adapter	McDonnell Aircraft Co., St Louis, Missouri

Preparations

Apart from the development of the hardware and the supporting infrastructure, there remained the matter of preparing the astronauts to participate in the Gemini programme. NASA had selected only one group of astronauts when Gemini was authorised, and to fulfil its requirements the cadre would need to be expanded. From the resulting group, the flight crews would need to be trained and selected for each mission.

THE SPACEFLIGHT LOGBOOK, 1961

By the end of 1961, when Gemini had received official approval, only two Americans had been in space, and neither of them in orbit. Alan Shepard had flown Mercury Redstone 3 in May, and Gus Grissom had flown Mercury Redstone 4 in July. Each flew only a 15-minute sub-orbital hop down the Atlantic missile range, just weeks after Russian cosmonaut Yuri Gagarin had completed one orbit of Earth in 108 minutes onboard Vostok.

Gagarin's pioneering spaceflight on 12 April had been followed in August by a flight by cosmonaut Gherman Titov, who spent a full 24 hours in space. He returned safely – although a little space sick – to tell the tale. The Americans had cancelled a third manned sub-orbital mission and decided to press for an orbital attempt by John Glenn early in 1962. Other Mercury orbital missions were scheduled, but none were expected to last much more than a day, although there were plans for a possible three-day flight at the end of the programme.

The end-of-year spaceflight logbook for 1961 therefore recorded approximately 27 hours and 18 orbits for the Soviets in two orbital missions, but just 30 minutes for the two American Mercury Redstone flights. In January 1962, therefore, NASA was without a manned orbital flight to its credit, and yet was planning a series of ten manned missions to begin within two years and be completed in less than four years. Each flight would have a crew of two in missions lasting up to two weeks, involving rendezvous and docking with a target vehicle, performing a wide range of scientific and technological experiments, and performing spacewalking activities. With the first manned Gemini flight planned to take place at the end of 1963 or early 1964, it soon became clear that seven Mercury astronauts would not be sufficient, and that NASA would need a new group of astronauts to train for the Gemini missions.

THE MERCURY SEVEN

In January 1962, when Mercury Mark II became officially known as Gemini, the first group of NASA astronauts (selected in April 1959) were all assigned to the remaining flights in the Mercury programme. Mercury Atlas 6 (MA-6) was planned as the first manned orbital flight of the series, with John Glenn as prime pilot. Scott Carpenter was assigned as his back-up (should Glenn be unable to make the flight) and was expecting to be assigned his own orbital flight – probably on MA-8.

These were still early days in the history of flight crew selection, and it soon became evident that even an assignment to a mission was no guarantee that the astronaut would actually fly it. The prime pilot for MA-7, Deke Slayton, had been named for the flight, but was shortly afterwards gorunded due to a heart murmur first revealed in 1959. Instead of being replaced by his back-up Walter Schirra, Scott Carpenter was given the mission, with Schirra taking the prime seat on MA-8 and Gordon Cooper serving as the replacement back-up.

For MA-9, Cooper was assigned to the mission, with Alan Shepard taking his second Mercury assignment – this time as back-up. At the time, there were rumours that MA-10 would be assigned as a three-day 'long-duration' mission in 1963. Shepard would take this flight, with Cooper serving as his back-up, with no prospect of making another Mercury flight. These final back-up assignments at the end of a series of flights, which could not rotate down the line to a prime position, were termed 'dead-end' assignments, as there was nowhere to go from them. The other five Mercury astronauts would be expected to have moved over to Gemini by the summer of 1963.

The first USAF 'space pilots'
In early 1961 (prior to the first Mercury flights) NASA indicated to the US Department of Defense that a further intake of astronauts would be required from the available piloting pool of the USAF and USN. During the year, the services had completed some preliminary medical examinations, and had compiled a list of potential candidates for NASA. Experience in the selection of the first NASA astronaut group in 1959, as well as the expansion of military man-in-space programmes, allowed the USAF to develop a syllabus for preparing military officers as flight crew members or programme managers.

In 1959, two instructors stationed at the USAF Flight Test Pilot School, William Schweikhard and Capt Edward Givens, realised that the Air Force would soon have an operational mission to carry out in space. With the jointly funded NASA/USAF/USN X-15 and the USAF X-20 Dyna Soar programmes in development, the officers believed that an aerospace course should be set up to offer some of the best pilots in America the chance to fly even higher.

Eight members of the ARPS II Class. These were potential candidates for the USAF manned space programmes, especially the MOL. (Courtesy USAF.)

At that time, the Commander of the Test pilot School was Maj Richard Lathrop, who was soon convinced that there was merit in such a course. Lathrop asked the two instructors to join forces with Maj Thomas McElmurry in order to move the idea through the rather slow chain of command. In 1960 a fourth officer, Maj Frank Borman, joined the team. At the time, the USAF had not even officially defined what its manned spaceflight goals should be, let alone allocate a budget.

By late 1960, approval was given for an 'in-house' training course to be developed. One of the problems that the course tutors encountered was in obtaining any sort of space training equipment, and when they did procure it it was spread across the country, forcing significant travel to use it. The lack of adequate funding meant that the number of personnel and the quantity of study material was limited. There were no qualified instructors, and no-one was really sure what form the syllabus should take. To overcome these difficulties, the instructors and early students taught themselves by trial and error.

Class I commenced on 5 June 1961, and consisted of just five students, all former

ARPS did not involve only flying, and there was much theory and academic study. On 20 April 1962 the second group was announced, and on 18 June the course began. (Courtesy USAF.)

graduates of the USAF Flight Test Pilot School: staff member Maj Frank Borman, staff member Maj Robert Buchanan, Capt James McDivitt from Test Operations, staff member Maj Thomas McElmurry (later to work at NASA as Chief of Flight Crew Support, Office of Manned Space Flight) and staff member William Schweikhard. Before the class graduated, the school was renamed the USAF Aerospace Research Pilot School (ARPS) on 12 October 1961, and the course was designated ARPS I.

When the first class graduated, the Air Force began to provide more support for the programme, with up-to-date aircraft (including the F-104 Starfighter) and, more importantly, additional funding. On 14 March 1962, the USAF also announced the names of six test pilots as Space Pilot candidates for the constantly changing X-20 Dyna Soar programme: civilian Neil Armstrong (an X-15 pilot), Maj Henry Gordon, Capt William Knight, Maj Russell Rogers, Milton Thompson and Maj James Wood (a candidate for the NASA Group 1 selection in 1959).

The ARPS II Class was announced on 20 April 1962, and began studies on 18 June that year. Eight candidates were named: Capt Charles Bock, Capt Albert Crews, Lt Cdr Lloyd Hoover USN (who subsequently withdrew from the course and was replaced by Lt Cdr Carl Birdwell, USN), Maj Byron Knoole, Capt Robert McIntosh, Capt Robert Smith, Maj Donald Sorlie, and Capt William Twinting. A ninth candidate, Maj Arthur Torosian, also joined the class after the announcement.

Hoover was subsequently selected as a USN backup X-15 pilot to Forrest Peterson, but never flew the aircraft. Bock became a B-52 carrier plane pilot for many of the later X-15 air launches. Crews replaced Armstrong on the X-20 programme later in 1962, while Sorlie became a lifting-body pilot. Twinting was later a Commander of Edwards AFB, and McIntosh achieved the rank of Brigadier General in the USAF Reserve.

The Next Nine

NASA had decided that any new intake of astronauts for its own programme would be open to both military and civilian test pilots alike, offering a broader scope for selecting the best pilots then available in America. The graduates of ARPS, and the two test pilot schools at USAF Edwards AFB in California and USN Patuxent River, Maryland, had produced test pilots for several years. Many of them had left the service, and were working for civilian aircraft companies, testing prototype aircraft prior to handing them over to the military.

The internal volume of the Gemini (and the improved helmet design) allowed more headroom than had Mercury, and with the even larger Apollo in mind, the candidate height restriction was raised to six feet. The upper age limit was lowered from 40 to 35, since it was expected that some of the astronauts would move over to Apollo following Gemini, probably having a ten-year astronaut career. A degree in biological sciences was an added option, but as before, only US citizens would be considered, and although every applicant was a volunteer, they also had to submit a recommendation from their parent employer or organisation.

On 11 April 1962, NASA officially announced the acceptance of applications up to 1 June. The agency was looking for between five and ten additional astronauts who, after a period of training, would join the Mercury astronauts to become crew-members of the Gemini series. Even with seven Mercury astronauts and ten new astronauts, NASA still expected that by 1964 a third group would be required to fill seats in the early Apollo missions leading to the first lunar landing.

By 1 June, a total of 253 applications had been received. From these, 32 applicants were aeromedically evaluated by the Aerospace Medical Science Division at Brooks Air Force Base, Texas, between May and August 1962. They consisted of thirteen Navy, nine USAF, five USMC and six civilian applicants, as follows (with date of their aeromedical evaluations in brackets):

Neil A. Armstrong, civilian (X-15 and Dyna Soar) (1 August)
Lt Cdr Roland E. Aslund, USN (27 July)
Lt Alan L. Bean, USN (2 August)
Lt Cdr Carl Birdwell Jr, USN (ARPS Class 2 student) (26 July)
Maj Frank Borman, USAF (ARPS Class 1 student) (11 July)
Capt Michael Collins, USAF (12 July)
Lt Charles Conrad Jr, USN (1 August)
Capt Roy S. Dickey, USAF (11 July)
Thomas E. Edmonds, civilian (23 July)
Capt William E. H. Fitch III, USMC (16 June)
John M. Fritz, civilian (2 August)
Capt William J. Geiger, USMC (17 July)
Lt David L. Glunt Jr, USN (18 July)
Lt Richard F. Gordon Jr, USN (17 July)
Orville C. Johnson, civilian (16 July)
Lt Cdr William P. Kelly Jr, USN (19 July)
Lt Cdr James A. Lovell Jr, USN (13 July)

Lt Marvin G. McCanna Jr, USN (10 July)
Capt James A. McDivitt, USAF (ARPS Class 1 student) (18 July)
Lt Cdr John R. C. Mitchell, USN (10 July)
Capt Francis G. Neubeck, USAF (20 July)
Lt Cdr William E. Ramsey, USN (24 July)
Elliot M. See Jr, civilian (24 July)
Capt Robert W. Smith, USAF (ARPS Class 2 student) (25 July)
Capt Robert E. Solliday, USMC (former Mercury candidate) (26 July)
Capt Thomas P. Stafford, USAF (9 July)
John L. Swigert Jr, civilian (25 July)
Capt Alfred H. Uhalt, Jr, USAF (12 July)
Capt Kenneth H. Weir, II USAF (30 July)
Capt Edward H. White II, USAF (30 July)
Lt Cdr Richard L. Wright, USN (31 July)
Lt Cdr John W. Young, USN (31 July)

The 32 were then subjected to written examinations of their scientific and engineering knowledge, and personal interviews by a selection board. The board included Deke Slayton, then Chief Co-ordinator of Astronaut Activities, Mercury astronaut Al Shepard, and former NASA test pilot Warren North (with Crew Systems Division).

There had been some thought given to simply re-evaluating the 25 finalists from the 1959 list who had not been selected, but in the event it was decided to also open up the next selection to new applicants. The first Director of the new Manned Spacecraft Center, Dr Robert A. Gilruth, finally announced the names of nine new astronaut candidates in the Cullen Auditorium of the University of Houston on 17 September 1962.

The nine former test pilots, who have been widely acknowledged as perhaps the best group of astronaut candidates the agency has ever chosen, consisted of four USAF, three USN, and two civilian candidates:

- Neil A. Armstrong, a 32-year-old former US Navy pilot and Korean War veteran who had worked as a NASA test pilot flying the X-1 and X-15 rocket research aircraft and had recently been short-listed for the X-20 Dyna Soar programme by the USAF.
- Maj Frank Borman, USAF, aged 34 and one of the first graduates of the USAF ARPS school.
- Lt Charles 'Pete' Conrad Jr, USN, aged 32 and one of the short-listed candidates from the 1959 Mercury selection.
- Lt Cdr James A. Lovell Jr,, USN, aged 34 and also a candidate for Mercury, although he was not selected due to a minor liver complaint at that time.
- Captain James A. McDivitt, USAF, aged 33, a fellow ARPS classmate of Borman who had been selected as a future X-15 rocket plane pilot.
- Elliot M. See Jr, a civilian test pilot from General Electric.
- Capt Thomas P. Stafford, USAF, selected on his 32nd birthday, who had

Members of the first and second NASA astronaut classes. (*Front, from left*) Group 1 (1959) – Cooper, Grissom, Carpenter, Schirra, Glenn, Shepard and Slayton; (*back, from left*) Group 2 (1962) – White, McDivitt, Young, See, Conrad, Borman, Armstrong, Stafford and Lovell.

been too tall for Mercury, but became a test pilot and authored two textbooks on test flying.
- Capt Edward H. White II, aged 31, who had been involved with all-weather testing at Wright-Patterson AFB, Ohio, and had narrowly missed qualifying as an athlete for the 1952 US Olympic track team.
- Lt Cdr John W. Young, aged 31, not eligible for Mercury as he was a student at test pilot school at the time. Earlier in 1962, however, he had set world time-to-climb records in the F-4 Phantom during 'Project High Jump'.

The group had been privately informed of their success on 14 September, but were instructed to wait for the public announcement three days later. Deke Slayton, by then medically grounded but still hoping for a flight assignment, was named Co-ordinator of Astronaut Activities on the same day, although he had been working in that role since losing MA-7 earlier in the year.

TRAINING THE NEXT NINE

On 1 October 1962 the nine new astronauts reported to MSC, and their first official assignment was to travel down to the Cape to witness the launch of Walter Schirra on Mercury Atlas 8 on 3 October. Gus Grissom had been working on the astronaut procedures training for some time, and was put in charge of the new group – referred to as the Next Nine by the press. To the general public they also became known as 'The Gemini Astronauts'.

Astronaut 'grubby school'

The first phase of the training for the new astronauts was principally academic. During the first six months or so, the nine rookies were joined by the seven Mercury astronauts for a programme of basic science studies. In the first year following their April 1959 selection the Mercury astronauts had received approximately 50 hours of space science lectures:

Elementary mechanics and aerodynamics	10 hours
Space physics	12 hours
Guidance and control	4 hours
Space navigation	6 hours
Communications	2 hours
Basic physiology	8 hours
Star recognition and celestial navigation	8 hours

These were primarily presented by senior members of Langley Research Center – except for the astronomy studies, which were held at the Morehead Planetarium at the University of North Carolina.

Now with the nine newest members, these classes were to be revised at MSC from the end of October and, as well as completing other aspects of training, would continue until February 1963.

Most of the courses lasted two hours each, except guidance and navigation, which lasted four hours each session, and orbital mechanics which lasted five hours each session. Each topic was broken into several sessions over the period (for example, Flight Mechanics I (29 Oct) II (5 Nov) III (12 Nov) IV (19 Nov), and so on) and overlapped other studies. Guidance and navigation occupied the entire academic timetable from October to February. All sixteen astronauts attended all the classes, which usually lasted for six hours each day, and Cooper and Shepard also had to accommodate their Mercury 9 training.

Group 1 and Group 2 academic studies, 29 October 1962–6 February 1963

Topic	Sessions	Total hours
Rocket propulsion	5	10
Flight mechanics	10	20
Astronomy	8	16
Computers	5	10
Guidance and navigation	22	88
Aerodynamics	3	6
Communications	6	12
Medical aspects of spaceflight	5	10
Upper atmosphere and space physics	4	8
Selenology (lunar geology)	5	10
Environmental control systems	2	4
Meteorology	4	8
Orbital mechanics	3	15
Total for each astronaut (13 topics)	82	217

Ed White examines a piece of lava at Meteor Crater, near Flagstaff, Arizona, during a geological field trip in 1964. Although in training for Gemini missions, the astronauts were expecting to progress to Apollo and perhaps a landing mission, and geology featured in their scientific training almost as soon as they joined the programme.

The nine new astronauts also conducted their first geological field trip to Meteor Crater and the surrounding area of Flagstaff, Arizona, over a two-day period (16–17 January), and rejoined the Mercury astronauts for a trip to Morehead Planetarium (29–30 January) for celestial recognition training.

The astronaut trainees formally concluded their academic training programme on 6 February 1963. Over the previous four months they had received basic science training for at least two days a week, and during the same period each man also received a series of familiarisation courses on the Gemini spacecraft and the Titan and Atlas Agena launch vehicle and docking targets, and visited contractors McDonnell Douglas, Martin, Aerojet and Lockheed. There was so much to study that they had to take home books to keep pace.

CB technical assignments
In addition to completing a basic astronaut training programme to prepare them for spaceflight, the nine also assumed additional technical assignments in support of the

developing Gemini and evolving Apollo programmes. These would eventually lead to flight assignments. The new technical assignments were announced on 26 January 1963. Of the Mercury astronauts, Gordon Cooper and Al Shepard were concentrating on the remaining flights in the Mercury programme (MA-9 and, for a while, MA-10). Gus Grissom was already specialising in Gemini; Glenn was to concentrate on the Apollo programme, with Scott Carpenter also working on Lunar Module training; and Walter Schirra (replacing Grissom) was responsible for Gemini and Apollo training procedure development. As co-ordinator of astronaut activities, Deke Slayton remained as overall supervisor of astronaut activities. These assignments also lined up Grissom and Schirra for early flight assignments on Gemini.

The technical assignments of the Next Nine were:

Armstrong	Training and simulators
Borman	Boosters (launch vehicles)
Conrad	Cockpit layout and systems integration
Lovell	Recovery systems
McDivitt	Guidance and navigation
See	Electrical, sequential and mission planning
Stafford	Communications, instrumentation and range integration
White	Flight control systems
Young	Environmental control systems, and personal survival equipment

In February the training programme expanded to include a series of eight three-hour and one four-hour space science and technology seminars over the next three months (11 February to 9 May 1963), totalling 28 hours. These included briefings on the unmanned Ranger and Surveyor lunar projects; solid rocket boosters; nuclear rocket propulsion; astronomy; and the four-hour circulation, respiration, hyperventilation, hypoxia and hypercapnia session.

The next twelve months would also see the group monitoring the MA-9 flight; conducting weightless flying in modified aircraft; completing their pressure suit indoctrination; taking a course in parachute jumping; a programme of wilderness survival training; even more detailed instructions in spacecraft sub-systems and launch support; a series of parasailing sessions towed behind boats as part of their water egress training; USN centrifuge training; rendezvous and docking practise; working with flight simulators and flight crew equipment such as cameras, waste management, and food systems; and practising their recently acquired skills in space navigation. They would also be working with PIs on proposed experiments and research investigations that they might need to conduct during their missions.

Many of the new astronauts found that one of the more difficult assignments was in learning how to deal with the press and meet the public, and in preparing their families for the attention of the world's media as a new celebrity. For many astronauts, this was by far the most difficult part of spaceflight training. Another aspect not covered in detail during the early years of the programme, and from which many suffered (if only temporarily), was in preparing to adjust to life after a spaceflight, both for themselves and their families. In some cases this was to prove much more difficult than preparing to make a flight into space, or the mission itself.

Group 2 astronauts and Deke Slayton at Stead AFB, Nevada, after a desert survival training session in 1963. (*Front row, from left*) Borman, Lovell, Young, Conrad, McDivitt, and White. (*Back row, from left*) Raymond Zedekar (astronaut training officer), Stafford, Slayton, Armstrong, and See.

Classrooms, deserts, jungles and a vomit Comet

On 20 May, the nine new astronauts completed zero-g indoctrination at Wright-Patterson AFB, Ohio, with the support of the 6750th Aerospace Medical Research Laboratory. Using a modified KC-135 aircraft, with the inside stripped of non-essential equipment and fittings and then adequately padded, the aircraft flew parabolic flight paths. At the peak of each climb, about 20 seconds of zero g was induced. The new astronauts undertook two flights involving twenty zero-g parabolas during each flight, with between 20 and 30 seconds of zero-g during each parabola. By the end of the two runs, each astronaut had experienced forty of these parabolic curves, logging about 15–20 minutes of zero g – and in some cases, with a demonstration of why the aircraft was also called the 'vomit Comet'!

The group also completed a two-week programme of Gemini systems briefings by the McDonnell training department based at MSC in late May, with the Mercury astronauts receiving the same briefings from late June. This was a more detailed familiarisation training procedure than had been previously conducted, and provided the astronauts with the opportunity to learn how the spacecraft operated, what it was like to be inside the crew compartment, and how much more they still needed to learn!

On 9 July, Phase 1 of the centrifuge programme began at the Naval Air Department Center, using the Aviation Medical Acceleration Laboratory centrifuge

At the Tropical Survival School, in the Panama Canal Zone, a native guide demonstrates how to convert a bamboo stalk into a container for catching and storing rainwater. (Courtesy British Interplanetary Society/NASA.)

which simulated the Command Pilot position (left seat) in the Gemini spacecraft. The engineering evaluation was divided into two parts, with the astronauts participating between 16 July and 17 August. In general, the astronauts were satisfied with the design and operation of the controls and displays, although some minor operational changes were recommended. Experience in the runs enabled the astronauts to determine that they would be able to cope with re-entry tasks without undue difficulty, even in the high acceleration conditions of extreme abort profiles. Centrifuge training was also conducted at the Ames Research Center in California, where high g-loads were also induced to match the stresses of the nominal powered launches on a Titan. This allowed the astronauts to expand their training in handling the controls despite the increased acceleration strains against them.

The first phase of the wilderness survival training was conducted between 5 August and 10 August, with a five-day desert survival course at Stead AFB, Nevada. All nine Group 2 astronauts (plus Slayton and Grissom) participated in the course, which was orientated towards Gemini mission recovery. It was divided into three phases, the first of them consisting of eleven days of academic preparations on the characteristics of world desert areas and suitable survival techniques. The second phase was a day of field demonstrations on the use and care of survival equipment and the use of the parachute for constructing clothing, shelters and signals. The final phase consisted of two days of remote site training, where the group (in two-man teams) was left alone in the desert at Carson Sink in Nevada, to apply what they had learned from the academic and demonstration phases of the programme. It was during this course that Young was stung on the ankle by a scorpion, but he refused any help from the accompanying medics, and attended to the wound himself. He reasoned that he was unlikely to find any convenient help in the middle of the Sahara if his spacecraft happened to land there!

Gene Cernan undergoes water survival training in the pool at the USN Pensacola Naval Air Station.

Water and land parachute training for the sixteen astronauts began on 8 September 1963. This was necessary because of the ejector seat crew escape system on Gemini. To simulate the effect, a towed 24-foot ringsail parachute carried each astronaut in turn to altitudes as high as 400 feet before the towline was released and the astronaut guided himself to a safe return, either on land or in the water.

ARPS Class III and Class IV

While NASA concentrated on preparing its astronauts for Gemini, the USAF continued with classes at the Aerospace Research Pilot School. On 22 October 1962, a third class was selected, and its members would receive specific X-20 training as part of their course:

Capt Alfred Atwell, USAF
Capt Charles A. Bassett, USAF
Maj Tommy Benefield, USAF
Capt Michael Collins, USAF

Water egress training also progressed to Gemini mock-ups, supported by pararescue divers, in the Gulf of Mexico.

Capt Joe H. Engle, USAF
Maj Neil Garland, USAF
Capt Edward Givens, USAF
Capt Gregory Neubeck, USAF
Capt James Roman, USAF
Capt Alfred Uhalt, USAF

An eleventh member subsequently joined the class:

Capt Ernst Volgenau, USAF

From this class, Neubeck was later selected for the USAF MOL programme, and Benefield became a test pilot for Rockwell working on the B-1 programme. Ed Givens, who had been instrumental in creating the course from which he was now a graduate, subsequently worked on USAF experiments for the Gemini programme (including the Astronaut Manoeuvring Unit) before joining NASA with the fifth group in April 1966 to train for the Apollo programme. Tragically, he was killed in an off-duty automobile accident in June 1967.

In May 1963, the fourth ARPS class began and became the final dedicated class. From 1 July 1963, the Experimental Test Pilot Course was termed Phase I, and the ARPS course Phase II, forming a single year-long training course, which began with Class 63A. The Class members of the fourth course were:

Capt Michael Adams, USAF
Capt Tommy Bell, USAF
Capt William Campbell, USAF
Capt Edward Dwight, USAF
Capt Frank Frazier, USAF
Capt Theodore Freeman, USAF

Capt James B. Irwin, USAF
Capt Frank Liethen, USAF
Capt Lachlan Macleay, USAF
Capt James McIntyre, USAF
Capt Robert Parsons, USAF
Capt Alexander Rupp, USAF
Capt David R. Scott, USAF
Capt Russell Scott, USAF
Lt Walter Smith, USN
Maj Kenneth Weir, USMC

Later, Macleay was selected for MOL, Liethen was killed in an F-100 on 12 October 1966, and Weir conducted test pilot work at Lockheed.

In September 1963, the USAF X-20 Dyna Soar programme was cancelled, but a new programme – the Manned Orbiting Laboratory – was under development, and received authorisation in 1965. Under this programme, from 1966, military astronauts would form two-man crews flying a modified Gemini spacecraft on 30-day surveillance missions in a rudimentary space station. One of the qualifying criteria for selection to MOL was to have graduated from ARPS (see Chapter 11).

The Fourteen
Although the new astronauts selected in 1962 were training primarily for Gemini missions, they were also receiving instruction on Apollo systems, and were providing support for technical aspects of Apollo development along with the seven original Mercury astronauts.

By September 1963, at the end of the first twelve months of the training for the second group, NASA had a cadre of just sixteen astronauts. It soon became clear to Deke Slayton that with two prime, two back-up, and about a dozen technical support duties on each mission, all the available astronauts would become totally involved in the early Gemini missions, and that there would be none spare to prepare for the more complex later Gemini missions involving rendezvous and docking, or for the early stages of Apollo. This was based on planning documents which showed that Gemini would consist of at least ten or possibly eleven manned missions, which would produce twenty flight seats, although some of the astronauts would probably fly two missions. There was no accounting for attrition, however, but it was already beginning to become apparent that the Mercury astronauts would not long remain as a group.

Deke Slayton was officially grounded, having lost MA-7, but he still hoped to regain flight status and fly a later Gemini mission. John Glenn had already begun to indicate his intention to leave both the USMC and NASA to enter private business, with a view towards a political career from 1964. Scott Carpenter was technically qualified to receive an early Gemini assignment as Command Pilot, but was never really in the running after his poor attempt at handling the attitude control system and maintaining the flight plan during MA-7. In charge of organising the Astronaut Office and selecting the candidates to fill future flight requirements, Slayton was fully aware that although Gemini was the next priority, he had to begin preparations for flying Apollo spacecraft to the Moon.

In 1963 there were plans to fly four three-man Apollo missions in Earth orbit, using the Saturn 1 launch vehicle, beginning in 1965. A further four missions were planned using the larger Saturn 1B launch vehicle in 1966, and then six manned Earth orbit and lunar orbit missions using the Saturn V would begin in 1967, leading to the first manned lunar landing attempt in 1968 or, most probably, 1969. If all of these missions were flown, a total of forty flight seats would have to be filled. Although Slayton did not believe that so many flights would really be needed, many still had to be included in the budget. Not wanting to be short of crew-members to fill whatever missions were required to complete the job of putting an American on the Moon by 1969, Slayton lent his support to a new selection process. The best estimates indicated that in 1963 he needed at least ten new astronauts – and possibly as many as twenty.

In selecting the new astronauts, Slayton devised a point system divided into three parts: academic, piloting performance, and character/motivation. There were ten points for each part, with the highest possible score being set at 30. Many were straightforward points for flying time and educational qualifications, and some were subjective, being based on personal interviews later in the selection process.

NASA issued the press release announcing a new astronaut selection on 6 June 1963. The criteria for qualification included being a citizen of the United States, no taller than 6 feet, and not over 34 years of age. A candidate also had to hold at least a BS degree in either the physical or biological sciences, and needed to have logged 1,000 hours flying time with NASA, the armed forces, or in the aircraft industry. The requirement for being a test pilot was dropped to accommodate a broader range of flying abilities, but each applicant again had to be recommended by his parent organisation.

NASA was expecting to announce the new group in October, and the cut-off date for civilian applicants was therefore set at 1 July, and for military applications pre-screened by their parent service, was 15 July. By the deadline, NASA had received 271 applications, of which 200 were civilian (three from women) and 71 military (26 USAF, 34 USN, 10 USMC and one from the US Army). Two of the military applicants were the first black jet pilots to be considered.

During July and August, 34 candidates underwent the exhausting week-long medical examinations that the previous groups had endured. In his 1974 autobiography, *Carrying the Fire*, Michael Collins graphically described the week at Brooks AFB, San Antonio, Texas, where they were greeted by a pleasant laboratory technician who was 'part Dracula and part leech'. There appeared to be great pleasure gained by this technician in extracting what appeared to be a quart of blood from every candidate, although this never seemed to prevent the body being prodded with even more needles at regular intervals. As Collins recalled, 'Inconvenience is piled on top of uncertainly on top of indignity,' as the body seemed to be prodded, poked, pummelled and pierced in every conceivable (and some inconceivable) ways. There was never any privacy, as every orifice was excruciatingly examined time and time again.

While at Brooks, each candidate endured 45 hours of tests and examinations in three categories of personal health, physical fitness and emotion. The official record does not explain what these 45 hours really meant to the man who was on the receiving end. For example, cold water was poured into one ear, and the conflicting information being relayed to the brain from a warm and a cold inner ear at the same time caused the eyeball to gyrate wildly!

In other tests, instrumentation taped to the body recorded the effects of the equivalent of running up a mountain on a treadmill; and strapped to a flat table, the candidate was tilted quickly at all angles to measure responses to immediate changes in the gravity vector. Then, in the cause of Space Age medical science – but in reality following the tradition of ancient torture chambers of the dark ages – 'steel eels' were inserted in the back passage to test for signs of cancer or other diseases. If they survived all of this, there was also the ink-blot test, eye pressure tests, enemas, prolonged breathing exercises, vibration tests, and many other tests. And this was even before the candidate was ever selected to train to become an astronaut!

The 34 (with dates of medical evaluation in brackets) were:

Capt Michael J. Adams, USAF (ARPS IV student) (1 August)
Maj Edwin E. 'Buzz' Aldrin Jr, USAF (13 August)
Capt William A. Anders, USAF (5 August)
Capt Charles A. Bassett II, USAF (ARPS III student) (15 August)
Lt Alan L. Bean, USN (Group 2 applicant) (14 August)
Capt Tommy I. Bell, USAF (ARPS IV student) (1 August)
Vance D. Brand, civilian (8 August)
Lt Eugene A. Cernan, USN (6 August)
Lt Roger B. Chaffee, USN (7 August)
Capt John K. Cochran, USMC (13 August)
Capt Michael Collins, USAF (Group 2 applicant) (15 August)
Darrell E. Cornell, civilian (9 August)
Ronnie Walter Cunningham, civilian (9 August)
Capt Reginald R. Davis USAF (6 August)
Donald G. Ebbert, civilian (8 August)
Capt Donn F. Eisele, USAF (5 August)
Lt Ronald E. Evans, USN (8 August)
Capt Theodore C. Freeman, USAF (ARPS IV student) (1 August)
Lt George M. Furlong, Jr, USN (5 August)
Lt Richard F. Gordon Jr, USN (Group 2 applicant) (31 Jul)
Capt Samuel M. Guild Jr, USAF (7 August)
Capt James B. Irwin, USAF (ARPS IV student) (2 August)
James E. Kirkpatrick, civilian (14 August)
Capt Charles L. Phillips USMC (12 August)
Lt William E. Ramsey, USN (1962 applicant) (31 Jul)
Capt Alexander K. Rupp, USAF (ARPS IV student) (2 August)
Russell L. Schweickart, civilian (13 August)
Capt David R. Scott, USAF (ARPS IV student) (2 August)

Lt Robert H. Shumaker, USN (6 August)
Lester R. Smith, civilian (9 August)
John L. Swigert Jr, civilian (Group 2 applicant) (14 August)
Capt Robert J. Vanden-Huevel, USAF (7 August)
Capt Clifton C. Williams Jr, USMC (12 August)
Lt Cdr John D. Yamnicky, USN (12 August)

Two days after the completion of the medical tests, 28 of the applicants were asked to report to MSC in Houston. where, in the first week of September, they would undergo a further three days of interviews, evaluations and briefings. Following this final cut, the successful candidates were selected, and then all of the finalists were informed of the result. The successful candidates were told to report to MSC on 17 October, and on 18 October there was a press conference announcing their selection as America's latest astronauts.

At the press conference at MSC, the fourteen members of the third NASA astronaut group were presented to the world's media, bringing the total selected by NASA since 1959 to 30. The group consisted of seven USAF, four USN, one USMC and two civilian candidates, and they were to report for duty at NASA between 1 January and 15 January 1964, to allow completion of their current assignments and to plan the moving of their families to Houston. Official training would begin on 2 February.

The Group 3 members were:

- Maj Edwin E. 'Buzz' Aldrin Jr, USAF, (33), a Korean War veteran with a PhD in orbital rendezvous.
- Capt William A. Anders, USAF, (29), with a degree in nuclear engineering.
- Capt Charles A. Bassett II, USAF, (31).
- Lt Alan L. Bean, USN, (31).
- Lt Eugene A. Cernan, USN, (29).
- Lt Roger B. Chaffee, USN, (28), who had flown U2 photoreconnaissance missions over Cuba during the 1962 missile crisis.
- Capt Michael Collins, USAF, (32), who had been told in 1962 that he had not been selected because he needed a further year of experience.
- R. Walter Cunningham, (31), a former USMC pilot and now a research scientist whose experiments had flown on satellites.
- Capt Donn F. Eisele, USAF, (33).
- Capt Theodore C. Freeman, USAF, (33).
- Lt Cdr Richard F. Gordon, USN, (34), one of the best pilots in the US at that time.
- Russell L. Schweickart, (27), a civilian research scientist and former USAF pilot currently serving in the Air National Guard.
- Capt David R. Scott, USAF, (31).
- Capt Clifton C. Williams, USMC, (31).

For Bean, Collins, and Gordon, it had been their second attempt at becoming astronauts. Of those from the final 34 who were not selected, Adams would be

The fourteen members of the third NASA selection of 1963. (*Front row, from left*) Aldrin, Anders, Bassett, Bean, Cernan and Chaffee. Rear from left: Collins, Cunningham, Eisele, Freeman, Gordon, Schweickart, Scott and Williams.

selected to join MOL Group 1 in 1965, before transferring to the X-15 programme. He was killed flying the #3 aircraft on 15 November 1967. Brand, Evans, Irwin, and Swigert would all finally make it to NASA as members of Group 5 in 1966.

In November 1963 Slayton – no longer cleared for solo flying – resigned his USAF commission as Major, but remained with NASA in a civilian capacity. He became the Director of Flight Crew Operations, and was responsible for the selection of flight crews, the assignment of astronauts to technical duties, and management of the NASA Flight Crew Support Division, the Astronaut Office (later receiving the mail code CB) and Aircraft Operations. Also in November, Shepard was finally grounded due to inner ear problems, and lost the chance to fly a Gemini mission. He became the Chief Astronaut for the next decade, and he and Slayton formed the most powerful team in American manned spaceflight. They chose all the American astronauts who flew into space between 1963 and 1982.

Slayton held these administrative roles until 1981, when he finally retired from NASA. Restored to flight status in 1972, he flew as a US crew-member on the 1975 ASTP mission, after waiting sixteen years. Shepard was also finally restored to flight status after corrective surgery in 1969. He flew to the Moon on Apollo 14 in 1971, and retired from NASA in 1974.

The first black astronaut
During the selection for the third astronaut group in 1963 there were political moves to include the first black astronaut candidate in the programme. Senator Robert Kennedy, brother of President John F. Kennedy, was particularly eager to find a suitable black pilot who could become an astronaut to show the equality of the programme and how it was open to all.

After a thorough search of military records, only one name came anywhere close to possible consideration – Capt Edward Dwight, a USAF bomber pilot. As a multi-engine pilot, his lack of jet pilot experience meant that he was passed over in the NASA selections for 1959 and 1962, as he would also be for 1963. He did not have an engineering degree, had not attended test pilot school, and was not eligible for the ARPS school (then headed by Chuck Yeager, the first man to break the sound barrier).

But the lack of even the basic selection criteria would not stop Robert Kennedy in his quest, and he personally asked General LeMay, Chief of Staff, to ensure that Dwight was enrolled in the ARPS – which, it was proposed, could lead to a qualification to put him forward for consideration by NASA as an astronaut candidate.

As a pilot, Dwight was considered average, with an average educational background. He was fully qualified to fly in an operational squadron, but not to compete against the best pilots in America in the space course. His lack of experience and qualifications were a disadvantage in the rankings of those who were eligible to attend the class, and initially he was included well down the list. There were far more qualified men ahead of him who would still not have been accepted into the school. At that time, there was a limit of eight places, and with more than a dozen qualifiers, bypassing the system would present more problems than it would solve.

By way of a compromise, it was decided that Dwight would be included only if all those who ranked above him were also included, and ARPS Group IV therefore included sixteen candidates rather than eight. Some of those participating in the course were average but not outstanding, and ironically at that time, NASA was not only considering applicants from ARPS or the USAF for astronaut training, as there were many suitably qualified pilots from the USN, USMC, NASA, and the aircraft industry. NASA also reviewed those former jet pilots who by then were working as research scientists. Dwight, however, was not keen to become an astronaut, and he just wanted to advance through the ranks in the Air Force. He was not short-listed for selection to NASA, and shortly afterwards he left the Air Force and moved to Colorado, where he became a successful sculptor.

Astronaut training, 1964

The new Group 3 astronaut trainees embarked on the third round of academic studies, as had the sixteen astronauts before them. This time the courses for the astronauts differed in that the geology programme and Gemini computer sessions were for all of the group (1959, 1962 and 1963 selections), while everything else was just for the new Group 3 astronauts and anyone who wanted to sit in. With the first manned Gemini missions fast approaching, many of the first two groups of astronauts were becoming involved in preparations for their assignment to early flight crews, and were involved in Gemini mission training. This would become even more involved as the mission to which they were assigned drew ever closer.

The programme continued from 3 February to late June, and again followed the usual two hours per topic, with an occasional double session combined on the same day. Between four and six hours' study was completed during each academic day (at least twice a week, as with the previous group).

Science training schedule for the Group 3 astronauts (February–June 1964)

Topic	Attendance	Sessions	Total hours
Geology Series I	All astronauts	25	50
Mineralogy and petrology	All astronauts	13	26
Flight mechanics	Group 3	18	36
Astronomy (three-hour sessions)	Group 3	5	15
Digital computers	Group 3	6	12
Gemini onboard computers	All astronauts	11	22
Rocket propulsion systems	Group 3	6	12
Upper atmosphere and space physics	Group 3	6	12
Aerodynamics	Group 3	2	4
Guidance and navigation	Group 3	15	30
Communications	Group 3	4	8
Medical aspects of spaceflight	Group 3	6	12
Meteorology	Group 3	2	4
Total for 13 topics (Group 3 only)		119	243

All astronauts were undergoing what was termed Series I of the geology programme, which involved a series of field trips. However, due to the nature of the overall training programme, not all of the astronauts could attend together, and they were often split into groups. This part of their training began with a one-hour geology course on 2 March 1964. The group was split for the Grand Canyon field trip, which was conducted in two parts, on 5–6 March (eighteen astronauts) and 12–13 March (nine astronauts). Then, on 2–3 April 1964, eighteen astronauts conducted a further geological field trip to Marathon Basin and Big Bend in West Texas. Another nine astronauts completed the same field programme on 16–17 April. Between 30 April and 2 May, seventeen astronauts conducted a geology field trip to Flagstaff, Arizona, and Kitt Peak Observatory, with another twelve making the trip on 20–22 May. Finally, twenty astronauts made the last field trip of this series, on 3–6 June 1964, to Philmont Ranch, New Mexico.

Series II of the geology programme for the 22 astronauts who were not assigned to Gemini began on 14 September 1964, and continued until 24 May 1965. On 21 June 1965, eleven astronauts began geology Series III, which continued until 27–29 December 1965. Most of this training was related to the forthcoming Apollo lunar missions. The reduced number available to attend reflected assignments in the Gemini programme as the series began flying and as astronauts were assigned to later missions planned for 1966.

Group 3 completed celestial recognition training at Morehead Planetarium on 26–27 March, and also completed jungle survival training programme in Panama in July 1964 and desert survival training in Nevada during August 1964.

In addition to completing their basic training programme and receiving Gemini familiarisation training and briefings, the fourteen new astronauts took more assignments in Apollo than had the previous groups, as many of them were expected to progress to Apollo missions.

Filling the seats

By 1964, both Glenn and Carpenter were out of consideration for flight assignments, leaving 28 active astronauts in the office (although technically, both Slayton and Shepard were medically disqualified). By this time, a significant change in the testing of Apollo hardware had had an effect on future crew assignments and seat availability. In November 1963, the Head of Manned Spaceflight, George Mueller, decided to alter the methods of testing the Saturn boosters. Wernher von Braun had planned to test one stage at a time, adding and testing later stages only when the previous stage had been flight-proven. Mueller thought that this was a waste of precious time and money, and opted for 'all-up' testing, so that all elements would be tested at the same time. Repeat flights would retest faulty equipment. This eliminated the need for large assembly lines, but also eliminated up to four manned Saturn 1 missions. These were cancelled on 30 October 1964, and with them went twelve flight seats.

By 1964, Slayton had a fair idea that he would have to train ten Gemini crews and probably eight Apollo crews before the first landing mission was attempted. It proved a fairly accurate prediction, as there were ten manned Gemini missions flown, although there were only five pre-landing Apollo missions.

Virgil Grissom received one of the earliest assignments to the Gemini programme in 1963. Here, a year later, he examines the Command Pilot station. Note the retro section, with its four solid-propellant rockets. (Courtesy British Interplanetary Society/NASA).

The reasoning behind the choice of one person over another to fly a particular mission (and with whom, and why) remained a mystery even to the astronauts themselves. In his 1994 autobiography, Slayton explained that he had decided that he needed to build up a pool of flight-experienced astronauts who had gained skills in long-duration flight, rendezvous and docking, and EVA during Gemini. From this, they could move on to early Apollo missions, and from there to the first landing crews. To achieve this his guidelines included:

1   Everyone was considered qualified and acceptable for any mission when selected by NASA. Slayton reasoned that if he had hired them and they stayed, they were eligible to fly.
2   Some were more qualified than others for specific seats on certain missions. Those with command, management, or test pilot backgrounds would be more likely to handle more challenging assignments.
3   Slayton attempted to match people in a crew based on individual talents and, wherever possible, personal compatibility, although this was not as important as skills. According to Slayton, becoming an astronaut was a demonstration of talent and motivation, or the candidate would not have been selected. He assumed that they would be compatible with each no matter how they were paired, but it soon became clear that this would not be the case as personalities started to clash.
4   He always kept future requirements and training plans in mind, and used a long-term plan that was constantly updated to determine how one assignment would have consequences for other plans.
5   He assumed an annual attrition rate of about 10% for accidents and resignations, and although he admitted this was a 'wild guess', it happened to be quite accurate.

Initial Gemini crew assignments

In 1963, Slayton began to form the first crews for Gemini. From Group 1, Shepard, Grissom, and Schirra were still available. Glenn was scheduled to retire in January 1964, Carpenter was still unacceptable to management as a result of his performance on his Mercury mission, and Cooper was still available, although as he had alienated a few of those in the higher administration at NASA during Mercury, his future in Gemini was by no means assured. Slayton still had all the Group 2 members, but the Group 3 class would not be available for about a year. From thirty names, he could confidently select thirteen that would crew the first few Gemini missions: Armstrong, Borman, Conrad, Cooper, Grissom, Lovell, McDivitt, Schirra, See, Shepard, Stafford, Young and White.

With each flight, four men (working in pairs) would be required: the prime crew of Command Pilot and Pilot, and their back-ups. Other astronauts could fill in support and CapCom duties in preparation for their own missions, including those from Group 3 later in their training.

The first four Gemini missions had unique requirements, whereas the last six were almost identical, and Slayton therefore chose the first four Commanders primarily from the Mercury astronauts:

Gemini 3 The first mission, planned for 18 orbits or one day in space. Al Shepard, as back-up on the previous mission (Mercury 9), was, at the time, the most eligible and capable to take the first Gemini.

Gemini 4 Planned to last seven days, and possibly to include the first spacewalking experiment, during which the pilot would stand up in the hatch. Slayton thought the first of the second astronaut group could gain valuable early experience as Command Pilot of this mission.

Gemini 5 The first rendezvous and docking flight with an unmanned Agena target vehicle. It would not be a long flight – possibly just one or two days. Slayton knew that Schirra was not keen on a long mission, and he was therefore assigned to Gemini 5.

Gemini 6 A marathon 14-day mission. Grissom had worked on Gemini systems since completing his Mercury flight in 1961, and was an ideal candidate to fly a really long mission. Slayton also thought that he could use Cooper as back-up Command Pilot on this flight.

It was with these assignments that Slayton also evolved the three-mission rotation system, which continued to the end of the Apollo programme in 1972. Upon

During the flight of Gemini 6 in December 1965, Al Shepard, Chief Astronaut – who should have been Commander of the first Gemini mission – chomps vigorously on a cigar during what he described as 'one of my weaker moments'.

completion of Gemini 3, the four astronauts would be available for reassignment. It seemed logical to promote the back-up crew of Gemini 3 into the prime crew for Gemini 6 (with the crews of Gemini 4 and Gemini 5 still in training). One or both of the Gemini 3 prime crew could then be the new Gemini 6 back-up crew, and it would also be possible to include a new astronaut (possibly from Group 3 by then) by promoting the Gemini 3 pilot to back-up Command Pilot. He would then be aimed at training for a later mission (which in this system would be Gemini 9) as prime Command Pilot.

The next step was to fill in the remaining assignments with Group 2 astronauts. According to Slayton lore, 'All astronauts were equal – just some were more equal than others.' In making the final decision, he sought the opinions of fellow Mercury astronauts Shepard, Schirra, and Grissom, as they would be flying with them.

During a pilots' meeting in the Astronaut Office in July 1963, Slayton told the team that Al Shepard and Tom Stafford would be the first crew to fly Gemini in space. They would be backed up by Gus Grissom and Frank Borman, who would move on to fly the 14-day Gemini 6 mission. Slayton thought that Borman would be tenacious enough to endure the long flight.

For the seven-day Gemini 4 mission, with the possibility of an EVA and limited rendezvous and station keeping practise with the second stage of the booster, McDivitt (thought to be the best of the second intake) was nominated as Commander. His pilot was Ed White, a former classmate at Michigan University. Although not considered to be one of the best from the group, Slayton thought that, due to the restricted volume of the Gemini spacecraft, compatibility was an important factor in crew selection for long missions. An added bonus was that White – a former athlete of Olympic potential – was also physically strong, which would be an advantage in attempting the first EVA. The back-up crew consisted of two friends from the USN Test Pilot School – Pete Conrad and Jim Lovell, who would then expect to rotate to Gemini 7.

For Gemini 5, the first rendezvous and docking mission, Slayton assigned John Young to Schirra. The back-ups were the final pair of Group 2 astronauts, Neil Armstrong and Elliot See (who was thought by Slayton to be the weakest pilot in the group), who would expect to fly as prime on Gemini 8.

The first crewing for Gemini was therefore as follows:

Mission	Prime	Back-up
Gemini 3	Shepard	Grissom
	Stafford	Borman
Gemini 4	McDivitt	Conrad
	White	Lovell
Gemini 5	Schirra	Armstrong
	Young	See
Gemini 6	Grissom	Shepard or Stafford
	Borman	Stafford, or first Group 3 Pilot
Gemini 7	Conrad	White
	Lovell	Group 3 Pilot
Gemini 8	Armstrong	Young
	See	Group 3 Pilot

This appeared fine on paper, but it was only weeks before it all had to change. The first casualty was the launch schedule, as it soon became apparent that the Agena would be ready in time for Gemini 6 rather than Gemini 5, the latter of which became the 14-day long-duration mission.

This meant that in order to fly the first Agena mission, Schirra and Young would need to become the new Gemini 3 back-up crew. Grissom and Borman were reassigned to Gemini 5 prime crew without first serving on a back-up crew.

Shepard grounded

Shepard and Stafford had been in training for only about six weeks when, during a party, Shepard confided to Stafford that he was having medical problems with dizziness, and was not certain whether anything could be done to correct it. By May 1963, Shepard had been preliminarily diagnosed as suffering from Ménière's disease – a problem with balance in the inner ear. He retained his flight status, and for a while kept Gemini 3, but by October the diagnosis results confirmed the suspicions, and he was taken out of flight consideration, which was a bitter pill to swallow. In November 1963 he was made Chief Astronaut – a post he was to hold for the next

A unusual photograph of the first manned flight training group. The Gemini 3 prime crew – Young and Grissom – and the back-up crew – Schirra and Stafford – are wearing the early (c.1964) silver-layered versions of the Gemini pressure suits.

decade. In this he ensured that although he could not fly himself, he would ensure that those who did were the best that NASA had to offer, while all the time secretly hoping for a chance to regain flight status and fly on Apollo.

Tom Stafford learned about Shepard's grounding from one of the NASA doctors and not from Shepard himself, nor from Slayton during the regular Pilot meeting. The result was a complete adjustment of crews by Slayton, although they were still not officially or publicly announced. At the same time, Gemini 3 was cut from 18 to just three orbits, Gemini 4 from seven days to four days, and Gemini 5 from fourteen to seven days. Gemini 6 was assigned the first Agena mission, and Gemini 7 became the 14-day marathon.

To replace Shepard, Slayton reassigned Grissom as Command Pilot of Gemini 3, and to ensure that Schirra would fly on Gemini 6, he became Grissom's back-up. Apparently, however, Grissom and Borman were not an ideal match, and Borman doubted that he could have worked with Grissom over several months, let alone fourteen days inside a Gemini spacecraft. Slayton therefore moved Borman out of the four-man unit, and moved up John Young, with a more compatible personality, to Gemini 3. Stafford was then without a Commander, and as he was the best of the group on rendezvous issues, it made sense for him to be in line for Gemini 6.

To save training time, and utilising previous experience on such a complicated mission, Grissom and Young would fill out the back-up Gemini 6 crew positions as a pair before splitting the crew, moving Grissom to Apollo and Young to command a later Gemini mission.

This sequence left Borman floating, and as he was one of the best choices of the Group 2 members for an early command, he kept the long-duration Gemini 7 mission, and Lovell became his Pilot. Conrad was moved to back-up McDivitt as Pilot on Gemini 4, and this allowed Slayton to bring in Cooper, who had been unassigned, for the prime crew of Gemini 5.

The back-up crew for Gemini 5 was Armstrong (another member strong on rendezvous techniques) and See as Pilot. (Slayton assumed that as See was weaker, he was balanced by a stronger Armstrong, who was aimed at Gemini 8). The revised assignments and missions therefore became:

Mission objective		Prime	Back-up
Gemini 3	Three orbits, first manned	Grissom–Young	Schirra–Stafford
Gemini 4	Four days, EVA, RV	McDivitt–White	Borman–Lovell
Gemini 5	Seven days	Cooper–Conrad	Armstrong–See
Gemini 6	First Agena docking	Schirra–Stafford	Grissom–Young
Gemini 7	Fourteen days	Borman–Lovell	
Gemini 8	Agena docking, EVA	Armstrong–See	

The announcement of the first assignments

On 13 April 1964 – only five days after the first unmanned Gemini had flown – Dr Robert Gilruth, Director of MSC, announced the names of the prime and back-up crews for Gemini 3. Gemini flight crews were at last in the final stages of mission training.

On 8 July, Slayton – by then the MSC Assistant Director for Flight Crew Operations – announced new technical assignments for the astronaut group. Al Shepard took Slayton's former job as Chief of the Astronaut Office – Code CB. The CB was now divided into three branches: Gemini, headed by Grissom, with Young, Schirra and Stafford joining him; Apollo, headed by Cooper, with McDivitt, White, Conrad and Borman; and the Operations and Training Branch, headed by Armstrong, assisted by See and Lovell. Carpenter was at that time on detachment to the USN Sealab Man-in-the-Sea project, and was not presented with a new MSC assignment. Carpenter remained with NASA until 1967, although he worked on the USN Sealab programme from 1964. He completed a variety of technical assignments in Gemini and Apollo Applications (Skylab) until leaving NASA, and retired from the USN in 1969.

The fourteen Group 3 astronauts were also assigned to the Operations and Training Branch, and were each given technical assignments in that office in addition to completing their basic training and their Gemini and Apollo preparations.

Aldrin	Mission planning (including trajectory analysis and flight plans – his PhD in rendezvous no doubt being helpful)
Anders	Environmental control systems radiation and thermal protection (utilising his degree in nuclear engineering)
Bassett	Training and simulators
Bean	Recovery systems
Cernan	Spacecraft propulsion and Agena docking target
Chaffee	Communications; and for Apollo, the Deep Space Network
Collins	Pressure suits and EVA experiments
Cunningham	Electrical and sequential systems and, using his previous experience, the monitoring of unmanned experiments in other programmes that were related to current and future MSC programmes
Eisele	Attitude and translational control systems
Freeman	Boosters (launch vehicle) – Titan, Atlas and Saturn
Gordon	Cockpit integration
Schweickart	Future manned programmes (such as Apollo and Apollo Applications, and Space Station), and in-flight experiments for Gemini and Apollo
Scott	Guidance and navigation
Williams	Range operations and crew safety

On 27 July, McDivitt and White were officially named as prime crew for Gemini 4, with Borman and Lovell as back-ups. After this announcement, the four moved to the Gemini branch under Grissom, and as a result the branch offices were again adjusted, reflecting long-term plans that would affect the careers of several of the team. The Gemini branch now had the crews for Gemini 3 and Gemini 4, plus their back-ups who were scheduled to fly Gemini 6 and Gemini 7.

The Apollo branch (headed by Cooper, with Conrad, both of whom were pointed at the Gemini 5 crew) had Group 3 astronauts Anders, Cernan, Chaffee,

The Gemini 4 prime crew shortly after selection to the mission in July 1964. The model of Gemini spacecraft is fitted with land landing gear configuration which was subsequently abandoned.

Cunningham, Eisele, Freeman, Gordon and Schweickart. These astronauts were expected to take early Apollo or Extended Apollo (later called Apollo Applications, which evolved into Skylab) assignments, although some became reassigned to Gemini from 1966 to fill out vacant seats.

The Operations and Training Branch (headed by Armstrong with See – both in line for Gemini 5 back-up and Gemini 8 prime crews) was now joined by Group 3 astronauts Aldrin, Bassett, Bean, Collins, Scott and Williams, who were first in line as Pilots in the next round of Gemini crew assignments.

With 26 flight-active astronauts in the office, it appeared that there were enough men available to complete all Gemini requirements and initial Earth orbit test flights of Apollo hardware prior to committing to lunar missions. There were forces at work to include a group of scientist-astronauts in with the pilot-astronauts for Apollo and later missions, but it was the members of these three groups that would fill the requirements for Gemini (see Appendix). Not all of them would see the end of the programme.

A further 43 astronauts joined NASA between 1965 and 1969. Two groups of scientist-astronauts (totalling seventeen) were selected in 1965 and 1967, a further group of nineteen pilot-astronauts were brought in during 1966, and seven former USAF MOL astronauts (who had trained on Gemini systems) were transferred in 1969 when the Air Force programme was cancelled. From these ranks, many would fly later Apollo, Skylab, and early Shuttle missions, although several would wait many years for a single flight, while others would never leave the launch pad.

The first astronaut fatality

Since the first astronauts had been selected in April 1959, the office had been remarkably free of accidents, considering the risky business in which they were employed. On 30 October 1964, however, their luck ran out when the first tragedy hit the office.

Group 3 astronaut Theodore Freeman – tipped as one of the first to receive a flight assignment – took a Northrop T-38 training jet out over the Gulf of Mexico, and then headed back to Ellington Field, near MSC, south of Houston. It was a normal proficiency flight to continue the monthly flying hours requirement which all astronauts have to maintain.

As a result of other air traffic in the area, Freeman had received a wave-off on his first approach. As he climbed and banked to line up for a second attempt, a snow goose was sucked into the engine intakes. For the next few seconds Freeman had sufficient power and lift to level out, and as the engines flamed out and the nose dipped, he ejected at about 1,500 feet. But the aircraft was diving as he ejected, and the action was forward instead of the normal upward profile, causing his seat parachute to not open fully. Freeman was found dead – still in his ejector seat, and with the parachute unopened – not far from the wreckage of his aircraft. The office mourned its first victim, and it was even more tragic that the loss was not in space but during routine flying. For the first of many occasions, a T-38 fly-over in the missing-man formation was featured at MSC. Freeman was buried at Arlington Military Cemetery.

Long-duration crews

On 8 January 1965, the first assignments for the new year were announced. Cooper and Conrad were named as prime crew for Gemini 5, with Armstrong and See as their back-ups. At that time, Slayton was considering Cooper as back-up Command Pilot for Gemini 9 and then as prime on Gemini 12. By the time those missions flew, many of the early Gemini Commanders would have started to move across to Apollo – but there would still be a need for an experienced Commander to end the Gemini series. Despite some reservations from higher management, Cooper had flown a good mission as Pilot of Mercury 9 in 1963, and was the most experienced American astronaut up to Gemini 4.

Following the three-orbit flight of Gemini 3 on 23 March, Schirra and Stafford were able to concentrate on their own impending docking mission, Gemini 6. In November 1964, Schirra had been told by Slayton that as back-up to Gemini 3 he would be assigned to Gemini 6, and in February he confirmed to the press that he and Stafford would be the first Americans to dock in space. Therefore, when the official announcement was made on 5 April it was largely academic, as it had been known by the media for several weeks. With Young and Grissom assigned as the back-up crew, the training for this demanding mission was eased by the experienced group.

On 1 July the Gemini 4 back-up crew, Borman and Lovell, were named for the prime positions on the 14-day marathon, Gemini 7. The back-up crew was announced as being Ed White (promoted to Command Pilot, with McDivitt moving on to Apollo and taking no further part in Gemini), and the first of the Group 3 astronauts to receive a flight assignment – Michael Collins. This crew would be in a position to rotate to prime on Gemini 10 after completing their Gemini 7 assignments.

On 28 June 1965, six scientist-astronauts were selected for the astronaut programme, to begin Pilot training the following month.

Rendezvous and docking crews

Shortly after Gemini 5 had returned from its eight-day marathon in August 1965, Slayton gathered some of the Group 3 astronauts from the Apollo Branch on a geological field trip to Oregon, where he took the opportunity to privately inform them of their first assignments to Gemini or Apollo. Dave Scott would be given the Pilot seat on Gemini 8, Slayton informed them, with Dick Gordon as his back-up.

At Mission Control, on 15 December 1965, astronaut CC Williams (*left*) discusses the scrubbing of Gemini 6 with Charles Bassett (*centre*) and Elliot See (*right*). See and Bassett were in training as the prime crew of Gemini 9, but ten weeks later were killed in the crash of their T-38. Williams was to serve as back-up on Gemini 10 before assignment to Conrad's Apollo crew, but in October 1967 he too was killed in a jet crash. His Gemini 10 back-up crew Commander, Al Bean, replaced him on Conrad's Apollo crew.

On 20 September the official announcement was issued, with Armstrong and Scott as prime crew and Conrad and Gordon as back-up crew. Slayton had broken up the Armstrong–See crew after they had completed their assignments as the Gemini 5 back-up crew. Slayton was beginning to feel that See could not withstand the physical demands of the EVA planned for Gemini 8, and he therefore replaced him with Scott. Although he was about to assign See as Command Pilot on Gemini 9, Slayton was not comfortable with See's piloting abilities, and paired him with Charles Bassett who, according to Slayton, was, 'one of the better guys – strong enough to support him and make up the deficiencies.'

For reasons of compatibility, Gordon was assigned with Conrad as back-up on Gemini 8, rather than as back-up on Gemini 9, for which he was originally intended. In addition, with Bassett receiving technical assignments in Gemini, and Gordon in Apollo, it seemed logical to fly Bassett first.

For Cooper there was a new assignment on 23 September, as he replaced Grissom as Gemini Branch Chief in the CB. There was, however, apparently no flight assignment for him in Gemini, and Slayton was then not planning to move him to an early Apollo mission, as he did not consider that he could handle a more complex mission. In the same announcement, McDivitt was named Apollo Branch Chief, with Buzz Aldrin as Operations and Training Branch Chief.

The Gemini 9 crews became official on 9 November, with See and Bassett as prime crew and Stafford and Cernan as their back-ups. Stafford would not be ready for full assignment to Gemini 9 preparations until Gemini 6 had flown in December, and so Cernan took a more active role in supporting the two prime crew-members. Cernan had carried out good work in support roles on Gemini 6, but (according to Stafford) he was merely assigned to Stafford, who did not request him. After Gemini 9, it was expected that Cernan would be rotated to fill in the dead-end assignment as back-up Pilot on Gemini 12 – the last mission of the series – while Stafford would receive an Apollo assignment.

Following the Gemini 7/6 flight in December, a group of six veteran Gemini astronauts were assigned to prepare for the first Apollo missions in Earth orbit. With Grissom as Apollo Branch Chief, they were Borman, Cooper, McDivitt, Schirra and White. From 3 February, Stafford became the new CB Gemini Branch Chief, until at least the summer of 1966.

Gemini veterans move to Apollo
In the early weeks of 1966, Grissom and the other Gemini veterans began full-time involvement in the Apollo programme, along with several Group 3 astronauts from the Apollo Branch Office. McDivitt had been working on Apollo since leaving Gemini 4 the previous June, and would soon be joined by Dave Scott when he completed his Gemini 8 flight in March. Bassett was assigned to follow him after Gemini 9 in May. This resulted in the announcement of the first crews:

Mission	Prime	Back-up
Apollo 1	Grissom–Eisele–Chaffee	McDivitt–Scott–Schweickart
Apollo 2	Schirra–White–Cunningham	Borman–Bassett–Anders

McDivitt's crew was already preparing to take the first manned flight of the LM after serving as Apollo 1 back-ups. This early crewing was altered just prior to Christmas when Eisele injured his shoulder and was exchanged with White from Schirra's crew. Anders and Schweickart had been assigned as future Lunar Module Pilots since 1965, as they were the best of the 1963 group that had not received Gemini assignments. Slayton had some reservations about their skills as operational pilots, which is why the more experienced pilots – Aldrin, Bean and Williams – were assigned to Gemini and not to Apollo. For the same reason, Eisele and Cunningham were also to receive Apollo assignments instead of Gemini flight seats. Although crew rotations resulted in Slayton assigning Anders as Gemini 11 back-up (see below), he moved back to Apollo immediately afterwards.

The loss of See and Bassett
John Young and Michael Collins received the official Gemini 10 assignment on 25 January, and were backed-up by Lovell and Aldrin, who should then have rotated to fly Gemini 13 – except that there would be no Gemini 13.

On 28 February, Gemini 9 astronauts See and Bassett flew a T-38 to the McDonnell Gemini plant in St Louis. Following in a second aircraft was their back-up crew, Stafford and Cernan. They were to conduct some training in the simulator, and check on the progress of their Gemini spacecraft, which was under construction at the St Louis plant.

The weather reports for St Louis had not been that good, with rain and fog in the landing area, and during the flight from Houston the weather deteriorated. By the time they approached Lambert Field next to the McDonnell plant, light snow flurries were hampering visibility. On their first landing attempt, Stafford followed a missed approach procedure and flew around for a second attempt. See, however, elected to land, but as he approached, the McDonnell building was shrouded in fog. When the building suddenly came into view, See tried to pull up the nose of the plane, but misjudged it. Probably realising that his sink rate was too high, he cut his after burner and tried to bank sharply to the right; but as he did so, the aircraft struck Building 101 and dislodged an internal support frame from the roof, which crashed down onto the workers inside.

The T-38 slid across the roof, and then fell and bounced in the car park beyond, exploding on impact and instantly killing both See and Bassett. On the ground, fourteen others were injured, though not seriously. The Gemini 9 spacecraft was unscathed. Unaware of the tragic accident, Stafford and Cernan landed safely, only to be told of the loss of their colleagues. Within hours, they were informed that they would take the places of the lost crew, and fly Gemini 9. On 3 March 1966, See and Bassett were buried at Arlington Military Cemetery, and once again the missing-man formation flew over MSC.

It was planned that after Gemini 9, See would back-up Gemini 12 and Bassett would move to Borman's Apollo crew. The consequences of their loss were reflected in the announcement on 21 March. These revised assignments completed the Gemini programme. Stafford and Cernan moved up to take their prime crew positions, and the previously announced Gemini 10 back-up crew of Lovell and Aldrin – the best

The Gemini 10 crews: (*from left*) Command Pilot John Young, Pilot Michael Collins, back-up Command Pilot Al Bean, and back-up Pilot CC Williams. Bean and Williams had replaced the original Gemini 10 back-ups Lovell and Aldrin, who were recycled to back up Gemini 9 after the death of See and Bassett.

available to support in such a short period of time – moved to back-up Gemini 9, and into line for Gemini 12. The loss of his friend Charles Bassett gave Aldrin his first seat into space. When he was officially told of the reassignment, Aldrin went to tell Bassett's widow, who informed him that Bassett had always felt that Gemini 12 belonged to Aldrin.

To fill in the dead-end back-up crew positions on Gemini 10, Slayton assigned Bean before he was due to head the Apollo Applications Programme. He had confidence that Bean would do a good job, and would be a good Command Pilot if for some reason Young could not make the flight. Despite a dead-end first flight assignment, Bean was the first of the Group 3 to receive a Command seat on a crew. CC Williams was assigned as back-up pilot for Gemini 10 after Slayton had originally lined him up as back-up pilot for Gemini 11 and Gemini 12 before moving him to Apollo.

On 19 March, Conrad and Gordon were confirmed as prime crew for Gemini 11. Armstrong was the most experienced back-up available, having just completed Gemini 8, and Anders was brought in to fill the back-up Pilot assignment. He would rotate back to Apollo, and afterwards to a new crew. Following Gemini 9, Slayton assigned Stafford to replace Bassett on the Borman crew, as an experienced Command Module Pilot. Collins would replace Anders on the same mission after Gemini 10, and Scott moved to McDivitt's Apollo crew after Gemini 8.

On 27 June 1966, the final assignments in the Gemini programme were

The Gemini 11 crews: (*front*) prime crew Dick Gordon and Pete Conrad, and (*rear*) back-up crew Bill Anders and Neil Armstrong. Armstrong had recently completed Gemini 8, and Anders had moved from Apollo training to back up Gemini 11. (Courtesy British Interplanetary Society/NASA.)

announced, with Lovell and Aldrin as prime crew for Gemini 12. Cooper (marking time) stepped up to fill in the last Gemini back-up Command Pilot assignment alongside Cernan, who had just completed Gemini 9 and was the best trained Gemini Pilot available prior to moving to Apollo.

TRAINING FOR A GEMINI MISSION

The completion of any spaceflight requires hundreds of hours of preparation for every hour spent in orbit. Most of the preparation for astronaut participation in a mission begins on the day on which they join the agency, in a programme of basic training skills to supplement their previous experience and knowledge. When an astronaut is assigned to a specific mission in a flight crew or back-up role, then training becomes more tailored to the demands of that particular mission. This can involve many briefings, meetings, visits to contractors or Principle Investigators, and

The last Gemini crew assignments, the back-ups for Gemini 12, Gene Cernan and right Gordon Cooper. Cernan had just completed Gemini 9 assignments, and Cooper was brought in to fulfil this dead-end assignment. (Courtesy British Interplanetary Society/ NASA.)

a personal training programme including physical training, home study, speciality training, and training in machines and devices that can accurately reproduce the space environment or elements of the flight.

Simulators offer, as accurately as possible, the conditions that the astronauts will encounter on the mission, allowing them to practise certain techniques or modes of the mission (from normal to emergency conditions) many times. The most accurate environment to prepare for a new spaceflight is space itself, but apart from utilising previous flight experience, this is impracticable. In the mid-1960s, training for Gemini involved all the state-of-the-art training devices and computer technology that was available.

Astronauts who used Gemini training devices later reported (having flown their mission) that the simulators were a high-fidelity representation of their mission, and accurately produced the conditions of the actual flight. There were, inevitability, a few minor discrepancies, such as the response of the controls, the displays in the simulator compared to those in the actual spacecraft and, of course, the ever present 1 g force on Earth as opposed to microgravity in orbit. In addition to the high-fidelity simulators, the astronauts also used several mock-ups or part-task trainers that were representative of hardware, or could perform only some of the functions of the real spacecraft. The major simulators related to Gemini training were launch, in-flight, and re-entry and landing.

Launch

The first training sequence for simulating the launch of a Gemini spacecraft by a Titan II was the *Dynamic Crew Procedure Simulator*, located at MSC. This was originally based at Ling-Temco-Vought Inc, Dallas, Texas, and was called the *Moving Base Abort Simulator*. It was used to provide launch profile simulations of normal flights and aborts as well as spacecraft/booster systems malfunctions. The simulator was able to reproduce a wide range of the motions which such malfunctions could inflict on the ascending spacecraft. As flights progressed, this (as with other simulators) was revised using actual flight data. After completing initial practise runs, the crews would begin their launch phase in this simulator as part of their Gemini mission simulator session.

The *Centrifuge* at the Naval Air Development Center Aviation Medical Acceleration Lab, Johnsville, Pennsylvania, was employed to simulate both launch and re-entry stresses on the crew. The first phase took place in July–August 1963 when, in a shirtsleeve environment, the astronauts evaluated Gemini displays and were spun on launch and re-entry acceleration profiles. For prime and back-up flight crew training, these profiles were run with the astronauts wearing full pressure suits.

In-flight

Initially, the *Gemini Mission Simulator* at McDonnell, in St. Louis, was the only flight simulator that matched the inside of the spacecraft. It provided an environment that represented the range of visual displays, sound cues, and even shaking, that the astronauts could expect to experience from launch to landing.

Each simulator interior was later adjusted for every crew mission and the physical requirements, based on Grissom's early work on this device. One of the first assignments on Gemini, it was given to Grissom due to his Mercury experience, working very closely with McDonnell engineers and technicians. The first cockpit was designed around him, giving him the best field of view, instruments, and windows, and was dubbed the 'Gusmobile'. The problem was that Grissom was also one of the smallest astronauts. Young (who was only two inches taller) had to have his seat compressed, and Stafford, being six feet tall, required both seat and hatch adjustments to fit in. In July 1963, measurements indicated that fourteen of the sixteen astronauts could not actually fit in the cockpit as designed, and it needed redesigning to allow them to fit in and close the hatch! The simulator was improved when it was moved to Houston in July 1964. A second unit was available at the Cape, but due to configuration and set-up work, it was not ready for use until November 1964.

The crew would first conduct simulations in their shirtsleeves before progressing to full pressure suits. As the training programme progressed, the Mission Simulator was linked to the Mission Control Center, with Flight Controllers manning the consoles and managing the duplication of events intended for the mission. They also had to solve in-flight emergencies or failures set by the training instructors. During the latter stages of training, the crew wore training or flight pressure garments, with a final run on the Dynamic Crew Procedures simulator only three weeks before the mission.

The Gemini mission simulator at MSC. The training instructor consoles are at the rear.

During post-flight debriefings, many astronauts reported on the accuracy of these simulators compared with the real events. The major problem was in keeping the simulator software up to date with the developing flight software. The Gemini 3 crew received their re-entry flight software to train on only two weeks before flying the mission, and the Gemini 4 crew received theirs only a month before launch. However, by the time the Gemini 8 crew was in training, accurate software was being used for their entire training programme – indicating the real-time learning curve that was being incorporated into the Gemini training programme.

For rendezvous, the astronauts began on the *Hybrid Simulator*, which included accurate flight controls and displays of both the Guidance and Control System and the Propulsion System. The rest of the crew compartment was only a mock-up. These runs were conducted in shirtsleeves, and often with instructors leaning over the side of the simulator, offering advice and comments as the run progressed. The crews conducted rendezvous runs in normal, back-up, and failure modes, with a random star-field background provided as an initial reference visual display. It was also found that data on the use of the attitude and manoeuvre consoles recorded in this simulator were accurate when compared with actual flight data.

The next stage of mission training took the crew to the Cape to use the Gemini Mission Simulator, in which a total spacecraft configuration could be used. Once again, shirtsleeve runs were performed before the crew progressed to wearing pressure suits; but in this device, 20% of simulator runs during the later stages of rendezvous training were completed with the astronauts wearing training pressure suits (which could take more wear), and then the flight suits.

The crews were also able to practise the time-consuming unstowing of cabin gear and crew compartment configuration and management for both the first orbit (M = 1 – up to 90 minutes after launch) and third orbit (M = 3 – up to 4 hours 30 minutes after launch) rendezvous profiles. During the missions, every minute of crew time was critical in achieving these two rendezvous profiles so early in the mission.

With the use of star-fields at the point of rendezvous based on the day and time of launch, the astronauts reported that the simulator optical system was accurate to the flight conditions, except for the magnitude and sharpness of the lights on the Agena. From Gemini 6, the Mission Simulator and MCC were integrated for rendezvous simulations, but it was not until Gemini 9 that a successful rendezvous simulation could be achieved on a target generated from Mission Control. Again, full pressure suits were used in these simulations, which resulted in numerous changes to cabin stowage and sequencing, based on simulator experience. Several failure modes in rendezvous, guidance and navigation were also inserted into the simulators for both the crew and the Flight Controllers to determine how they would overcome the problems – which did not involve just one failure, but could include multiple failures at the same time.

Each mission carried a number of experiments (see p. 331), and accurate training models were reproduced so that the crews could train on identical hardware in the Gemini spacecraft mock-ups and simulators at McDonnell and MSC. The astronauts took cameras, film, and some of the experiment hardware on T-33 and T-38 aircraft flights, which provided them with useful experience in handling this equipment in a confined space. The main difficulty encountered in supplying experiment training devices was in providing accurate representations of the equipment in a time-scale that allowed the astronaut to train with the hardware which they were to operate in space. In some cases this was just a few days before launch.

In missions involving docking with the Agena (initially Gemini 6, and then Gemini 8 through Gemini 12), the Gemini Mission Simulator provided a visual target that responded both to crew input and to the simulator instructor station, providing crew-controlled movements from the Gemini simulator, and Agena movements from the instructor. The crew could experience both docked and undocked modes, which allowed attitude manoeuvres and the simulation of primary and secondary propulsion systems, and reproduced accurate time delays between control input of commands and the execution of that command. Target-vehicle failures were also factored into this training period.

At MSC there was also a *Gemini Docking Trainer*, at MSC, which provided experience of physically docking the Gemini to an Agena mock-up. All control modes of the Gemini and Agena could be reproduced, and variable lighting could also be used to simulate the exact conditions of illumination on each flight. The most difficult task was to simulate a dynamic 6° of freedom motion capability on the ground to precisely equal that in orbit, and as a result, the crews indicated that the final docking manoeuvre was somewhat easier on the mission than in the simulator.

On the final two missions, the Dynamic Crew Procedure Simulator at MSC was used for accurate simulations of *Tether Dynamics*. In using this facility, the two

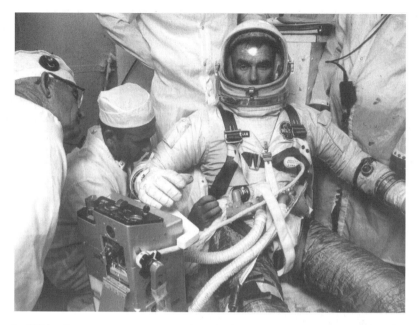

In the White Room at LC 19, Gene Cernan undergoes a 1-g walk-through of EVA operations for Gemini 9. Suit technicians check the connections of the ELSS that attaches to the chest of the suit.

crews were able to develop the basic timelines and control tasks for the experiment, and demonstrated the value of simulator training in flight by how accurately they completed the experiment on orbit.

All crews completed a series of *system operations* for both normal and emergency situations, following briefing at MSC and the Cape and using the mission simulators in those locations. Less emphasis was placed on emergency situations, allowing more time for planned mission operations. When MCC-H was linked to these simulations they became known as *network simulations.*

EVA training (see p. 269) was initially divided into two phases, and then three phases towards the end of the programme: 1-g training, zero-g training, and underwater training.

The first phase was a series of *1-g walk-throughs*, allowing the astronauts to familiarise themselves with checklists, hardware and storage, as well as the EVA tasks themselves. Initially in shirtsleeves, training suits were later used, and pressure suits were also used for altitude chamber runs, air-breathing platform simulations, and the evaluation of body harnesses and restraints.

The EVA astronauts then progressed to *zero-g training* onboard a USAF KC-135 aircraft, following the ballistic trajectory of parabolic curves to induce the short periods of weightlessness. Inside the padded crew compartment of the aircraft were mock-ups of different parts of the spacecraft. Depending on the mission EVA requirements, these could include the hatch and Pilot crew station; the Re-entry

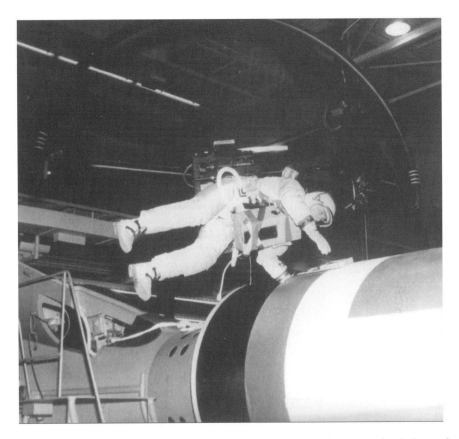

Using a 'Peter Pan' rig to support the weight of the astronaut, zero-g simulations of
EVA techniques – such as this translation from Gemini to Agena – could be practised.

Module (for AMU donning); partial spacecraft nose section docked to a target
docking adapter (for experiment retrieval, deployment, or attachment); a quarter
section of an Adapter Module for experiment evaluation (D016 Power Tool
experiment), or a second including a nitrogen connection to attach the 50-foot
umbilical; or an adapter with all internal EVA provisions. Here, it was normal for
both the prime and back-up Pilot who would attempt the EVA to complete a series
of five training sessions in the aircraft, each with forty parabolas per session. As well
as the Pilots, the Command Pilots participated in and observed these training runs,
allowing for post-simulation debriefing and improvements where necessary.

The final phase was introduced from July 1966, as part of a Langley Research
Center sponsored demonstration of *water immersion EVA* training techniques. These
were filmed, and were then evaluated with the objective of providing water tanks to
simulate EVAs on future missions. Simulations were initially conducted in support
of Gemini 10, using contractor personnel at the Environmental Research Associates
Laboratory in Maryland. Gemini and Agena mock-ups were constructed, and

Inside the KC-135 'vomit Comet', zero-g activities could be practised for about 20 seconds at a time in forty or more parabolic profiles. Here the astronaut simulates exit and entry of the Gemini hatch during EVA.

Buzz Aldrin, working at the docking collar of the Agena, completes underwater simulations of the EVA which he planned to undertake on Gemini 12. This was the first time that the water tank was used for training for EVA activities.

Gemini EVA suits and support equipment were provided for the underwater tests. Further evaluations were conducted with a re-enactment of the Gemini 9 EVA, using the pilot as a test subject to evaluate the accuracy of the tank compared with flight conditions. The recommendations derived from this simulation were used on the impending Gemini 11 EVA, and finally in training the Gemini 12 Pilot.

Re-entry and landing
The crew initially practised the re-entry phase of their mission at MSC, and then later at the Cape. Using the Mission Simulators, they could reproduce two reset points for training – one immediately prior to retro-fire and the other at 400,000 feet. This capability enabled them to concentrate on repeated retro-fire and re-entry sequences independent of the rest of the mission simulation.

Performed in shirtsleeves as well as in their suits, the simulation used accurate constellation positions, allowing the crew to work in accurate conditions linked to Mission Control for the simulations, and to repeatedly practise accurate procedures and timelines.

The crews also had to simulate *water egress* following splash-down. Initially, there was a bench review of the process at McDonnell before using the water tank set-up at Ellington AFB. Training was conducted both with and without suits, and either floating or submerged. They then practised emergency egress from a model of Gemini in the Gulf of Mexico, and finally practised while fully suited and using lift rafts. They also received a refresher course in parachute landings.

Practise makes perfect
In NASA's training plans, the first mission preparations for Grissom alone indicated how much training was involved:

Cockpit hours (as a crew)	36 simulations at St Louis
Spacecraft check-out hours	14 hours; nine inside the altitude chamber
Cape simulator	77 hours
Ling abort simulator	225 runs
Spacecraft at Cape time	19 hours
Launch aborts	20 normal and 46 aborted launches
Insertions into orbit	13 normal speed, five over speed, four under speed
Platform alignments	8
Flight plan enactment	9
Retro-fires	107
Re-entries	64
Total flight profiles	51
Systems malfunctions	211 (57 sequential, 34 electrical and communications, 17 attitude control and manoeuvring electronics, 30 orbital attitude and manoeuvres, 16 re-entry control, 36 guidance and control, and 21 environmental control
Centrifuge	9 hours 30 minutes
Mission flight plan simulators	77 hours

For every hour in space, an astronaut spends dozens more hours in simulators. Here, Aldrin and Lovell participate in systems tests and simulations on the vehicle (Gemini 12) that they were to fly in space. (Courtesy British Interplanetary Society/NASA.)

Planned mission training hours

	GT-6	GT-8	GT-10–12
Examples from 1965–1966			
Spacecraft testing	150	150	100
Gemini mission simulator	75	75	100
Experiments	–	30	75
McDonnell engineering simulation	30	20	75
Systems briefing	50	50	66
Operations review	30	30	30
Flight plan review	30	–	24
Zero-g aircraft (including EVA)	8	24	24
Planetarium	8	24	20
Spacecraft mock-up	16	16	16
Translation and docking training	8	16	16
Egress and recovery	12	24	12
Dynamic crew procedures simulator	12	12	10
Part-task trainer	0	0	10
Air bearing (EVA)	0	15	10

Grissom also spent an average of 25 hours a month flying the T-38 between centres, and 200 hours in briefings and reviews of mission plans. And besides all this there were many more hours of physical training, medical examinations, suit fittings, experiments training, celestial recognition training, and water egress training – for a three-orbit, 4-hour 30-minute mission with few experiments, no rendezvous or docking, no EVA, and very little extended-duration habitation training. Every mission benefited from previous experience gained by both the back-up and the flight crews, and training changed as the programme developed.

An example of CB support assignments and deployment during a mission in progress (Gemini 4, flown in June 1965; prime crew, of McDivitt and White) is shown below. Borman and Lovell (the back-up crew) were at the Cape until launch, and then flew to Houston to monitor the flight.

Aldrin	Observer at the Cape and then at Houston
Anders	CapCom (Guaymas, Mexico)
Armstrong	Observer at the Cape (back-up Command Pilot, Gemini 5)
Bassett	Observer at Houston
Bean	Observer at Houston
Cernan	CapCom 3 at MCC-H
Chaffee	CapCom 2 at MCC- H
Collins	Observer at the Cape and then at Houston
Conrad	Observer at the Cape and then at Houston (prime Pilot, Gemini 5)
Cooper	Observer at the Cape and then at Houston (prime Command Pilot, Gemini 5)
Cunningham	CapCom (Kauai, Hawaii)
Eisele	Observer at Houston
Gordon	Observer at the Cape and then at Houston
Grissom	CapCom 1 at MCC-H (Command Pilot, Gemini 3; back-up Command Pilot, Gemini 6)
Schirra	Observer at Houston (Command Pilot, Gemini 6)
Schweickart	Blockhouse ('Stoney') at the Cape
Scott	CapCom (Carnarvon, Australia)
See	Observer at the Cape (Back-up Pilot, Gemini 5)
Shepard	CapCom at the Cape (Chief Astronaut)
Slayton	Observer at the Cape and then at Houston (Director, FCOD)
Stafford	Observer at Houston (Pilot, Gemini 6)
Williams	Booster (Titan) fuel tank monitor at the Cape
Young	Observer at Houston (Pilot, Gemini 3; back-up Pilot, Gemini 6)

(Carpenter assigned USN Sealab programme; Freeman deceased; Glenn resigned)

For Gemini, flight control specialists took over the manning of remote stations, replacing the astronauts who had done so during Mercury missions. Launch control at the Cape and at MCC Houston was manned by astronauts in the CapCom role. The original MCC Houston was first used during Gemini 4, and was last used during STS-70 in 1995, prior to a new control centre being activated in the same extended building. Up to Gemini 4, several astronauts were still deployed to some remote stations as the new network came on line.

Prime and back-up crew assignments

Mission	Prime (Commander–Pilot)	Back-up (Commander–Pilot)
3 (i)	Shepard–Stafford	Grissom–Borman
	Shepard grounded; crews reassigned as	
3 (ii)	Grissom–Young	Schirra–Stafford
4	McDivitt–White	Borman–Lovell
5	Cooper–Conrad	Armstrong–See
6	Schirra–Stafford	Grissom–Young
7	Borman–Lovell	White–Collins
8	Armstrong–Scott	Conrad–Gordon
9 (i)	See–Bassett	Stafford–Cernan
	See and Bassett killed in air crash; crews reassigned as	
9 (ii)	Stafford– Cernan	Lovell–Aldrin
10	Young–Collins	(i) Lovell–Aldrin (original)
		(ii) Bean–Williams
11	Conrad–Gordon	Armstrong–Anders
12	Lovell–Aldrin	Cooper–Cernan

CapCom assignments

Mission	Cape	Houston
3	Cooper	Chaffee (monitor only)
4	Williams	Grissom
5	Grissom	McDivitt, Aldrin, Armstrong
7/6	Bean	See, Cernan, Bassett
8	Cunningham	Lovell
9	Aldrin	Armstrong, Lovell, Gordon, Aldrin
10	Cooper	Cooper, Aldrin
11	Williams	Young, Bean
12	Roosa (first Group 5 1966 assignment)	Conrad, Anders

Flight operations

Before the separate objectives accomplished during the Gemini programme are reviewed, it is worth recalling the flight activities of the period between April 1964 and November 1966. Two unmanned and ten manned missions were flown over a 31-month period – an average of one mission every 2.6 months. However, the ten manned flights which took place between March 1965 and November 1966 were completed in just twenty months – an average of one launch every eight weeks. This is a frequency that even the Shuttle has difficulty in matching or surpassing a generation later.

PLANNING FOR THE FLIGHT

With authorisation to proceed with the programme and the definition of mission objectives under discussion, it was not long before hardware was being fabricated and plans were being laid to expand the astronaut corps. Another element of preparation for flight operation was the planning of the missions themselves.

Mercury Mark II flight plans

Within the planning documents for Mercury Mark II in 1961 were the origins of what became the Gemini flight plans. A Master Plan for Orbital Operations, dated 19 July 1961, indicated that four 18-orbit Mercury flights in 1963 would be followed by eight one-man Mercury Mark II missions, flown at two-monthly intervals beginning in October 1963 and completed by December 1964. It was during these eight Mark II missions that one of the objectives of the programme – the development of the techniques of orbital rendezvous and docking – would first be attempted.

By 14 August 1961, a preliminary development plan refined operations further, and stated that there would be six major objectives achieved on the ten missions of the new spacecraft. In total, there would be three unmanned flights and seven manned missions. The first in the series would be flown in March 1963, as an unmanned engineering compatibility test between the spacecraft and the launch vehicle that would also carry biological payloads. The orbit of the vehicle was planned for a high ellipsis, ranging between 100 and 870 miles. This would take the spacecraft through the Van Allen radiation belts to gather information on the radiation levels, and thus meet the second of the programme's objectives, besides supplying information relevant to the Apollo lunar missions that were to follow Mercury Mark II.

The subsequent missions were still to be attempted every two months until September 1964. The second Mercury Mark II would also be the first with a crew on board, and would attempt an 18-orbit test of the spacecraft and systems. Astronauts would follow this on flights 3 and 4, on missions of extended duration that could last up to seven days each. These three missions would partly achieve the programme's primary objective of extended-duration spaceflight. There would then be even longer flights of up to fourteen days on Mission 6 and then Mission 8 (5 and 7 had a different objective). However, there would be no astronauts onboard, and they would instead carry animals and other biological specimens.

There was concern at the time that the spacecraft systems were still unreliable for such a long duration in space, and that manual input by the astronauts after two weeks in space might jeopardise safe re-entry. In 1961, only one man had been in space for 24 hours – cosmonaut Gherman Titov – and had experienced alarming symptoms of disorientation after only 17 orbits. To send astronauts into space for fourteen days before the hazards were investigated more thoroughly was considered dangerous. Manned long-duration flights for Mercury Mark II at this time were therefore limited to no more than seven days.

The third objective was a controlled landing – preferably on land rather than in water – and this was to be the target of several manned flights. The fourth objective was to rendezvous and dock with a separately launched Agena B target vehicle, and this was planned for missions 5, 7, 9 and 10. The fifth programme objective was listed as astronaut training, and this would only be seen as successful by the productivity of the programme as it unfolded.

There was a sixth objective, but this would only be attempted upon the successful conclusion of all the previous five. Here, four add-on Mercury Mark II missions were planned that would extend the programme into May 1965. Under these additional flight plans, a Mercury Mark II would rendezvous and dock with a liquid-fuelled Centaur upper stage. This, in theory, could propel the combination to lunar orbit in order to supplement the goal of Apollo on a mission of up to seven days. Initially, two deep-space high elliptical missions would be achieved, before sending two circumlunar flights to fly around (but not orbit) the Moon in March and May 1965.

The report also suggested an alternative, accelerated plan of nine flights, which would see the 18-orbit unmanned qualification and radiation test flight followed by an 18 orbit-manned qualification, and just one manned seven-day flight. Flights 4 and 5 would conduct rendezvous and docking development tests with Agena B targets, and flights 6 and 7 would attempt the deep-space orbital missions using the Centaur upper stage. The programme would be completed by missions 8 and 9, taking American astronauts around the far side of the Moon in a Mercury Mark II/

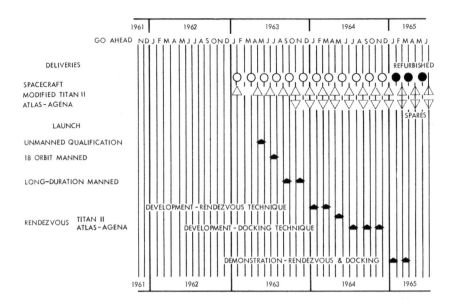

This launch schedule, dated 27 October 1961, was issued as part of the final version of the Mark II Project Development Plan.

Centaur combination as early as May 1964 – which would have been 4½ years before Apollo 8 actually achieved the feat!

Just one week later, a release dated 21 August omitted any reference to the Mercury Mark II lunar missions. The only comments stated that *if* Centaur was employed as a target vehicle, then the added velocity potential would 'allow for expanded investigations'. Exactly what these 'expanded investigations' might be was not revealed. The suggestion of sending Mercury Mark II to the Moon was also soon dropped, as it might seriously affect the budget request for Apollo. However, other studies were implemented (see Military Gemini, p. 361).

A new forecast

A revised development plan, dated 27 October 1961, reflected a new approach in flying twelve missions between May 1963 and March 1965, instead of the ten previously planned. The new objectives for the series were listed as:

Flight	*Objective*
1	Unmanned qualification test of a spacecraft and launch vehicle up to 18 orbits
2	Manned 18-orbit qualification test
3	Long duration of seven days and possibly up to fourteen days
4	Long duration up to fourteen days
5, 6, 7	Rendezvous development techniques
8, 9, 10	Docking development techniques
11, 12	Demonstration of rendezvous and docking techniques

In this forecast, all references to the lunar and deep space missions, as well as the animal flights, had disappeared. Since the spacecraft would not fly into or through the Van Allen belts, and because ground-based studies suggested that fourteen days was achievable in stages, the astronauts would attempt two-week missions. But with a series of Apollo Earth-orbital missions also planned at this time, further duration studies on mission of up to fourteen days could be achieved on those flights, leaving Mercury Mark II to tackle a much more demanding objective.

The whole emphasis of the programme was to overcome the recognised difficulties of rendezvous and docking, and to investigate elements of the technique, rather than use it in an operational role, as was planned for Apollo. Increasing the number of missions from ten to twelve also allowed NASA to double the high-priority rendezvous and docking missions from four to eight.

Missions would now commence in May 1963, and were still aimed at every eight weeks between launches, completing the programme before the Apollo lunar missions began. Mercury Mark II had also expanded its costs from $347.8 million (1961) to $529.45 million (1961), which included a new allocation of $29 million (1961) for a paraglider landing system for the ground landing capability, and provision for spare spacecraft and boosters.

Earlier hardware planning envisage the use of just eight spacecraft and ten boosters, with two of the spacecraft being refurbished and re-used. This new plan included provision for refurbishing three flown spacecraft – but only as spares, and not as flight vehicles. The plan also encompassed fifteen Titan IIs (twelve flight vehicles and three spares, instead of the ten with three spares), and eleven Atlas Agenas (instead of only four). The programme would then have eight flight-qualified targets and three spare vehicles.

A new year and a new name
Following the official approval for the new programme in December 1961, and the naming of McDonnell Douglas as primary contractor for the new vehicle, the initial plan was to deliver twelve spacecraft, fifteen launch vehicle adapters, eleven target vehicle adapters, static test articles and ancillary hardware. The first spacecraft would be delivered for March 1963, and then the eleven others every 60 days.

Two weeks after the naming of the new programme, the late official acceptance of the programme saw the first flight slip to the third quarter of 1963, with six weeks between each of the first three missions before reverting back to two months between the remainder. The programme was scheduled to end in the second quarter of 1965. The first Agena mission was planned for the first quarter of 1964.

By the release of the second analysis, dated 14 March 1962, the schedule reflected launch vehicle delivery as well as spacecraft delivery. Agena was not included in this forecast, as it was still too early in its involvement in Gemini, and so effective analysis was limited. One way in which Gemini benefited over Mercury was in the use of several test articles. Although it slowed the production line of flight hardware, it assisted in providing valuable data and experience that could easily be incorporated into later spacecraft and compensate for any delays. The report indicated that the first unmanned qualification mission was manifested for the third quarter of 1963. The second flight, and the first to carry a crew, slipped to the fourth quarter of the year, and the first Agena flight slipped to the second quarter of 1964. The rest of the missions remained on target for every two months, ending with the twelfth flight in March 1965.

TEST HARDWARE

In addition to the flight units, the contract with McDonnell envisaged a series of mock-ups and boilerplates (full-size mock-ups) for systems testing and astronaut training.

Boilerplates and test articles

A third analysis of projected flight operations, dated 4 May 1962, included a ground test programme, and also commented that construction of flight hardware had commenced. Two boilerplate spacecraft had been added to the programme for ground testing. At this time, flight 2 – the first to carry astronauts – was also planned to demonstrate the paraglider; but this was already encountering delays, and it was thought that it would slip beyond flight 2, although a close eye was being kept on its scheduling. The manufacturing schedules were also being affected by the late delivery of key components by various vendors. Spacecraft 1 (unmanned) was still awaiting instrument and recording systems and the inertial platform, and on Spacecraft 2 (first manned), communications and electrical systems components were also delayed. NASA's optimism in predicting an on-schedule launch was dispelled with another launch slip for the first flight from August to September 1963 (announced on 28 August 1962) due to these supply delays. But it was still expected that the first manned mission would meet its November 1963 launch date.

Just under two months later, on 19 October, MSC Assistant Director for Administration, Wesley L. Hjornevik, presented senior MSC management with a 1963 budget cut to of $27 million from an already tight Gemini budget. Discussions indicated that the programme could survive on this target only if something was cut from the objectives.

The long-duration flights required the least expense for development, but would need more mission support during the flight. The rendezvous and docking experiments and the Agena target vehicle required a significant slice of the Gemini programme budget, but cancellation or a cutback in either might have had significant consequences for the Apollo lunar programme. The paraglider ground-landing programme seemed to be the most likely candidate for deletion from the programme, together with a rethink concerning the Agena docking programme to determine whether some missions could combine objectives instead of duplicating previous mission activities.

There was also the series of mock-ups, boilerplates and simulators, and it appeared that some cuts would be made with these non-flight models of the spacecraft. However, although construction time for these spacecraft would slow production, McDonnell felt that time would be saved later in the flight programme, as problems would be remedied early in the development cycle. Using a series of static and dynamic test articles (facilities not available during the Mercury programme), McDonnell proposed four metal models of the spacecraft for use in escape and recovery testing (boilerplates), and another four articles for tests on the structural integrity of the spacecraft. There was also an extra flight spacecraft and target-docking adapter to be used for an extended series of laboratory-simulated missions called Project Orbit. Several mock-ups and partial spacecraft models, and two flight trainers, were also proposed.

To compensate for the budget reductions, McDonnell agreed to limit the amount of development in several areas of systems testing, in both spacecraft and ground equipment. In addition, they also reduced the spares list, the quality assurance and reliability programme, and the massive paper trail, to help alleviate the pressure on the test programmes. This was very helpful, but did not immediately resolve the budget difficulties.

As a result of the budget restrictions, the first flight was subsequently targeted at December 1963, with the first manned flight in March 1964 and the other ten at two-monthly intervals. The first Agena was planned for the fifth mission. This, of course, also affected the launch vehicle production programme at the Martin (Titan), General Dynamics (Atlas) and Lockheed (Agena) facilities, which were already suffering their own setbacks in their development programme for Gemini.

Project Orbit

While the budget issue was constantly reviewed during the ensuing months, the spacecraft test programme continued. In October 1962, McDonnell had decided that one STA could be deleted from the programme, and the Gemini programme office agreed. The GPO also suggested that Project Orbit might be cancelled, but McDonnell's response was not favourable. As a compromise it was decided to modify Project Orbit to simply test the spacecraft's heat balance, and so the test programme was renamed Spacecraft Thermal Qualification Test.

On 20 February 1963, the GPO also decided, that Spacecraft 3 would rejoin the flight programme and would be replaced by a 'beefed-up Structural Test Article'. McDonnell had planned to use Spacecraft 3 in a series of simulated manned orbital flights in a vacuum chamber, as part of the original Project Orbit programme. It would have been used in laboratory simulations to investigate any encountered problems, and to evaluate engineering changes during the life of the programme. Structural Test Article Number 1 was intended for load tests on the paraglider, ejection seats, hatch mechanism, and cabin pressurisation. In the new plan, STA-1 would be redesignated Spacecraft 3A, and would replace Spacecraft 3 in the vacuum tests, and so allow that vehicle to be used in the flight programme. In view of the money crisis, McDonnell conducted a full review of the Structural Test Article programme, which resulted in the elimination of STA-1 objectives, instead assigning 3A to handle the launch, re-entry, abort, and parachute load tests. STA-4 would now be used for seat hatch and pressurisation loads, and for a programme of dynamic response tests.

Boilerplate 3A sits in the production area of the McDonnell plant in St Louis, prior to shipment to Weber Aircraft for qualification of the ejection seat system.

Escalation of costs

When NASA revised its preliminary budget for Fiscal Year 1964, announced on 8 March 1963, the estimated costs for Gemini had almost doubled the amount first approved in December 1961, and it was now in excess of $1 billion. As well as problems in the development of the Titan II launch vehicle and the slow pace of spacecraft delivery, a critical review by programme management into the flight schedule resulted in a further change to the launch manifest.

The first change was the replacement of one of the manned flights with a second unmanned flight to further qualify the Titan II for manned flight. The flight schedule, approved on 29 April 1963 then consisted of:

Mission 1 Launch, December 1963. A qualification of the Titan with the Gemini as an instrument carrier, which would not separate from the second stage nor be recovered.

Mission 2 Launch, July 1964. A sub-orbital ballistic test flight to qualify the Titan and the heat shield for manned flight.

Mission 3 Launch, October 1964. This would be the first manned flight. Originally an 18-orbit mission (24 hours) it was decided, on 2 June 1963, that a three-orbit mission was more suitable for the first manned flight.

Mission 4 Launch, January 1965. A seven-day and first rendezvous mission with an evaluation pod. The three months between launches also allowed more time to evaluate the previous flight and prepare the next flight.

Mission 5 Launch, April 1965. The first rendezvous flight with an Agena.

Mission 6 Launch, July 1965. A 14-day long-duration flight, allowing time to learn from the results of the seven-day mission.

Missions 7–12 Each of these would feature rendezvous and docking with Agena, and would last for about three days, with the final flight occurring in January 1967. The first land landing was expected in October 1965, from the seventh mission onwards. Up to then, each mission would descend by parachute into the ocean.

In July, the GPO explored the possibility of providing a back-up spacecraft for the first unmanned test flight. These would be constructed from a boilerplate re-entry module and a production article adapter section. By August this was approved, and it received the designation Gemini–Titan 1A. This was to be constructed by local contractors to MSC (not at McDonnell), and would be modified after delivery in September 1963. The plan was to use the mock-up for flotation tests at MSC and in the Gulf of Mexico, and to prepare the vehicle for a flight only if Gemini 1 failed in its primary objectives, whereupon Gemini GT-1A would fly a duplicate mission.

Objectives and manifests

The combination of hardware, testing and budget difficulties gradually delayed the launch programme into 1964. Difficulties with the development of the Titan and the elements of the spacecraft components, the threat of the USAF taking over the Gemini programme, restricted budgets and rising costs, and a troubled test and qualification programme, not only conspired to delay the flight programme, but also threatened to affect the programme objectives.

The qualification of Titan II, the Agena target vehicle, the paraglider, and spacecraft systems, all contributed to delays and changes to the flight programme (discussed later). Before reviewing the operational objectives of Gemini in detail a discussion of the pre-launch activities, objectives and achievements for each of the twelve missions follows.

THE GEMINI PRE-LAUNCH SEQUENCE

Although every Gemini mission was different in terms of mission objectives and flight activities, each followed a basic profile of preparation and launch. This began several months in advance of the flight, with the construction and testing of the components at the contractors, and then, after acceptance testing, shipping to the Cape for preparation for flight.

Preparation At McDonnell, the spacecraft were fabricated and the equipment was installed, with the two modules (Re-entry and Adapter) joined together. Everything was then tested in a programme of System Assurance Tests and simulated flight tests, and a session in the altitude chamber, before shipment to the Cape. They were then take to LC 19 for mechanical mating to the Titan and a programme of combined tests, a final systems test, and simulated launches. Several of these tests

A Gemini spacecraft is prepared for a programme of thermal tests at MSC, Houston.

and simulations were conducted with either McDonnell personnel, or with astronauts from the back-up or prime crew.

Titan II The Titan was prepared at the Martin plant at Denver, where major welding was completed, and the tanks were then delivered to Baltimore for assembly and tests. Horizontal tests were followed by vertical tests, at first without power, and then with power applied. Sub-system Function Verification and Combined Systems tests were completed before an inspection by a Vehicle Acceptance Team to DD-250 Standards. Following shipment to the Eastern Test Range (ETR), the Titans were erected at LC 19. The first stage was raised vertically, and was then joined by the second stage to begin a programme of on-the-pad tests with power applied. Sub-system and combined system tests were conducted along with a tanking exercise, which was then followed by the mating of the spacecraft on the top of the launch vehicle. Next came a Combined Systems Test and practise count-down exercises – called the Wet Mock Simulation Launch. Preparations were completed with a Final Status Simulated Flight Test.

All the final tests on the individual and combined modules or sections of the hardware included radio and radar circuits, simulated flights linked to computers,

Workers cover the stages of the Titan II with canvas, and the first stage is then loaded onto a modified Boeing Stratocruiser ('pregnant guppy') for shipment to the Cape.

A Gemini spacecraft is hoisted to the top of the gantry at LC 19, prior to mating to the Titan launch vehicle. Tests have been carried out independently on both vehicle, and after mating they will be combined in a series of integrated tests leading to the final preparations for launch.

simulated launches with the flight crew on board, servicing and fuelling of the launch vehicle with a full check-out of the engine systems, a combination check of the Gemini–Titan interfaces, a 300-minute dress rehearsal count-down, and a final two-day service and check-out. It was only at the completion of these intricate tests that the count-down for launch commenced, ending with the launch of the vehicle and ascent into orbit.

The target vehicle The Atlas was adapted to carry the Agena and was fitted with Gemini-specific equipment before a systems test and a composite test were completed. After a data review and a final acceptance review, the vehicle was delivered to ETR, and was then erected on LC 14 for a sub-systems test. The Agena target was modified from the basic Agena stage, and received the Gemini docking equipment before a combined systems acceptance test and an inspection by the Vehicle Acceptance team. The target vehicle was a production-line vehicle modified from the standard Agena upper stage for Gemini as part of a Government Furnished Equipment programme. When delivered to the vendor, it was considered to be flight-ready, and was converted to accept Gemini docking hardware. The same tests were used to verify its readiness and certification as a Gemini Target Launch Vehicle.

Preliminary tests were held in Hanger E at the ETR, with interface tests and systems tests completed before the mating to the Atlas on the pad at LC 14, where further system tests were conducted on each stage and on the combined vehicle. In the docking phases of the Gemini programme, it was the Agena that was launched to orbit (or an attempt was made to launch it to orbit) before the Titan and the manned Gemini vehicle left the pad.

THE GEMINI MISSIONS, 1964–1966

All twelve missions flown under the Gemini programme were assigned a series of primary objectives that were designed to obtain the maximum return from the mission and secure important flight and operational data both for later Gemini missions and for Apollo and other programmes. How well these primary objectives were attained would classify the completed mission as a total success, a partial success, or a failure. Secondary objectives were also assigned, supplementary to the primary goals of the mission, and were performed only if the security and safety of the mission and achievement of the primary objectives were not compromised. These were not included in the assessment of the success of the mission.

In the twelve missions, classification of flight activities were not readily defined, as several missions that incorporated rendezvous and docking also included EVA or duration attempts. Broadly, however, the first three missions (Gemini 1, Gemini 2 and Gemini 3) focused on engineering flight tests, and the following three (Gemini 4, Gemini 5, and Gemini 7) were to investigate duration flight. The remaining missions (Gemini 6, and Gemini 8 through Gemini 12) were assigned to the development of rendezvous and docking procedures. It is in these three areas that I have summarised the flight operations, with the last six missions sub-divided into two groups. The first

The Gemini programme emblem and the official mission emblems worn by the crews of Gemini 5 to Gemini 12.

(Gemini 6, Gemini 8 and Gemini 9) attempted to achieve the first docking with target vehicles, while the latter three missions (Gemini 10, Gemini 11 and Gemini 12) applied this skill further to develop additional methods of rendezvous and re-rendezvous that had direct application to the Apollo programme. (A summary of all Gemini missions is presented in the table on p. 161)

FLIGHT TESTS: GEMINI 1, 2 AND 3, 1964–1965

Gemini 1 (GT-1)
Launched 8 April 1964; unmanned; not recovered.

Primary objectives To demonstrate Gemini launch vehicle performance and to flight-qualify the vehicle and its sub-systems for future missions. A programme of structural integrity tests on the spacecraft and the launch vehicle, and compatibility between the two, was to be completed from launch to orbital insertion, all of which were achieved. The primary flight programme also successfully demonstrated the launch vehicle and ground guidance systems, and monitored the malfunction detection system.

Secondary objectives To evaluate the spacecraft and launch vehicle operational procedures during count-down, launch, and ascent to orbital insertion, and to verify the launch and tracking networks, training for flight controllers, and to demonstrate the operational capability of the pre-launch and launch facilities. Again, all of these were achieved.

Gemini Titan 1 on the launch pad, with the launch vehicle erector lowered, during electrical and electronic interface tests. (Courtesy British Interplanetary Society/NASA.)

Launch vehicle preparations Major welding of the first Gemini Launch Vehicle (GLV) began in September 1962, with assembly completed on 21 May 1963. Testing on the vehicle occurred between May and October 1963. It was not accepted after the first Combined System Acceptance Test (CSAT) on 6 September or the Vehicle Acceptance Team (VAT) inspection on 11 September, and had to be tested again. The launch vehicle was delivered to the ETR on 26 October 1963, and was installed on the pad three days later. On 13 November 1963, power was applied for the first time. A series of tests was completed in January 1964, with the Sub-system and Combined Systems test and a Sequence Compatibility Firing (SCF). The Wet Mock Simulated Launch (WMSL) was conducted on 2 April 1964, and the Final Status Simulated Flight Test (FSSFT) on 5 April.

Spacecraft preparations Spacecraft 1 arrived at the ETR on 4 October 1963, and was mated to its launch vehicle on 5 March 1964. A month later, on 3 April, the spacecraft completed its Wet Simulated Launch/Simulated Launch Demonstration test for a final Flight Simulation Flight Test on 6 April. Each of these was completed a day after the similar tests on the Titan.

Emblem As this was an unmanned test flight there was no emblem adopted for this mission.

Call-sign None, as this was an unmanned mission.

Mission Flight operations of the Gemini programme began with an unmanned mission, 2 years 4 months 1 day after the official approval of the programme had been announced. Launch occurred at 11.00 am Eastern Standard Time (EST) on 8 April 1964, from LC 19. With no plans to separate the spacecraft from the second stage of the Titan or to recover the spacecraft, the launch vehicle placed the spacecraft in an orbit of 86.6 x 173 miles, six minutes after launch. Officially the mission lasted only three orbits, with all objectives met by GET 4 hours 50 minutes. However, tracking of the spacecraft by the Goddard Spaceflight Center network continued until it re-entered the atmosphere over the South Atlantic on 12 April and burned up during its 64th orbit.

Gemini 2 (GT-2)
Launched 19 January 1965; unmanned sub-orbital mission.

Primary objectives The second mission was targeted at demonstrating the efficiency of the re-entry heat-shield during a maximum-heating-rate re-entry. This time, the structural integrity of the spacecraft could be tested from launch to recovery for the first time, and the evaluation of the onboard systems and back-up guidance steering signals could continue during launch. Once again, all primary objectives were met satisfactorily.

Secondary objectives These included tests on the cryogenics, fuel cells and reactant supply system. These were all achieved, except for the fuel cell, which had been deactivated prior to launch. Further tests and demonstrations of the launch count-down, mission profile, tracking communications and flight controller training were all achieved.

Launch vehicle preparations Major welding on the GLV-2 began in September 1962, with assembly completed by 9 January 1964. A programme of unpowered and powered tests occupied the next three months before it was accepted and delivered to ETR on 11 July 1964. By 11 September the vehicle was erected on LC 19, and its sub-system and combined systems tests were completed by 20 October. GT-2 was mated to its launch vehicle on 5 November, to begin a series of tests that took most of that month, culminating in the Wet Mock Simulated Launch on 24 November. On 14 January 1965, just five days before launch, the FSSFT was conducted.

Spacecraft preparations GT-2 was shipped to the Cape on 21 September and on 5 November was mated to its launch vehicle for the series of pre-launch tests that lasted most of that month. A final simulated flight test occurred on 14 January 1965.

Emblem None, as it was another unmanned mission.

Call-sign None, as it was another unmanned mission.

Gemini 2 – weighing approximately 6,900 lbs – is carefully lowered for mating to the top of the Titan II second stage at LC 19. The absence of the Equipment Module's protective covering provides a clear view of the detail inside the module. (British Interplanetary Society/NASA.)

The launch of Gemini 2 from LC 19 on 19 January 1965. In the background, a helicopter hovers to record the launch on film!

Mission Launched on a sub-orbital ballistic test flight at 09.04 am EST, 19 January 1965. Gemini 2 reached a maximum altitude of 92.4 miles, with retro-fire initiated at GET 6 minutes 54 seconds. It landed in the Atlantic Ocean, 1,848 nautical miles south-east of Cape Kennedy, after a flight of 18 minutes 16 seconds. The success of this mission qualified the spacecraft and launch vehicle to carry the first astronaut crew on the next mission. This capsule was subsequently transferred to the USAF MOL programme, and after refurbishment it was relaunched, on 3 November 1966, on a second sub-orbital heat-shield qualification test (See Military Gemini, p. 361).

Gemini 3 (GT-3)
Launched 23 March 1965; Gus Grissom and John Young; three orbits.

Primary objectives To demonstrate manned orbital flight in the Gemini spacecraft and qualify it for further use, and to evaluate the two-man design and its effect on flight crew performance. To evaluate the global tracking network and demonstrate the capability of manoeuvring the spacecraft in orbit for the first time. In addition to all of these being achieved, the mission also demonstrated OAMS capability for back up retro-fire, and continued to demonstrate the performance of systems, checks and procedures from pre-launch to post-landing. One primary objective not fully achieved was the ability to control the re-entry flight path and the intended landing point. The landing was subsequently not as accurate as expected.

Secondary objectives To evaluate flight crew equipment, biomedical instrumentation and the personal hygiene system, all of which were achieved. The crew partially achieved the three experiments assigned and took some general photographs, despite the wrong lens on the 16-mm camera. They also successfully evaluated the effects of low-level longitudinal oscillations (the 'pogo' effect) of the Titan on the flight crew.

Launch vehicle preparations Major welding of the propellant tank commenced in June 1963, and the vehicle was completed on 6 June 1964. Like GLV-1, the GLV-3 Titan failed its first CSAT on 7 August and the VAT inspection on 17 August, but was finally accepted by 7 October. Delivery to ETR occurred on 23 January 1965, just two months before launch. Two days later it was on LC 19, with power applied on 29 January. The on-the-pad systems testing occupied most of February, and the Wet Mock Simulated Launch was completed on 8 March. The FSSFT was completed on 17 March, less than a week before lift-off.

Spacecraft preparations Installation of equipment for the first manned spacecraft began on 12 September 1964, with mating of the Re-entry Module and the Adapter Module occurring two weeks later. An altitude chamber test was completed on 18 November, followed by a simulated flight test on 21 December 1964. GT-3 arrived at ETR on 4 January 1965, and was mated to the launch vehicle on 17 February. A series of pre-launch tests culminated in the final simulated flight test on the spacecraft, held on 18 March 1965.

Emblem At this time, NASA astronauts had not begun designing mission emblems for their flights, but as a veteran of Mercury, Grissom had named his spacecraft (see below). There was a 'souvenir commemorative emblem' marketed commercially for Gemini 3 – a round design with a representation of the capsule in the centre. The mission designation of Gemini 3 was at the top with the names of the crew, and at the bottom, the name they assigned to the spacecraft. Years later, during training for his Shuttle missions (in the early 1980s), Young wore a NASA flight suit with a representation of each of his mission emblems, including one for Gemini 3 which revealed a Gemini capsule floating in the ocean after splash-down – again reflecting the name that they had assigned to the spacecraft.

Call-sign 'Molly Brown'. Grissom and Young wanted to give their spacecraft a name following the tradition set during the Mercury manned missions. Grissom used the name Liberty Bell 7 for his Mercury flight in 1961. That spacecraft sank due to a faulty circuit prematurely blowing the hatch and allowing in the sea water, so he wanted to name his next spacecraft 'Molly Brown', after the unsinkable heroine of a popular Broadway stage show. Although Grissom may have found the name a logical choice, NASA headquarters did not think it was a dignified name for a US spacecraft, and asked the astronauts for an alternative. Grissom proposed Titanic – which made Molly Brown a much better choice. The name was never official, and all subsequent Gemini missions adopted the flight number as their official 'call-sign'. Indeed, call-signs did not return until March 1969, when Apollo 9 flew with two separate spacecraft (the Command Module and the Lunar Module), requiring a method of radio identification – a tradition that continued to the end of Apollo in 1972, and which was reactivated to identify each Shuttle orbiter in the late 1970s.

Mission Astronauts Grissom and Young became the first to ride Gemini into orbit, with Grissom being the first American astronaut to fly in space twice. Gemini was launched at 09.24 am EST, 23 March 1965. During the flight, Grissom successfully performed three orbital manoeuvres, ensuring that Gemini 3 entered the record books as the first manned spacecraft to manoeuvre to a new orbit and not just translate around its axis. The changes were minute, but successfully demonstrated the capability of the spacecraft for later missions involved in rendezvous and docking. The maximum apogee was 121 miles, and perigee was 87 miles. The landing occurred at 14.16 pm the same day, after a flight duration of 4 hours 52 minutes and three orbits, successfully completing the test programme and allowing Gemini to progress to the first of the long-duration missions.

EXTENDING THE DURATION: GEMINI 4, 5 AND 7, 1965

Gemini 4 (GT-4, or GT-IV)
Launched 3 June 1965; Jim McDivitt and Ed White; four days.

Primary objectives To demonstrate the performance of spacecraft systems for a period of approximately four days, and to evaluate previously developed procedures for crew work–rest cycles, eating schedules, and real-time planning for long-duration flights. All objectives were achieved, although the computer-controlled re-entry was not flown due to an inadvertent alteration to the computer memory.

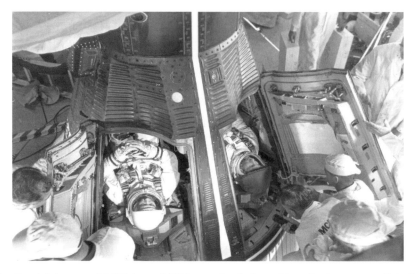

The Gemini 4 crew – McDivitt and White – take their positions in the spacecraft inside the White Room at LC 19. This was to be home for the next four days.

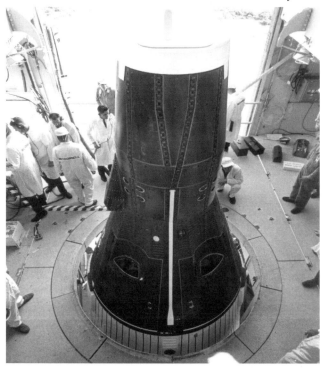

When everything is secure, the hatch of Gemini 4 is closed as the count-down progresses towards zero on 3 June 1965. Note the work platform surrounding the spacecraft, allowing foot access. McDivitt and White are inside the spacecraft, participating in pre-launch check-outs.

Secondary objectives To demonstrate EVA in space and evaluate of attitude and position control using a hand-hold propulsion unit or safety tether lines, which was achieved. Conduct station keeping and rendezvous manoeuvres with the expanded second stage of the Titan was only partially achieved due to excessive use of propellants in station keeping. Further evaluation of systems and manoeuvring ability was also achieved, together with eleven experiments.

Launch vehicle preparations Major welding on GLV-4 commenced in November 1963, and the assembly was completed by 4 September 1964. The series of system tests and verification programmes were completed over the next two months. The VAT inspection was delayed from mid-December until 19 March 1965, to allow modifications to be completed at the Baltimore plant instead of at the Cape. The Titan arrived at the ETR on 21 March, and was erected on the Pad by 29 March. A further series of vehicle checks was completed during April before the Wet Mock Simulated Launch on 13 May. The FSSFT was conducted on 29 May 1965.

Spacecraft preparations Installation of equipment commenced on 31 January 1965, and after the programme of assurance, flight tests and altitude chamber tests, the spacecraft was shipped to ETR on 4 April for mating to the launch vehicle on 23 April. Joint system tests began on 30 April, with the final simulated flight test occurring on 29 May.

Emblem The only emblem from this mission is a round souvenir emblem, with a depiction of a Gemini spacecraft and a tethered astronaut on EVA The crew names and flight designation are included, together with the phrase First Space Walk – although this was the second EVA (after cosmonaut Alexei Leonov from Voskhod 2 on 18 March) but the first US EVA. The crew wore the Stars and Stripes on their flight suits for the first time.

Call-sign There was no official call-sign for this mission, apart from 'Gemini Four'. A few newsmen covering the flight called the mission 'Little Eva' to commemorate the spacewalk, and the crew had thought of calling the spacecraft 'American Eagle', or 'Phoenix', or even 'Lady Bird' (after their First Lady, Lady Bird Johnson), but this was not adopted.

Mission This was the first of three highly successful long-duration missions, this one lasting for four days. Launched on 3 June 1965, the maxim orbital parameters achieved were 159.9 x 86.1 miles, and station keeping with the spent second stage was achieved during the first orbit. Ed White became the first American and second person (after cosmonaut Alexei Leonov) to walk in space, spending 20 minutes outside the spacecraft in a 36-minute period with the hatch open. Splash-down occurred after 97 hours 56 minutes and 62 orbits (setting a new US record) at 12.12.12 pm EST, 7 June 1965.

Gemini 5 (GT-5, or GT-V)
Launched 21 August 1965: Gordon Cooper and Pete Conrad; 8 days.

Primary objectives To evaluate the performance of the rendezvous and guidance systems using a rendezvous evaluation pod (REP). This was not achieved because of a decision to power down the spacecraft. However, the demonstration of manned orbital flight for approximately eight days, exposing the two-man crew to long periods of weightlessness, was successfully achieved.

Secondary objectives Demonstration of a controlled re-entry to a predetermined point was not achieved due to incorrect co-ordinates being transmitted to the spacecraft from the ground network. Evaluation of fuel cell performance under flight electrical loads was achieved, as was the demonstration of all phases of guidance and control system operation required to support rendezvous missions, and the evaluation of the rendezvous radar. Most of the seventeen experiments were conducted successfully, but the evaluation of the ability of either pilot to manoeuvre the spacecraft in orbit to close proximity with another object could not be achieved.

Launch vehicle preparations GLV-5's major welding began in January 1964, and assembly was completed by 9 December that year. The programme of testing occupied most of the next three months, with the final VAT inspection undertaken on 26 April 1965. The launch vehicle was transferred to the Cape on 18 May, but due to the preparations for Gemini 4 it was not installed on the pad until 7 June. The combined systems tests, verifications and tanking tests were conducted over the next eight weeks, and ended with the FSST on 12 August.

The crew performed a simulated dual count-down (including scheduled and unscheduled holds) on the launch pad on 22 July, with the astronauts in Gemini 5 on LC 19 and an Atlas Agena on LC 14. At the end of the test – which lasted 867 minutes instead of the planned 505 minutes – the lowered launch erector tower could not be raised, and the crew had to be 'rescued' by using a cabin at the end of a crane called a 'cherry picker', similar to that used on Mercury. Ironically, it was Cooper who had insisted that the crane be available on Gemini for such an eventuality, and he was the first to use the system operationally.

Spacecraft preparations Equipment installation began on 14 March 1965, with mating of the Re-entry Module and the Adapter Module two weeks later on 1 April. Over the next two months, the series of tests was completed before the spacecraft was shipped to ETR on 19 June. Gemini 5 was mated to its launch vehicle on 8 July, to complete a series of launch tests and checks by 13 August.

Emblem On 14 August 1965, a memo from NASA Administrator James Webb stated that from Gemini 5, the astronauts would be allowed to wear an identification emblem on the right chest area of their suit beneath their name tag. In addition, the Commander or senior Pilot would designate or recommend a design for his flight patch, subject to approval by the Director of MSC and the Associate Administrator for Manned Spaceflight at NASA HQ. This tradition has since continued for more than 35 years into the current Shuttle and ISS programmes.

The first emblem designed by the Gemini 5 crew emphasised the pioneering nature of their mission. Set at eight days, this was to be the longest flight to date, and was the key to the later 14-day mission and the Apollo lunar missions. It was therefore important to achieve the full duration – which also would surpass the Soviet record of five days held by cosmonaut Bykovsky since June 1963. It was Conrad's father-in-law's whittling of a model of a covered wagon of the Old West, and the slogan 'California or bust', that inspired the design of the patch, and the theme of early pioneering of a new frontier. Thus, a covered wagon was central to the design of the Gemini 5 patch; and the crew added the phrase '8 days or bust', a natural progression. When this design was presented to Webb, he rejected the slogan on the grounds that if the mission was terminated early there would be many who would say 'it busted', which would not be a favourable image for NASA. The slogan was therefore covered for flight with a piece of parachute cloth (which was reportedly ripped off after landing, the mission having surpassed the 8-day objective).

Call-sign 'Gemini Five'.

Mission Flown during the period 21–29 August 1965, this was a demonstration of an eight-day duration mission that was under planning for Apollo missions to the Moon. Launched at 09.00 am EST, 21 August, the early stages of the flight went according to plan. However, during station keeping with the REP, a rapid drop in pressure in the cryogenic storage tanks supplying oxygen to the fuel cells resulted in many planned activities being cancelled or abbreviated. This was the first flight of fuel cells (rather than the previously used chemical batteries) on a US manned spacecraft, and they significantly prolonged the extent of the mission. The crew performed a rendezvous with a 'phantom Agena' during FD3, which ground tracking indicated would have placed the spacecraft to within 0.3 of a mile of a target spacecraft. The crew also demonstrated the effectiveness of visual observations from orbit. During the mission, Gemini 5 attained a maximum orbit of 188.9 x 87.4 miles, with splash-down achieved at 07.55.14 am EST, 29 August. The flight time was a new world endurance record of 190 hours 55 minutes and 120 orbits surpassing Bykovsky's Vostok 5 record.

Gemini 7 (GT-7, or GT-VII)
Launched 4 December 1965; Frank Borman and Jim Lovell; 14 days.
   Gemini 6 was to have performed the first docking with an Agena target, but the target vehicle was lost 6 minutes 16 seconds into the launch on 25 October 1965, as a result of a propulsion system failure. The unmanned target vehicle broke up and led to the postponement to Gemini 6, to be re-flown during the Gemini 7 mission.

Primary objectives To demonstrate the capability of supporting a spacecraft and crew on a 14-day mission, and to investigate the effects of the duration spaceflight on the crew. These were successfully achieved.

Secondary objectives To provide a rendezvous target for Gemini 6, and to conduct a programme of station keeping with the second spacecraft, which were also successful. They also demonstrated the ability to station keep with the second stage of their launch vehicle. In addition, twenty experiments were performed, and evaluations of a lightweight spacesuit, the spacecraft entry guidance capability, and a programme of systems checks were also successfully accomplished during the mission.

Launch vehicle preparation The major welding of the propellant tank on GLV-7 was conducted in May 1964, with the tanks delivered to the Baltimore plant on 25 February 1965. The vehicle assembly was completed by 20 May, and the series of system and verification tests were conducted during the summer, finishing with the VAT inspection on 28 September. GLV-7 arrived at ETR on 19 October, and was erected on Pad 19 on 29 October, after GLV-6 had been removed the previous day.

Spacecraft preparations Equipment had been installed in Gemini 7 by 29 June 1965, and the two main elements of the spacecraft were mated by 26 July. The series of tests began with the System Assurance tests on 12 August, and were completed with an Altitude Chamber test on 17 September. Gemini 7 was shipped to the Cape on 9 October, and was mated to the launch vehicle by 22 November. The final simulated flight test took place on 27 November.

Emblem The main feature of this circular design was an Olympic torch, symbolising the medical and endurance aspects of the 14-day mission. The mission numerical (in Roman figures) was added to the patch, but the astronauts' names were not included, as they did not wish to detract from the emphasis on the mission rather than the crew (although their names are present on souvenir emblems.) NASA artists completed the artwork, and the curling flame from the torch indicated motion – again representative of the long mission.

Call-sign 'Gemini Seven'.

Mission Gemini 7 set a world endurance record that lasted until June 1970, when it was surpassed by the 18-day Soyuz 9 mission. It held the US record until the first manned mission to Skylab – 28 days – in June 1973. Indeed, the mission remained the fourth longest US manned spaceflight (behind the three manned Skylab missions) for more than 26 years, until it was surpassed by the STS-50 extended-duration orbiter mission in June 1992. Gemini 7 remains a significant milestone in the development of extended human spaceflight. Flying a mission approximately twice that expected of the Apollo lunar missions, they supported the rendezvous of Gemini 6, performed twenty experiments, and evaluated a new lightweight pressurised suit for added comfort. Gemini 7 was launched at 14.30 pm EST, 4 December 1965, and attained a maximum orbital altitude of 177.1 x 87.2 miles. After 330 hours 35 minutes and 206 orbits, Gemini 7 splashed down in the Western Atlantic recovery area, only 6.4 miles from the planned point, at 09.05.34 am EST, 18 December 1965.

With the recovery of the Gemini 7 crew, the major objective of sustaining a crew for two weeks had been achieved, a demonstration of EVA had been accomplished, and the first experiments in rendezvous and station keeping had been completed. It was now Gemini's task to achieve the physical docking of two spacecraft in orbit, and to further evaluate EVA tasks.

INITIAL RENDEZVOUS AND DOCKING OPERATIONS: GEMINI 6, 8 AND 9, 1965–1966

Gemini 6 (GT-6 or GT-VI-A)

Launched 15 December 1965; Walter Schirra and Tom Stafford, one day.

Primary objectives Originally intended to demonstrate Gemini–Agena docking, this was not achieved due to the loss of the Agena target vehicle, and the primary objective of the mission was therefore changed to a rendezvous with Gemini 7.

Secondary objectives The initial secondary objective associated with the docking of the Agena was not accomplished due to loss of the target vehicle. However, the revised secondary objective of closed-loop rendezvous with Gemini 7 was achieved, as were station keeping, and visibility tests using Gemini 7 as a rendezvous target. The re-entry guidance capability of the spacecraft was also proven.

Launch vehicle preparations Major welding commenced in April 1964, with assembly completed by 25 February 1965. Vehicle tests occupied April, May, June and most of July, with the launch vehicle finally being shipped to ETR on 2 August. The first erection of the vehicle on the Pad occurred on 31 August, and the series of on-pad tests was completed by mating GT-6 to the launcher on 17 September. A Final Status Simulated Flight Test was conducted on 19 October, but the 25 October launch was scrubbed after the loss of the Agena. The vehicle was destacked on 28

The astronauts are helped out of their overshoes prior to sliding into the spacecraft.

October to allow preparation for the launch of GT-7. The day after the GT-7 launch on 4 December, GT-6 was reinstalled on the pad, and a second FSSF test was completed. The mission was aborted on 12 December and rescheduled for 15 December.

Target vehicle preparations The basic Gemini Agena Target Vehicle (GATV) was designated 5002 and was completed on 17 December 1964. Modification and final assembly was achieved on 18 May, followed by a Combined Systems Acceptance Test in June and a VAT inspection in July. The vehicle was delivered to ETR on 25 July. A series of tests began in Hanger E on 23 August and was completed by 30 September, with the Agena mated to the Atlas on 10 October.

The Atlas for the first target launch (Target Launch Vehicle – TLV-5301) had been at the Cape since 4 December 1964, and was erected on the pad in October 1965. The target vehicle was added to the top of the stack on 10 October, and following launch tests, was scheduled for a 25 October launch. When the vehicle was lost seconds after separation of the Atlas and Agena stages, the Gemini 6 mission was scrubbed.

Spacecraft preparations Installation of equipment was completed by 4 May 1965, and the Re-entry Module and Adapter Module were mated eight days later. Systems checks were completed in June, with an altitude chamber test occurring on 21 July. The spacecraft was shipped to ETR on 4 August, and was mated to the Titan on 18 September. Tests continued on the combined vehicles during the next month. When the 25 October launch was scrubbed, the spacecraft was destacked on 28 October until after Gemini 7 launch. On 5 December it was restacked on its launch vehicle on LC 19, and was re-tested. After the 12 December abort, the spacecraft remained on top of the launch vehicle for the rescheduled 15 December launch attempt.

Emblem For his mission emblem design, Schirra wanted to reflect the night sky as it would appear during rendezvous with the Agena on the original mission. The prediction was that rendezvous should occur with the constellation Orion in the background, and this was reflected on the emblem. Although the mission was changed to rendezvous with a Gemini (7) and not dock to an Agena, the emblem design remained the same. A representative Agena was included on the six-sided emblem, which also included the mission designation, despite being the seventh Gemini launched. The lemon-yellow outline of the stars was to represent the twinkling effect of stars as seen through the Earth's atmosphere from the ground. The mission designation '6' linked through the stars and the Agena, joining the navigation and the target vehicle. The mission designation and the names of the astronauts completed the design.

Call-sign 'Gemini Six'.

Mission After the loss of their target vehicle, the Gemini 6 crew prepared to meet their colleagues in orbit on 12 December. However, although ignition of the Titan was achieved at 09.54.06 am, the engines automatically shut down 1.2 seconds later, prior to lift-off. Schirra correctly assessed the situation, and determined that it was safe to remain in the vehicle and not eject. It was later discovered that a small electrical plug in the tail of the Titan had dropped out prematurely, and a small plastic dust cover had obstructed the oxidiser inlet line of a gas generator. Either of these problems would have caused the lift-off to be aborted. By not initiating ejection from the craft, Schirra protected the option of launching the Gemini on a subsequent flight day. The third launch attempt of 15 December at 08.37.26 was achieved successfully, allowing rendezvous with Gemini 7 on the fourth orbit. The spacecraft orbited at 168.1 x 86.9 miles, and Schirra completed a programme of manoeuvres enabling him to bring Gemini 6 to within a few feet of Gemini 7. Then the spacecraft approached to within one foot of Gemini 7, at 5 hours 56 minutes into the flight. The two spacecraft remained in close proximity for the next 5 hours 18 minutes. Splashdown in the Western Atlantic occurred after 25 hours 51 min and 16 orbits, and was within seven miles of the primary recovery vessel.

Gemini 8 (GT-8 or GT-VIII)
Launched 16 March 1966; Neil Armstrong and Dave Scott.

Primary objectives To dock with an Agena target vehicle and to conduct EVA. The first was achieved, but the second was not.

Secondary objectives Despite the early termination of the mission, preventing the EVA, most of the secondary objectives were at least partially achieved during the abbreviated flight. These included a fourth revolution rendezvous and docking, completing a systems evaluation, conducting ten experiments (the latter two both partially achieved), evaluation of an auxiliary tape memory unit, and parking the target vehicle in a 220-nautical-mile circular orbit, allowing further re-use. The docking practise and re-rendezvous were not achieved, and neither was the objective of using the Agena for the orbital manoeuvres.

Launch vehicle preparations Major welding of the GLV-8 vehicle was completed in September 1964. Following the series of tests between September and November 1965, the Titan was delivered to the ETR on 6 January 1966. It was erected on LC 19 by 13 January, and then completed the programme of on-the-pad tests and checkouts, which lasted into March. The spacecraft was mated to the top of the Titan on 5 March, and the simulated final flight test followed four days later.

Target vehicle preparations GTV-5003's basic vehicle had been completed by 20 July 1965, and was modified into the Gemini target by 14 October. After completing a series of tests early in the new year, by 21 January the target was shipped to the Cape to undergo further tests and check-outs prior to the mating with the Atlas launch vehicle on 1 March.

The Atlas (TLV-5302) had been completed on 2 April 1965, and had undergone tests and installation of Gemini-specific flight equipment during the following three months. It was delivered to the ETR on 11 August 1965, and on 5 January was installed on LC 14 to undergo further systems checks prior to target vehicle mating on 1 March. A joint Flight Acceptance Composite Test was conducted on 7 March, followed by a simultaneous launch exercise on 9 March, in preparation for the actual launch on 16 March.

Spacecraft preparation Equipment installation was completed by 17 September 1965, with mating of the Re-entry Module and the Adapter Module three days later. The test programme continued over the next few weeks, ending with the altitude chamber test on 13 December. The spacecraft was shipped to ETR on 8 January and was temporarily mated to its launcher on 25 February before final mating on 6 March. Further tests on the pad led up to the final simulated flight test on 10 March 1966.

Emblem Another circular design, the Gemini 8 emblem featured the crew names one above the other in the upper left quadrant. To the right was a prism that split the light of the stars Castor and Pollux – the brightest stars in the constellation Gemini – into a spectrum of light, forming the zodiacal symbol of the programme and the Roman numerals of the flight. As the mission objectives of this flight encompassed most of the Gemini programme (EVA, rendezvous and docking, and experiments), the seven spectral colours reflected the notion of this flight, covering the spectrum of the Gemini programme objectives.

Gemini 9 is mated to its Agena in this docking compatibility test at the Cape, prior to both spacecraft being installed on top of their respective launch vehicles. (British Interplanetary Society/NASA.)

Call-sign 'Gemini Eight'.

Mission The Agena was launched at 10.00 am EST, 16 March 1966, followed 1 hour 42 minutes later by Gemini 8. Six hours after launch, the Gemini had rendezvoused with the Agena for the first time, and at GET 6 hours 33 minutes the Gemini docked with the Agena, completing the first space docking in history. After 27 minutes, however, the docked configuration encountered a greater then expected yaw and roll rate. The crew tried to command the Agena to bring the combination back under control, but it soon became apparent that it could be the Gemini and not the target vehicle that had the faulty thrusters. The rates continued to increase to a point where the crew felt the structural integrity of the spacecraft was threatened in the combined configuration. Having successfully slowed the rate sufficiently to undock from the Agena, the roll and yaw rates of the Gemini increased significantly, indicating that it was indeed their own spacecraft's attitude control system that was at fault. As the rate increased to an alarming one full revolution a second, the crew were forced to disarm the attitude control system and initiate the re-entry control system. This was a back-up measure, but it called into play a mission rule of immediate return to Earth. Gemini 8 landed in a secondary recovery area in the Pacific, east of Okinawa, after only 10 hours 41 minutes and seven orbits in flight, but just 1.1 miles south of the planned area – the most accurate recovery to date. Landing occurred at 22.22.28 pm EST.

Dick Gordon, back-up Pilot on Gemini 8, participates in check-out procedures at the radar boresite tower during Agena docking simulations at the Cape. (British Interplanetary Society/NASA.)

Gemini 9 (GT-9, or GT-IX-A)
Launched 3 June 1966; Tom Stafford and Gene Cernan; three days.

The original Gemini 9 crew – Elliott See and Charles Bassett – were killed on 28 February 1966 in the crash of their T-38 jet at the McDonnell factory, St Louis, and their back-up crew – Tom Stafford and Gene Cernan – were subsequently named to the flight. The Agena for the mission was launched at 10.15.03 am EST, 17 May 1966, as the new prime crew was in the spacecraft awaiting confirmation of orbital insertion of their target. It was a call they would not receive. At 2 minutes 1 second after Atlas launch, booster engine number 2 pitch control was lost, and the vehicle and the Agena were destroyed. The crew left their spacecraft and had their mission rescheduled for 1 June, using the Augmented Target Docking Adapter as a target vehicle – a back-up for Agena in this eventuality.

Primary objectives The primary activity of docking to the Agena was lost with the destruction of the vehicle, and docking with the ATDA could be only partially achieved, as the launch shroud was still attached. The original EVA was lost when the original mission was terminated, but the revised EVA was achieved. However, the test of the AMU had to be abandoned due to overheating of the suit and fogging of the visor.

Secondary objectives Most secondary objectives on the original mission involved the lost Agena, so these could not be achieved. The revised secondary objectives featured rendezvous with the ATDA, and although docking could not be achieved, the secondary objectives, including EVA, most of the seven experiments, and a controlled re-entry, were completed.

Launch vehicle preparations Major welding of the propellant tanks commenced in February 1965, and the assembly was completed by 10 November. Over the following three months the system and verification and acceptance tests were conducted before delivery to the Cape on 10 March 1966. The vehicle was erected on LC 19 on 24 March, and after further tests the spacecraft was added to the top of the Titan on 8 May. The final status Simulated Flight Test (SFT) took place on 11 May. When the 17 May launch was postponed due to the failure of the target to orbit, a second SFT was conducted on 26 May, only to have the launch again scrubbed because of a spacecraft computer problem.

Target vehicle preparations The basic Agena target vehicle (5004) was completed by 25 October 1965, and its modification for Gemini use was completed by 26 January 1966. Following the test programme, it was delivered to ETR on 4 March to begin a further series of tests, before mating with the Atlas on 2 May. Systems tests at LC 14 were completed on 10 May. The Atlas (TLV-5303) had been delivered to the Cape on 13 February 1966, and, following the mating to the target vehicle and systems tests, the launch occurred on 17 May.

When the original vehicle was lost, the Augmented Target Docking Adapter was prepared for flight. The TLV-5304 was assigned to the mission. This vehicle had been at the Cape since 8 May, and during 11–17 May was in storage, before modification to accept the ATDA payload between 18 and 20 May. The vehicle was erected on the Pad on 21 May, with the Agena mated to it on 25 May. Launch was scheduled for 1 June.

Spacecraft preparations The Re-entry Module was mated to the Adapter Module on 22 November, and was followed by the installation of equipment by 7 December. Systems tests began on 30 December, and were completed with the altitude test on 10 February 1966. Gemini 9 arrived at the Cape on 2 March, and completed a soft mate to the launch vehicle on 3 May, being hard-mated on 8 May. Tests on the combination were completed by the final SFT on 11 May. The mission was scrubbed after the loss of the Agena on 17 May, and was followed by a systems re-test, mate and repeat of the FST on 26 May. The mission launch was again scrubbed and recycled on 1 June due to computer software problems.

Emblem This design was a departure from the previous circular emblems, being a shield shape, depicting large Roman numerals IX for the mission designation. Above these were representations of a Gemini and an Agena with an astronaut figure on EVA – the original mission objectives of docking to the Agena and EVA by the Pilot. The astronaut's tether formed an Arabic number 9 and the patch was once again devoid of the two astronaut names (which were added only to souvenir emblems).

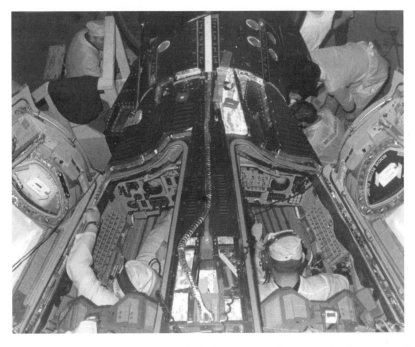

Another view of the Gemini 8 spacecraft during pre-launch tests at the Cape, revealing the paucity of space in the crew compartments. Note there is no additional 'flight stowage' inserted at this time. (British Interplanetary Society/NASA.)

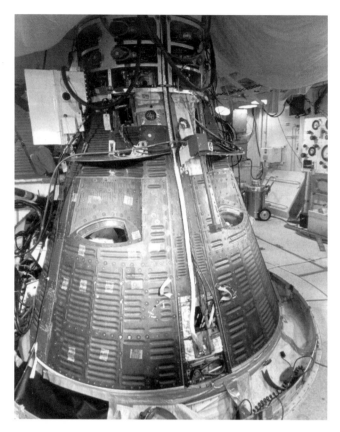

Gemini 8 is checked out by the flight crew (Armstrong and Scott) during a simulated count-down at LC 19 a week before launch. A check-list can be seen through Armstrong's window at left. Note the connections and umbilical around the spacecraft for the test.

Call-sign 'Gemini Nine'.

Mission The ATDA was successfully launched at 10.00.023 am EST, 1 June 1966, and was placed in a 161 x 159-mile orbit. However, lifting the crew off the pad proved more problematic. At T minus 1 minute 40 seconds, a hold was initiated, and the count recycled to T minus three minutes. Launch Control reported that the guidance system update of the spacecraft computer was not able to transfer data from ground equipment to the spacecraft. This was followed by two further holds due to the same problem, until the mission was postponed for two days.

At 08.39.33 am EDT, 3 June, Gemini 9 finally left the pad for orbit. The plan was to rendezvous with the target during the third revolution, to re-rendezvous during the fourth, and to rendezvous from above on orbit 12. This was achieved, but docking and docking practise was abandoned upon discovery that the launch shroud had not jettisoned as it should have done, giving the target an 'angry alligator' appearance. Gemini rendezvoused with Agena at 4 hours 15 minutes into the mission, and completed 46 minutes of station keeping exercises. The second rendezvous at 6 hours 36 minutes was followed by a further 39 minutes of station keeping. The third RV from above was the most difficult, because of the terrain moving in the background. This was completed at GET 21 hours 42 minutes, followed by 1 hour 17 minutes of station keeping.

Cernan opened the hatch at GET 49 hours 23 minutes, and completed a 2-hour 7-minute EVA. But he was unable to complete his EVA tasks due to suit fogging and physical exhaustion. He did, however, strap on the AMU and power up, but he did not release it. After much effort, the hatch was locked at GET 51 hours and 30 minutes. Gemini 9 orbited at 168.2 x 85.7 miles, with splash-down occurring at 09.00.33 am EST, 6 June, after a mission of 72 hours 20 minutes and 45 orbits.

PUSHING THE ENVELOPE: GEMINI 10, 11 AND 12, 1966

The final three Gemini missions were all to feature both rendezvous and docking with an Agena target, and EVA operations. None were duration missions, but they were highly successful in demonstrating the primary techniques that would later apply to Apollo.

Gemini 10 (GT-10, or GT-X)
Launched 18 July 1966; John Young and Michael Collins; three days.

Primary objectives Gemini 10 had only one primary objective – to rendezvous and dock with the Agena target, which was achieved.

Secondary objectives These included rendezvous and docking with Agena on the fourth orbit, checking onboard navigation, using the large propulsion system of the Agena in primary and secondary modes, and conducting EVA, all of which were achieved. Most of the fourteen experiments were also achieved, but docking practice could not be carried out due to insufficient fuel.

Launch vehicle preparations The major welding of propellant tanks on the GLV-10 launch vehicle was completed by April 1965. Assembly was completed by 28 February, and following the series of acceptance tests, it was delivered to the ETR on 20 May. GLV-10 was erected on LC 19 by 8 June, and a series of tests continued, culminating in a final SFT on 13 July.

Target vehicle preparations The basic Agena vehicle (5005) was completed on 2 February 1966 and received its modifications to be used as a Gemini target by 12 April. After the test programme was completed, the vehicle arrived at ETR on 16 May to begin a second series of pre-launch tests. The Agena was mated to the Atlas on 1 July, followed by a systems test on LC 14 on 12 July.

The Atlas (TLV-5305) was delivered on 1 November 1965 before undergoing systems integration to support a Gemini-related mission on 18 February 1966. Early in June, a series of tests was conducted before it was shipped to the ETR on 19 June. Installed on

LC 19 by 25 June, the Agena was mated on 1 July and the series of combined systems tests led to the final systems test on 12 July, in time for a 16 July launch.

Spacecraft preparations Equipment installation was completed by 29 January 1966, and was followed by the mating of the Re-entry Module and the Adapter Module on 4 February. Systems tests began on 2 March and continued until the end of April, with an altitude chamber test conducted on 28 April. Spacecraft 10 was shipped to the Cape on 13 May, and was mated to its launch vehicle by 5 July. The final SFT was completed on the spacecraft on 13 July, three days before launch.

Emblem The emblem for Gemini 10 was designed by John Young's wife Barbara, and was based upon suggestions from both astronauts. Again round in shape, it featured a large red Roman numeral X – symbolic of the designation of the flight. Representations of Gemini and Agena orbited the numeral, duplicating the chase manoeuvres prior to the docking. The two stars of Gemini and (not for the first time) the crew names were not included on the emblem. Collins called this one of the best Gemini designs, being clean without names and machines cluttering up the design. Later souvenir emblems included the two astronauts' names.

Call-sign 'Gemini Ten'.

Mission The Agena for this mission was successfully launched at 15.39.46 pm EDT, 18 July 1966. Less than two hours later, Young and Collins lifted off onboard Gemini 10 at 17.20.27 pm. As planned, the first rendezvous with the target came at 5 hours 21 minutes into the mission, with the docking achieved 31 minutes afterwards. For the next 38 hours 47 minutes, the two spacecraft remained docked. During this time, the crew completed six major manoeuvres – three using the Agena primary propulsion system, and three using the secondary propulsion system. The first of the series placed the combination in an elliptical orbit of 412.2 x 158.5 miles. During the docked period, Collins also performed the first of his two EVAs. At GET 23 hours 24 minutes he opened the hatch and performed his assigned tasks for the next 50 minutes during a stand up EVA. The SUEVA was terminated early when both crew-members experienced eye irritation.

At GET 44 hours 40 minutes the spacecraft undocked from Agena 10 and proceeded to rendezvous with Agena 8, which had been placed in a parking orbit on 16 March. Collins opened the hatch again at GET 48 hours and 41 minutes to perform his umbilical EVA. He exited the spacecraft and retrieved experiment packages from the Agena 8 target, which had been exposed to the space environment for four months. The EVA lasted 39 minutes, and a docking with Agena 8 was not attempted. The highest apogee of the mission was 412.2 miles, and the lowest perigee was 86.3 miles. Splash-down was achieved at 16.07.06 pm EST, 21 July, and was estimated as being 3.4 miles from the planned impact point. Mission duration was 70 hours 46 minutes, with 43 orbits completed.

Gemini 11 (GT-11, or GT-XI)
Launched 12 September 1966; Pete Conrad, Dick Gordon; three days.

Primary objectives Only one primary objective was assigned to this mission – a rendezvous during the first orbit, which was achieved.

Secondary objectives Gemini 11 was assigned secondary objectives that included docking practise, performing EVA, conducting eleven experiments, manoeuvring in a docked configuration to a high apogee orbit, conducting a tethered vehicle test demonstration of an automatic re-entry, and parking their Agena target. All objectives were achieved apart from one experiment that was not completed due to the early termination of the EVA.

Launch vehicle preparations Major welding of the GLV-11 began on 28 June 1965, with assembly completed by 5 April 1966. The programme of system and verification tests was conducted between April and June, with the launch vehicle delivered to the Cape by 12 July. Erection on LC 19 was completed by 22 July, to begin further testing. The spacecraft was mated on 24 August, and a final SFT was completed on 1 September.

All the training is completed, all the simulations have been performed, the spacecraft and the launch vehicle are ready to go, and so are the crew. Here Gemini 11 astronauts Pete Conrad and Dick Gordon leave the Ready Room at LC 16 after suiting up for their mission. (Note the mobility of the Gemini suits.) Behind them is Chief Astronaut Al Shepard.

Target vehicle preparations Agena Target 5006 was converted for Gemini on 6 June 1966. Following systems tests, it was delivered to the Cape on 15 July, and proceeded to undergo further systems checks and verification before being mated to the Atlas on 22 August. A combined systems test programme was completed on LC 14 by 31 August.

The Atlas TLV-5306 basic vehicle had been delivered on 14 January 1966, and was fitted with the Gemini specific equipment by 20 June. It arrived at ETR on 18 July, and was installed on LC 14 ten days later, with the Agena topping off the stack on 22 August. The joint Flight Acceptance Test conducted on 26 August was repeated on 1 September. There were two launch attempts, on 9 and 10 September, before the vehicle finally left the pad on 12 September.

Spacecraft preparations Equipment was installed from 13 April, five days after the two primary modules were mated. The test programme began on 29 April, and was completed with an altitude chamber test on 15 June. Spacecraft 11 was then delivered to the ETR on 7 July, and was mated to its launch vehicle by 24 August. The final Simulated Flight Test occurred on 2 September.

Emblem Again departing from the 'traditional' roundel, this emblem depicted the basic shape of the Gemini capsule, and as both prime crew were Naval officers, the colours of blue and gold represented their parent service. Across the patch, four stars represented the four elements of the mission: the star nearest the Earth represented the first orbit target rendezvous; the second, located near the docked spacecraft, was symbolic of that objective; the third was situated near the tethered EVA astronaut (also emphasising tethered operations with the docking vehicle); and the fourth star, at the top of the emblem, was representative of the 850-mile record altitude achieved on the mission. The Roman numerals XI was depicted 'rising from Earth' – symbolic of both the designation of the mission and also rising to the new altitude record. The astronauts' names were added to the base of the design in their flight positions – Conrad to the left, and Gordon to the right.

Call-sign 'Gemini Eleven'.

Mission At 08.05.02 am EST, 12 September 1966, the Atlas carrying the Agena lifted off successfully to place the vehicle in orbit. At 09.42.56 am the same day, the two astronauts were launched to begin their Gemini 11 mission. The crew achieved the first orbit rendezvous and docking with the Agena, and during the second day they fired the Agena engines to raise the orbit to a record 850 miles apogee. This remained a record for manned spacecraft until surpassed by Apollo 8 in December 1968 as it headed for the Moon. Indeed, apart from the nine Apollo lunar distance missions, this was by far the highest apogee attained by humans until the Hubble Space Telescope deployment mission on STS-31 in April 1990.

Gordon completed a tether fastening to the docking vehicle during his umbilical EVA; but he expended a great deal of energy, and the planned two-hour EVA was terminated after only 33 minutes because of Pilot fatigue. During the third day the crew achieved another first as they undocked and performed a successful tethered demonstration, completing two orbits of Earth with the two spacecraft tethered but undocked. Splash-down occurred just 1.5 miles from the prime recovery ship, at GET 71 hours 14 minutes during the 44th orbit.

Gemini 12 (GT-12, or GT-XII)
Launched 11 November 1966; Jim Lovell and Buzz Aldrin; four days.

Primary objectives To rendezvous and dock with a target vehicle and to perform at least three periods of EVA, all of which were achieved.

Secondary objectives To perform docking practise, accomplish tethered station keeping exercises using gravity gradient technique, perform fifteen experiments, and complete a controlled re-entry technique, as demonstrated by Gemini 11. All of these were achieved. The only objective not achieved was to perform manoeuvres with the Agena primary propulsion system to change orbit, because controllers noted a fluctuation in the Agena's propulsion system and cancelled the planned manoeuvre.

Launch vehicle preparations Gemini Launch Vehicle 12 – the last of the series – began major propellant welding on 22 November 1965, a year before the mission was flown. By 1 June 1966 assembly was completed, and the Titan arrived at ETR on 3 September, being erected on the LC 19 on 19 September. The spacecraft was mated to the Titan on 25 October during a programme of pad tests and verifications, and a final SFT was completed on 2 November. The launch on 8 November was postponed due to a malfunction in the secondary autopilot, and again on 9 November due to a malfunction in the new autopilot. Gemini 12 – the 19th and final mission of the Gemini programme – was launched on Veterans' Day 1966.

Target vehicle preparations Target Vehicle 5001 was completed as the basic vehicle by 30 April 1964, and was modified for Gemini by 24 September. After systems tests it arrived at the Cape on 29 May 1965, and was used in tests on the Pad on 26 July 1965. It was then moved and returned from ETR for refurbishment on 23 November, after which it was redesignated 5001R, having been completely disassembled and then rebuilt. On 4 September 1966 it was tested and then redelivered to the Cape as 5001R, and on 23 October it was mated to the Atlas for final systems tests on LC 14.

The Atlas basic vehicle (TLV-5307) was delivered on 11 April 1966, and was converted for use on Gemini on 22 August 1966. This Atlas was originally assigned to the unmanned Lunar Orbiter programme, but following the loss of the Gemini 9 target vehicle on 17 May it was necessary to procure an additional Atlas for the final mission. TLV-5307 was delivered to ETR on 19 September, and received the Agena target on 23 October. The simulated launch occurred on 1 November, and following the two aborted launches it was launched on 11 November.

Emblem Gemini 12 was originally scheduled for launch on or around 31 October 1966 – Halloween – and therefore the design of the crew emblem reflected Halloween colours, with a gold spacecraft outlined in orange. The Roman numerals XII were also included in orange. The black background was a depiction of both the mystic event and the deep black of space. The spacecraft represented the hour hand of the clock – pointing to the 'witching hour' of midnight – and again also represented the flight designation. The crescent Moon in gold was at the left of the emblem. This was chosen by the crew, not only for its association with Halloween, but also because this final Gemini flight pointed to the Apollo programme in the new year. Indeed, at one point there was discussion about a joint mission of Gemini 12 and Apollo 1. The emblem was designed by McDonnell artists, and the crew's idea of a witch on a broomstick was not incorporated.

Call-sign: 'Gemini Twelve'.

Mission The final launches of the programme occurred at 14.07.59 pm EST, 11 November 1966, for the Agena, and at 15.46.33 pm EST for Gemini 12. The last Gemini was an unqualified success and a fine conclusion to the programme. Aldrin had used underwater training for this EVA, and this combined with the provision of handrails, foot restraints and waist tethers, proved that EVA was effective, allowing him to complete all nineteen assigned tasks in the most successful EVA programme of the series. For Lovell, the flight set an individual flight record of 425 hours 9 minutes 31 seconds on just two missions, making him the most experienced space explorer in the world. Aldrin completed two SUEVAs of 2 hours 18 minutes and 1 hour 6 minutes, and one umbilical EVA of 2 hours 6 minutes, and gained the record for EVA at 5 hours 30 minutes.

Splash-down occurred at 14.21.04 pm, 15 November 1966, just 2.6 miles from the prime recovery point after a duration of 94 hours 34 minutes and 59 orbits, and ended the flight operations of the Gemini series.

ENTER APOLLO

With all twelve missions successfully completed it was time to move on to the Apollo programme. Beginning in February 1967, three astronauts – Gus Grissom, Ed White and Roger Chaffee – were to take the first manned Apollo on a 14-day test flight in Earth orbit. The mission, designated AS-204, would be using the Block 1 configuration for the first of a series of manned flights in Earth orbit to qualify the Apollo hardware and to test Block II lunar mission hardware before venturing to the Moon. Their mission was also known as Apollo 1.

Flight data

Mission	I	II	III	IV	V	VI-A
Spacecraft No.	1	2	3	4	5	6
Launch weight (lbs)	7,026	6,882	7,111	7,879	7,947	7,817
Command Pilot			Grissom	McDivitt	Cooper	Schirra
Pilot			Young	White	Conrad	Stafford
Back-up Command Pilot			Schirra	Borman	Armstrong	Grissom
Back-up Pilot			Stafford	Lovell	See	Young
Launch date	1964 Apr 8	1965 Jan 19	1965 Mar 23	1965 Jun 3	1965 Aug 21	1965 Dec 15
Launch time (EST)	11:00:01	09:03:59	09:24:00	10:15:59	08:59:59	08:37:26
Launch azimuth (deg.)	72	105	72	72	72	81.4
Launch vehicle	GLV-1	GLV-2	GLV-3	GLV-4	GLV-5	GLV-6
Target vehicle				Second stage	REP	GT-7
Target launch vehicle				GLV-4	GLV-5	GLV-7
Highest apogee (n. miles)	173.0	92.4	121.0	159.9	188.9	168.1
Lowest perigee (n. miles)	86.6	n/a	85.6	86.1	87.4	86.9
Inclination (deg.)	32.59	n/a	32.6	32.56	32.59	28.97
Period (min.)	89.3	n/a	88.3	88.90	89.50	87.92
Orbits	(64)	n/a	3	62	120	16
Duration (h:m:s)	n/a	00:18:16	04:52:31	97:56:12	190:55:14	25:51:24
Landing latitude	n/a	16° 36' N	22° 26' N	27° 44' N	29° 44' N	23° 35' N
Landing longitude	n/a	49° 46' W	70° 51' W	74° 11' W	69° 45' W	67° 50' W
Nautical miles from target	n/a	34	60	44	91	7
Recovery area	n/a	Mid-Atlantic	W. Atlantic	W. Atlantic	W. Atlantic	W. Atlantic
Prime recovery ship	n/a	Lake Champlain	Intrepid	Wasp	Lake Champlain	Wasp
Date	(1964 Apr 12)	1965 Jan 19	1965 Mar 23	1965 Jun 7	1965 Aug 29	1965 Dec 16

Mission	VII	VIII	IX-A	X	XI	XII
Spacecraft No.	7	8	9	10	11	12
Launch weight (lbs)	8,076	8,351	8,268	8,295	8,374	8,296
Command Pilot	Borman	Armstrong	Stafford	Young	Conrad	Lovell
Pilot	Lovell	Scott	Cernan	Collins	Gordon	Aldrin
Back-up Command Pilot	White	Conrad	Lovell	Bean	Armstrong	Cooper
Back-up Pilot	Collins	Gordon	Aldrin	Williams	Anders	Cernan
Launch date	1965 Dec 4	1966 Mar 16	1966 Jun 3	1966 Jul 18	1966 Sep 12	1966 Dec 12
Launch time (EST)	14:30:03	11:41:02	08:39:33	17:20:26	09:42:26	03:46:33
Launch azimuth (deg.)	83.6	99.9	87.4	98.8	99.9	100.6
Launch vehicle	GLV-7	GLV-8	GLV-9	GLV-10	GLV-11	GLV-12
Target vehicle	Second stage	5003	ATDA	5005	5006	5001R
Target launch vehicle	GLV-7	5302	5304	5305	5306	5307
Highest apogee (n. miles)	177.1	161.3	168.2	412.2	739.2	162.7
Lowest perigee (n. miles)	87.2	86.3	85.7	86.3	86.6	86.8
Inclination (deg.)	28.89	29.07	28.91	28.87	28.85	28.87
Period (min.)	89.39	88.83	88.78	88.79	88.99	88.87
Orbits	206	7	45	43	44	59
Duration (h:m:s)	330:35:01	10:41:26	72:20:50	70:46:39	71:17:08	94:34:31
Landing latitude	25° 25' N	25° 13' N	27° 52' N	26° 44' N	24° 15' N	24° 35' N
Landing longitude	70° 06' W	136° E	75° 00' W	71° 57' W	70° W	69° 57' W
Nautical miles from target	6.4	1.1	0.38	3.4	2.65	2.6
Recovery area	W. Atlantic	W. Pacific	W. Atlantic	W. Atlantic	W. Atlantic	W. Atlantic
Prime recovery ship	Wasp	Leonard Mason	Wasp	Guadalcanal	Guam	Wasp
Date	1965 Dec 18	1966 Mar 16	1966 Jun 6	1966 Jul 21	1966 Nov 15	1966 Dec 15

Flight tests

After the exhaustive ground testing programme as part of the development and construction phases, and before the Gemini missions could pursue the main objectives, the spacecraft needed to be tested in flight.

The Titan itself also needed to be qualified to carry an instrumented Gemini capsule before placing astronauts on board, and the system of crew escape needed to be evaluated before operational use on a manned launch. To accomplish this, a programme of Titan development launches was scheduled, a sequence of test ejections of the escape seats was planned, and three engineering test flights of the Gemini capsule – two unmanned and the third manned – were manifested.

TESTING THE EJECTOR SEAT SYSTEM

During June 1961, McDonnell had produced an Advanced Capsule Design that included easier methods of pilot escape than had been possible from Mercury. By 7 July it was proposed that the design should feature ejection seats rather than escape towers to pull the astronauts clear of an exploding booster, although some early designs of what became Mercury Mark II still retained an escape tower and ejection seats for emergency evacuation during descent. As the design evolved and the development of the paraglider continued, there were some concerns about how the crew could use ejection seats when a paraglider had been deployed.

The escape tower system was finally rejected when the rendezvous radar equipment was placed in the nose, and the Titan's non-explosive propellant merely burnt, allowing more reaction time for the malfunction detection system to activate. On 11 December 1961, NASA listed its guidelines for two-man spacecraft development, which would include ejection seats rather than an escape tower. By 14 March 1962, NASA had decided ejection had to be initiated manually, but with the option of a later automated ejection capability. Additionally, if either seat was triggered then this would simultaneously set off a dual ejection. The design featured the capability for ejection off-the-pad and up to 60,000 feet during powered ascent. In the event of paraglider (or parachute) failure, ejection could also be used to escape the plummeting spacecraft.

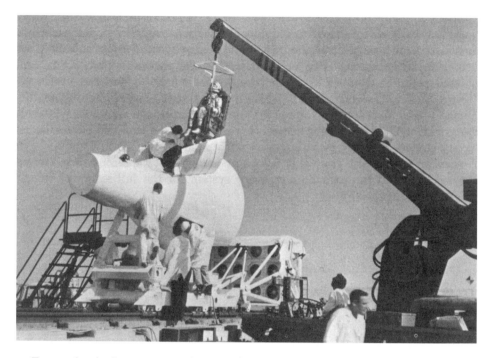

To test the ejection system, an instrumented mannequin is lowered into a boilerplate spacecraft prior to a run on the rocket-powered dynamic sled track.

To achieve this, the spacecraft hatch had to be opened before ejection initiation, and each astronaut had to be able to separate from the seat and land by parachute with all his survival equipment. Beginning on 29 March 1962, a series of regular meetings was established to evaluate the progress of the ejection seats on Gemini. Even though ejection systems had been used on military aircraft for some years, their application to a rocket-launched spacecraft was beyond the capability of existing systems. Contracts for the design of new seat (Weber Aircraft) and rocket catapult (Walter Kidde and Company Inc) systems were issued in April and May respectively.

On 29 May, the plans for a programme of development tests of the ejection seat were completed. The group included representatives from McDonnell, Weber Aircraft, NASA's Gemini Project Office and Life Systems Division and the Procurement Office, and the USN Ordnance Test Station, who met at China Lake, California, where the tests would be held. The test programme would begin on 2 July, with simulated off-the-pad aborts using a tower, followed by sled tests (beginning on 9 November) which would include simultaneous ejection with open hatches under dynamic pressure.

SOPE tests
The tower tests simulating off-the-pad ejection revealed the need for two important design changes: a drogue gun method to deploy the personal parachute, and a three-point restraint-harness-release system of the type used in military aircraft of the day.

Achieving suitable altitude in the tests was a fine balance in ensuring that the rocket catapult (rocat) line of thrust was straight, and that the shifting centre of gravity of the seat–man combination while the escape rocket was still firing did not take the seat off course. It was determined that solid-mass simulations would not be sufficient to do this, and so life-sized dummies were used in the test programme

In the first series of tests, five successful ejections were completed by early August, although each had its own small technical problems that needed to be resolved before the next test could proceed. These Simulated Off-the-pad Ejection (SOPE) tests revealed design faults in the seat and in the recovery parachutes, which twice failed to deploy. When the tests resumed on 30 August after bench tests and ground firing, a successful drop test of the seat and dummy revealed that the design changes had been successful, and so the off-the-pad tests resumed in September. Later that month, however, tests revealed problems with the rocket motor, and pushed this phase of testing back into 1963.

During the eighth SOPE test on 7 February, two dummies were ejected using a new personal recovery system for the first time. The ballute (*ball*oon and parach*ute*) was designed to stabilise an astronaut at extremely high altitude, where the air is too thin to support a conventional canopy parachute. Although the ejection seat and dummy separation worked satisfactorily, faults in the test equipment prevented full deployment of the parachutes, and the ballute either failed to inflate or did not release either dummy correctly. Subsequent improvements (producing fuller inflation at very low dynamic pressures) resulted in five highly successful similar tests in March.

Tests resumed in May 1963, and featured the design changes from past tests and the removal of the 'add-on' equipment that had marred the February tests. The series finished with two dual SOPE tests: number 10 on 2 July, and number 11 on 16 July. These were successful, except for the failure of a seat recovery parachute (not part of the ejection system) resulting in major damage to one of the test seats. Despite development problems, the overall off-the-pad test programme was successful, and completed the development phase of pad ejection testing.

'A hell of a headache, but a short one'
The next stage was to test the ejection seat on the rocket sled at China Lake. The system was tested on ground mock-ups, during which the spacecraft hatch was opened and then the ejector seat containing the dummy astronaut was expelled through the open hatchway. As CB point of contact for crew survival equipment, monitoring of these tests fell to John Young. During one such test, Young (soon to fly as Pilot on the first Gemini manned mission) took particular interest, as on this occasion the hatch was not opened and the seat containing the dummy ploughed straight through the closed hatch door. As Young keenly observed, if such an event occurred on a manned flight it would result in 'a hell of a headache, but a short one.'

This phase of testing featured a Gemini boilerplate (full-size mock-up) spacecraft carrying two ejection seats mounted side by side. This unit was carried along a rail by a rocket-propelled sled. Officially known as the Supersonic Naval Ordnance Research Track, in typical space age language it was shortened to its acronym and

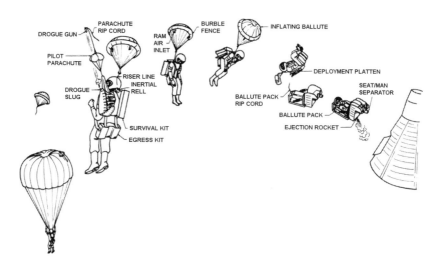

The deployment sequence of the proposed ballute stabilisation device.

was known as SNORT. This series was designed to simulate a variety of high-altitude abort situations evaluating the system under extremely high dynamic pressure, as would be encountered on real missions.

Ballute and SNORT

The sled runs commenced on 9 November 1962, with a qualification run without an ejection test. Unfortunately, the rocket motor came loose and smashed into the boilerplate, starting a fire and inflicting serious damage on both the boilerplate and the sled, which delayed further runs. The first dynamic dual ejection took place on 20 June 1963, and was successful, except that the sled did not reach the required velocity level. Refinements to the ejection system meant that the next test did not take place until 9 August, with the fourth on 16 January 1964. In the intervening months, two dummy drops using a modified egress kit were successfully completed. Test run 5, on 7 February, was a repeat of the second test and closed the development sled testing.

On 30 October 1963, in a meeting to discuss the ejection seat test results, it was determined that the ballute was a problem area. Qualification tests conducted in the wind tunnel at the Arnold Engineering Development Center revealed that the ballute failed at supersonic speed and did not open at subsonic speeds. To solve this, the riser lines were lengthened and their diameter was increased. A series of dummy and manned ballute descents were planned to fully test both the system and changes in the deployment method.

Between 22 November 1963 and 5 February 1964, a programme of 24 test drops designed to deploy the ballute stabilisation system (18 live and six with dummies) was conducted at the Naval Parachute Facility at El Centro, California. During the early test drops, excessive rates of rotation indicated that a three-foot ballute with dual-point suspension was not sufficient, and it was discovered that a single point of

A jump test of the 36-inch ballute with dual suspension, at the Naval Parachute Facility, El Centro, California. The jumper at right carries a helmet camera to record the events.

suspension on a 48-inch ballute was the optimum choice for Gemini. The test drops had been conducted with ballute designs of 28-inch, 36-inch and 42-inch from altitudes of 12,500–35,000 feet.

Further wind tunnel tests were completed by 9 April 1964 – two at subsonic and four at supersonic speed at design condition, followed by two further runs at 150% of the design maximum dynamic pressure. All showed that a 48-inch ballute would be a satisfactory stabilisation device. In January 1965, Gemini escape-system personal parachute qualifications began, with ten dummy drop tests conducted between 11 January and 16 February 1965. Despite some development difficulties, the series was successful. At the same time, a programme of twelve manned low-altitude jump tests was conducted between 28 January and 10 February, and this too was successful, leading to the first high-altitude live jump test on 17 February.

With the first manned Gemini launch set for March 1965, the long delayed twelfth

SOPE test was conducted on 16 January. The delay had been caused by the unavailability of pyrotechnics. Unfortunately, the seat hit the hatch and failed to leave the test vehicle. After modifications, the thirteenth (February 12) and fourteenth (6 March) tests were conducted successfully.

Meanwhile, on 28 January 1965 the High Altitude Ejection Test (HAET) resumed, with the first ejection in flight to demonstrate a Gemini personal recovery system. The seat and the dummy were ejected at 15,000 feet from an F-106 flying at Mach 0.72, and a second ejection test was completed on 12 February at 40,000 feet and Mach 1.7. Once again some small equipment failures led to redesigns, but on the whole the tests were highly successful and did not require re-flights.

Finally, on 17 February there began a series of live jumps from high altitude. Twenty-four low-altitude manned jumps were planned to qualify the system, but the ballute failed to deploy on the first manned jump due to the failure of an aneroid device responsible for ballute deployment initiation. A duplicate failure affected the second F-106 ejection test on 12 February, and the design was reviewed. The test programme was replaced with dummy drops from high altitude. This new programme featured ten dummy drops from 12,000–18,000 feet between 2 and 5 March, and five manned jumps from 15,000–31,000 feet between 8 and 13 March. All were successful except one live drop, where the ballute again failed to deploy. When the test subject had free fallen to 9,200 feet, he hit the manual override, activating the personal parachute.

This final series of tests completed the qualification of the personal parachute system for Gemini and completed all tests of the Gemini escape system. The final live jump occurred on 13 March 1965, and just ten days later Gemini 3 left the launch pad carrying astronauts Grissom and Young, each lightly holding the ejection handle on their recently qualified ejection seats, and hoping that if they were called upon to eject, the hatch would open first!

TROUBLE WITH THE TITAN

The Titan II launch vehicle was selected to launch the Gemini flights because of its simplified operation, greater lifting capability, and availability over other vehicles. Its other advantage was the shortened count-down required prior to launch. This would be an important factor in the later dual launch and rendezvous missions with the Atlas Agena.

The first static test firing of the Titan II took place at the Martin facility in Denver, Colorado, on 28 December 1961. As well as testing the engines, the exercise provided an evaluation of the equipment intended for launch operations. Some of the equipment intended for Titan II had previously been tested on Titan I launches from the Cape, where the Titan II would also be test flown. Preparations also included a 20-second static test firing on 9 March 1962, during which the first stage reached a thrust of 439,000 lbs without moving off-the-pad.

Just one week later, on 16 March, the Titan II was launched for the first time, and completed a successful test demonstration flight. The second test flight was not so successful, however; the 7 June flight recorded a lower than predicted second stage thrust, resulting in an early termination of the flight. A third launch, on 12 July, was successful; but on the fourth launch, on 25 July, the second stage again faltered and was prematurely shut down. This low thrust represented about 10% of its intended design figure which, converted to payload weight, meant that it would have to fly with a payload reduction of 700 lbs. This posed a serious threat to its use as a Gemini launch vehicle.

The very first flight of the Titan II had recorded the 'pogo' oscillations along its length, which seemed to indicate a major structural problem that would impart heavy g-loads on any occupant. NASA was so worried that the USAF Ballistic Systems Division (BSD), which was responsible for the development of the missile, set up a Committee for Investigation of Missile Oscillations to investigate the problem. Surprisingly, however, the problem proved easy to solve.

A Gemini Titan 1 sequence compatibility firing of the two (separated) stages at LC 19 on 21 January 1964.

Martin engineers decided to increase the pressure in the first-stage fuel tank, and on the fourth launch the pogo reduced by 50%. It was suggested that the phenomenon was similar to the chugging of water pipes (termed 'water hammer'), and that installing a surge suppression standpipe in the oxidiser line would further alleviate the problem. This system was installed for the eighth development launch on 6 December. However, instead of reducing pogo, it raised it to ± 5 g! In addition, the violent shaking which the pogo created caused the first-stage engines to shut down early. NASA optimistically concluded that fitting the oscillation chamber had in fact added to the problem, but in turn had also focused the investigation into the problem.

The next test flight, on 19 December, was more than satisfactory. It did not carry the standpipes, and featured aluminium piping instead of steel. This was followed on 10 January 1963 by test flight ten, which recorded ± 0.6 g from instruments placed where the Gemini would be located. Although still higher than the ± 0.45 g recorded on the Mercury flights, it was heading towards NASA's goal of ± 0.25 g. Although it appeared that the investigation into the pogo problem was on the right track, it also revealed the extent of the second problem. It had been assumed that the lack of full thrust on the second-stage engines was a direct result of the pogo phenomenon but on the tenth flight, even though the pogo oscillations had been reduced, the second stage thrust was still only about half of what was expected.

No risk to Gemini

This was a much more serious threat to flying Gemini on Titan, and to make matters worse for NASA, static test firing data from Aerojet General had revealed that the second stage could experience difficulty in supplying a steady burn profile after the 'shock' of ignition in flight. There was to be no test flight programme for the improved Titan booster for Gemini prior to the unmanned launch, and all data would be received from the ICBM test flights. On 29 January 1963, the USAF froze the missile design with the addition of several devices to increase first-stage pressure (including redesigned turbopump impellers and the addition of aluminium oxidiser feedline over steel). This reduced the effects of pogo on the airframe and structures below the requirements for the Titan II as an ICBM, but not for the higher demands set by NASA for Gemini.

In March, discussions on Titan II's status as a missile and Gemini launch vehicle were addressed by the Gemini Program Planning Board, during which it was explained to NASA that the pogo problem would soon be resolved by these new additions, and that the unstable operation of the second-stage engine was not a risk to Gemini. NASA was not convinced, and a week later added a second unmanned test flight to the programme, cutting the manned flights to just ten. Secretary for Defense, Robert McNamara, tried to convince Robert Seamans that the Titan II problem would be solved well before the first Gemini flight, but the Associate Administrator was still not totally convinced.

Addressing the problems

NASA insisted that ± 0.25 g was the highest vibration that the agency would accept.

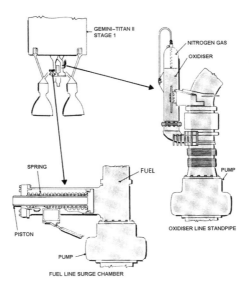

Pogo suppression equipment, which proved effective during the Titan II development programme.

The pogo and the second-stage fluctuations posed a serious threat to Gemini, and to resolve the problems, on 29 March 1963 an open meeting was held at Andrews AFB, Maryland. Those in attendance included representatives of all the major Titan contractors, NASA representatives, and top officials from BSD and Space Systems Division (SSD).

Director of BSD Titan System Program Office, Brig General John L. McCoy, addressed both the pogo and the combustion instability on the second stage. McCoy emphasised that his primary role was to develop a weapon system, and since this was not affected by either problem, he objected to the use of ICBM missile tests to meet Gemini standards. The contractors told NASA that results had indicated that both problems were close to being solved.

This was not the news that NASA wanted to hear, and during a closed session between NASA and USAF officials, it was clearly pointed out by NASA that the pogo problem was far from being understood. Without also solving the second-stage irregularity, the booster could not be termed 'man-rated'. These problems had to be solved before risking astronauts on top of the booster, and since the USAF was a major partner in Gemini, NASA felt that the service should also foot some of the bill for solving these problems. However, NASA also indicated that even if the agency had to pay for the modifications itself, then the solution had to be found for Gemini to fly on Titan II, whatever the cost.

The result was an agreement from the USAF to solve the pogo problem, refine the designs of both the first and second stage engines, and improve reliability as part of the ICBM development programme. The DoD would pay for the work from current emergency funds, with the understanding that missile development would not be delayed while waiting for Gemini Titan hardware, and that McCoy held final responsibly for deciding if and when to fly Gemini improvements on the ICBM test missiles. The problems of Titan became a matter of 'utmost urgency' to both NASA and the USAF.

More tests and more problems

The eighteenth Titan launch was the tenth successful flight, but the celebrations were short-lived. On the next test, just five days later, the missile, incorporating the latest refinements to resolve the pogo problem, left the pad while leaking fuel in its engine compartment. The resulting fire caused damage to the flight controls, and the missile pitched over, breaking up after only 52 seconds without providing clear results on whether the improvements had solved the pogo problem. The next test, on 20 June, was from Vandenberg on the California coast, and though the first stage performed well, the second stage again underperformed. Had Gemini been on board, the astronauts would have been forced to abort.

There had been twenty test launches out of a programme of 33, with only ten being anywhere near successful. Titan II was commissioned to become one of America's strategic deterrents, and once again ICBM requirements overrode the Gemini difficulties for the remaining launches. NASA conducted its own studies, and found that there remained three principle problems in clearing the missile for manned use: the pogo effect, the dynamic instability of the Stage 2 engines, and various design faults in the engines themselves. The effects of pogo on the astronaut crew were investigated in centrifuge tests at Ames, and the conclusions reaffirmed the earlier results in that, although a crew could endure a higher g level from the pogo, it would not be wise to exceed the $\pm$ 0.25-g limit.

The combustion instability and a whole range of minor engine problems had worried NASA to the degree that, after a visit to Aerojet facilities in Sacramento in July 1963, representatives recorded more than forty instances in which the design of engine parts could be improved. It was concluded that of the ten failures in the flight test programme, only one could be blamed on pogo alone, and none on dynamic instability. All the remaining failures were traced to small engine faults – such as a clogged injector, a failed weld, or a broken line. In a meeting of the Gemini Program Planning Board on 5 August, it was concluded that the reduction of pogo, the provision of a stable second-stage engine, and various improvements to the Titan engines, were only part of the problem. Improvements in the Aerojet effort would help resolve many of the problems encountered. There was still much work to be done – but at least the problems were identified, and a joint co-operative solution was underway.

Gemini 1A proposed

NASA had revised its early Gemini launch schedule at the beginning of 1963, to fly two unmanned missions (Gemini 1 and Gemini 2) and then a manned test flight before flying a further nine manned flights to achieve the programme objectives. When the Titan problems continued, the GPO began considering the insertion of a third unmanned test flight between the first and second missions. This would increase the flight programme from the planned twelve missions to thirteen missions, but would offer an extra safeguard against a failed first mission.

The new proposal would see a boilerplate spacecraft (not a production vehicle) fitted with instrumented pallets in the crew positions. The flight would be designated Gemini 1A and the cost of this additional mission was estimated at around the $2 million. It was therefore decided, at a Planning Board meeting on 6 September, that the mission would be used only as a back-up to a failed Gemini 1 (although it still required the development of the hardware and the availability of a launch vehicle). Its use would be determined only after the result of the Gemini 1 launch. The spacecraft was to be Boilerplate 1A, which was undergoing construction at a local Houston contractor, for flotation tests, and would be adapted to fly the 1A mission. On 24 September the structure was delivered to MSC for preparation to make it flightworthy, and on 13 December it was shipped on a flatbed truck to the Cape, where it arrived three days later. McDonnell were to fabricate the adapter section that connected the boilerplate to the top of the Titan second stage, and this arrived at the Cape on 24 January 1964. However, the reasons why this spacecraft and mission were included had all but been resolved by early 1964, rendering the mission redundant, and on 17 February 1964, Gemini 1A was officially cancelled from the programme.

Replacing the Titan with the Saturn 1B
Titan II test programme flight operations were resumed on 21 August 1963, and were a welcome success for everyone except NASA. This time, although the missile performed well, the Malfunction Detection System (MDS) it carried (a 'piggyback' test device to report, but not act upon, Titan performance data) suffered a short circuit 81 seconds after launch. Neither did the next launch, on 23 September, help to alleviate the gloom surrounding the missile tests, as guidance malfunctions took the vehicle out of its planned trajectory. Further talks between NASA and the BSD refined the requirement for Gemini, and after much discussion, a meeting on 11 October 1963 detailed all remaining test flights. Weekly status reports indicated to Seamans that progress was good, if slow, and although NASA was committed to pursuing the Titan II option, thought was given to substituting the NASA Saturn 1B for both the Titan and Atlas. A study conducted at MSFC Propulsion and Vehicle Engineering Laboratory showed this to be unfeasible.

Titan II success at last
On 1 November 1963, Titan II vehicle N-25 was launched from the Cape on the 23rd test flight. The vehicle carried the improvements installed to reduce pogo and data indicated that only ± 0.11 g had been recorded. For the first time, the NASA upper limit of ± 0.25 g had been surpassed. It was the beginning of five months of success for Titan, which indicated that the problems of both pogo and the engines had been corrected. Although the result fell within NASA standards, further discussion between NASA and the USAF sought to completely eliminate the problems.

In review, NASA praised the hard work of the USAF and Aerojet General at both engineering and management levels to resolve the problems of pogo, stability and engine design. There had been 32 test flights in the Titan II programme, and of these, ten (31.25%) would have failed to orbit a Gemini spacecraft. But in the summer of 1963, when there had been ten failures in twenty launches, the 50% average appeared far worse for the programme. The final twelve launches were a string of successes, and cleared the way towards the two unmanned launches designed to qualify the Titan for man-rating on the third mission. At last, Gemini was ready to fly.

GEMINI TEST FLIGHTS

With the completion of Titan test flights and the qualification of the astronaut escape system, NASA looked towards the first Gemini test flight. While other aspects of the Gemini programme (such as the Agena target, the fuel cells and, most seriously, the paraglider recovery hardware) were encountering their own problems, progress continued towards the first three missions to prove that the system worked.

In the revised flight schedule issued in April 1963, the first manned launch had been manifested for October 1964. However, by late 1963 the problems with the Titan, delays in the delivery of the second spacecraft, and difficulties in the

Installation of the ballast seat and instrumentation in the 'crew compartment' of Gemini 1, to record launch and flight data.

installation of test and check-out equipment at the Cape, were already threatening to delay the first launch and push the first manned flight into 1965. A review of test and check-out procedures led to a reduction of the programme at the Cape, which was merely duplicating work already carried out at McDonnell. Once the spacecraft had arrived at the Cape, there seemed to be little point in stripping it down and testing it once again. The programme of systems check-outs would therefore only be conducted on the complete spacecraft. It was also decided to eliminate static flight-test firing of the vehicles on the pad from the first three missions, which again streamlined the pre-launch preparations and helped to reduce some of the launch delays which the programme was facing.

Gemini 1 was originally scheduled to arrive at the Cape in August 1963 for a December 1963 launch, but the delays had pushed the launch back to February and then April 1964. Spacecraft 1 arrived at the Cape on 4 October 1963, but problems with Titan ICBM meant that some of the Martin engineers who would also work on the Gemini Titan launches were also assigned to resolve the Titan II ICBM test flight difficulties.

The plan was to fly Gemini 1 on an orbital trajectory, but without a planned recovery, to check the spacecraft/Titan interfaces. Gemini 2 would follow a sub-orbital test profile, to evaluate maximum heating rates at re-entry and to demonstrate recovery techniques for the spacecraft during ocean retrieval. The third mission would carry astronauts Gus Grissom and John Young on a three-orbit mission to verify the performance of spacecraft systems and the ability of the astronauts to manoeuvre the vehicle, and to demonstrate the capability for manned spaceflight operations.

'Something wrong with the range clock'

The first few weeks of 1964 proved to be fraught, with work teams changing from two eight-hour shifts to two twelve-hour shifts to prepare elements of the hardware. Compatibility tests proved stubborn to resolve, while procedural errors voided a propellant tank test, and malfunctioning hardware plagued one test element after another. At the third attempt on 21 January, a wet (fuelled) simulated flight test resulted in a 30-second firing of the first-stage engines (though still with problems), followed by a 30-second firing of the second stage, not yet mated to the first stage. All engines recorded the correct thrust and gimbals, as required for flight. This was the first and only static firing of the Gemini launch vehicle.

It was still a battle through March to prepare the spacecraft for launch, set for 28 March 1964, and progress was slow, with minor problems threatening the launch date. These included a small dent some 0.0015-inch deep in the 0.64-inch thick dome of the oxidiser tank, caused by a worker dropping a wrench. Tests proved the structural integrity of the tank to be intact, however. Further procedural errors and test equipment faults once more delayed the launch. The tests took two weeks longer than planned, and a new launch date of 7 April was again moved (to 8 April) when a transformer and switch motor on the pad burned out and took time to replace.

The final count-down, started in the early hours of 7 April, included built-in holds, propellant loading and systems checks. Finally, at one second past 11.00 am

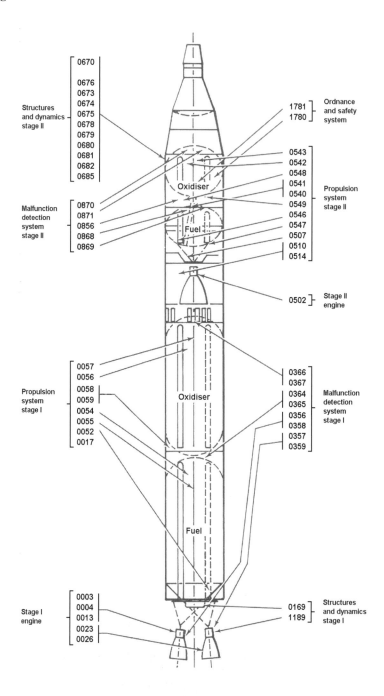

The location of sensors on GLV-1 used for the first mission data collection in order to compile a post-flight Mission Analysis Report.

on Wednesday, 8 April 1964, the first-stage engines on the Titan roared into life. Four seconds later, the first Gemini mission left LC 19 and opened Gemini flight operations. In a subsequent press conference, veteran launch director Walt Williams (on his final NASA mission before leaving the agency and being replaced by Chris Kraft) joked that the one-second delay in igniting the engine was not caused by the vehicle, but rather, 'there must be something wrong with the range clock.'

'A storybook sort of flight'
The Titan's engines had built up to 77% of full thrust when the vehicle was finally released. According to the official post-flight mission report, 'Lift-off is defined as the time at which the pad disconnects separate after the vehicle had lifted 1.5 inches.' The official lift-off time was recorded as 11:00:01.69 am EST. For 23 seconds Gemini 1 rose vertically, and during this period the roll programme was initiated to change from a 'launch stand' azimuth of 85.2 to a flight azimuth of 71.9. From the ground, it quickly disappeared into a blue Florida sky far beyond the unaided visual range of the onlookers, and from the array of telemetry signals, the launch seemed to be perfect.

Gemini flight operations begin with the launch of Gemini 1 on 8 April 1964.

At Ground Elapsed Time (GET) 153.98 seconds, the staging sequence was initiated, which followed the sequence of Stage I shut-down, Stage II ignition, and separation of the stages by firing the explosive bolts holding them together. Telemetry indicated that all went according to the flight plan, and Gemini 1 headed towards orbit. At 20 seconds after shut-down of the second stage, Gemini 1 was 621.4 miles down range, 99.42 miles high, and travelling at 25,879 feet per second. The spacecraft, still attached to the now spent Titan II second stage, was in orbit.

The second stage shut-down signal was 3.73 seconds later than predicted, and data indicated that Stage 2 chamber pressure was lower than nominal. This accounted for some of the extended burn time, which caused a 24-fps excess of trajectory speed and resulted in an orbit that was about eleven miles higher than planned. However, the launch had clearly demonstrated both that the Titan could indeed launch a Gemini into orbit, and that the combination was of a sound design. So relieved were both NASA and the USAF SSD that the launch was quickly termed 'a storybook sort of flight'. The recent problems with Titan seemed far behind them.

Meanwhile, Gemini 1 continued in orbit. The official mission would last three orbits, or about 4 hours 50 minutes, when the mission was terminated but the tracking continued. It was expected that the combination would remain in orbit for about 3.5 days, but because of the higher orbit at insertion it stayed up about four days. Radar data indicated that the combination was tumbling (which was expected) at about 2 rpm. On 12 April, on its 64th orbital pass, Gemini 1 re-entered Earth's atmosphere and burned up over the south Atlantic, halfway between South America and Africa. The first flight of Gemini was over, but the important analysis was yet to come.

Flight 1 analysis
The Gemini 1 spacecraft carried no flight systems on board, although the C-band transponder and telemetry transmitters were Gemini flight sub-systems. Dummy equipment equivalent to the mass of flight hardware was installed, and the spacecraft was instrumented to collect data on heating, structural loading, vibration, sound pressure levels, temperature, and pressure during launch.

Recorded vibration was below predicted levels, with longitudinal loading less than 25% of the structural integrity of the spacecraft. The heating characteristics of the spacecraft were near predicted values, and the internal temperatures in the crew compartment were much lower than expected. Sensors also recorded lower than expected temperature levels from the two spacecraft windows. Acoustic pressure levels reached a peak near the point of maximum dynamic pressure (as expected), but were still several decibels less than those used in the design parameters.

The cabin pressure release valve also worked as predicted and observed pressure measures indicated a leak rate of about 43.0 cubic inches per minute, which was well within the permissible rate of 61.0 cubic inches per minute. Spacecraft instrumentation also performed satisfactorily

The Adapter Module connecting the Re-entry Module to the Titan recorded lower temperatures than predicted, and revealed ample margin for more severe cases on future missions. Despite tumbling of the vehicle, all ground stations were able to track the combination, with telemetry transmissions recorded as excellent.

The Titan was structurally sound during the flight, with flight loading recorded at 32% of design limits. Also below design levels were the vibrations of the structure, the engines, and the sloshing of the fuel load. It had been predicted that the external temperature on the forward skirt could reach 325° F, but it was recorded as being only 157° F. There would be a review of insulation in this area, to perhaps remove material and slightly lighten the launch mass. Engine performance and the propellant feeding system were satisfactory, with engine starts and shutdown occurring well within design tolerances. The flight control system operation was also acceptable.

The post-flight report stated that the Gemini 1 mission had demonstrated the flight compatibility of the spacecraft and launch vehicle, and that the structure of the GLV was adequate for the environment in which it had flown, with basic dynamic characteristics all well with design limits. The pogo was of low amplitude, and even the slight build-up just prior to staging was not considered significant. The MDS performed correctly, and would have supplied the correct information to the instrument panel in the spacecraft, had one been carried. Most importantly, the structure of the spacecraft maintained its integrity under flight conditions.

GEMINI 2 (GT-2)

Gemini 1 had tested the launch and orbital characteristics of the spacecraft/Titan II structures, and the second mission was planned as a sub-orbital test that would result in a high re-entry heating rate on the heat shield, qualify the parachute recovery system on an operational mission, and test the effectiveness of the ocean recovery infrastructure in retrieving the spacecraft.

The transportation of the spacecraft from the construction line to the launch pad presented many problems, and became the *major* problem of the second mission as the testing programme dragged on throughout 1964. For once, Titan progressed relatively smoothly through its preparations and was delivered to the pad in July; but the spacecraft would not be ready until September. On 17 August, as the ground teams prepared for the final ground testing, a severe thunderstorm reached the Cape area, and at about 23.30 pm, lightning struck LC 19. Subsequent inspections did not reveal any damage to the pad, the erector blockhouse or to the Titan, but could not reveal anything about the state of the electrical circuits. NASA termed the event an 'electro-magnetic incident', and called for a thorough systems check. This revealed a number of failed parts, but no indication of a direct lightning hit.

Battling with the elements

While preparation of the hardware continued, it was not just the systems with which the launch crew was concerned. After the lightning strike in August, Hurricane Cleo also headed for the Cape, and the launch crew had just enough time to lower the second stage and place it under cover. The first stage remained upright but lashed to the pad, the erector lowered, as the hurricane passed over on 27 August.

After the storm passed, testing and stacking resumed, and confirmed earlier

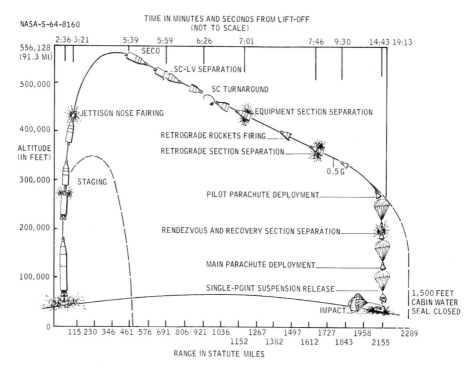

The Gemini 2 sub-orbital flight trajectory, 19 January 1965. (British Interplanetary Society/NASA.)

electrical problems that raised the question of whether the launch vehicle was fit to fly. NASA suggested replacing GLV-2 with GLV-3, but based on recommendations from Martin and Aerojet it was decided to go with GLV-2. Then, for a third time, the weather came into play. Although Hurricane Cleo had skirted the Cape area, the next hurricane, Dora, was heading there on a direct path.

Testing was halted yet again, and this time both stages were lowered and secured as Hurricane Dora headed towards the Cape on 8 September. But it never reached there, and by the next day it was no longer a threat – but indications that Hurricane Ethel would arrive over the weekend kept the Titan grounded until 14 September, after the danger had blown over. With no more spare time, the GLV-2 was back on the pad by the end of the day. Spacecraft 2 arrived on 21 September.

Three cosmonauts in one spacecraft

On 12 October 1964, the Soviet Union announced the launch of a 'new' multi-crewed spacecraft called Voskhod (Sunrise), carrying three cosmonauts in a 'shirtsleeve environment' without spacesuits. Originally, it was thought that this was a considerably enlarged and improved spacecraft over the one-person design of Vostok that had orbited the first cosmonauts between 1961 and 1963. Gemini was intended to launch two astronauts, and yet again the Soviets stole the thunder by orbiting three, two of whom were not even pilots. Engineer Konstantin Feoktistov and Dr Boris Yegorov joined Pilot Engineer Vladimir Komarov on a 24-hour spaceflight to test the new spacecraft design.

Voskhod was, however, no more than a stripped-down Vostok, and was designed to squeeze three couches inside and 'prove' that Soviet space technology was once again far in advance of anything that the Americans could orbit. The reason for not wearing suits was not improvement in design, but simply the fact that they could not fit three men inside if they had worn pressure garments. The design of Vostok had included an ejector seat (like Gemini) to evacuate the sole cosmonaut in the event of launch mishap, or prior to landing; but with three cosmonauts onboard, this was redundant, and as a proposed escape tower was not ready in time, Voskhod flew with no means of launch escape for the crew. They also had only a retro-rocket system in the parachute lines to soften the descent, as they could not eject prior to landing.

The Soviets had achieved the first three-person spaceflight and the first orbiting of non-pilots, but little more. Voskhod, like Vostok before it, still could not manoeuvre in space – but Gemini *could* manoeuvre, as Gemini 3 would soon demonstrate.

More delays

Delays had pushed the earliest launch date for Gemini 2 back to mid-November 1964, and Gemini 3 could therefore not fly before the end of January 1965 at the earliest. Systems testing then slowed the pace considerably, so that the launch of Gemini 2 slipped to 9 December. While progress towards the launch continued, Martin took the opportunity to perform tanking exercises to train the crews to load the Titan and to determine how accurate they could do so. This instigated a new series of tanking exercises that lasted throughout the Gemini programme and also created a new fraternity of highly respected and dedicated launch technicians – who became known as the Wednesday Evening Tanking Society (the WETS) and the Thursday Evening Tanking Society (the TETS).

The Gemini 2 count-down finally reached zero, and at 11.41 am on 9 December the first-stage engines of the Titan roared into life – and one second later promptly shut down again! Just 3.2 seconds prior to launch, the controllers noted a drop in hydraulic pressure in the primary control system and tripped the guidance and control system from the primary to the secondary system which, as the vehicle remained on the pad building up thrust, was effectively an automatic shutdown command.

Fifteen minutes later, the flight was officially cancelled, and the vehicle was secured to fly another day. The cause of the abort was quickly traced to an unexpected high pressure in a hydraulic line, which fractured an aluminium housing on a servo-valve and allowed fluid to spill out. It was later determined that early in the design process, the thickness of the aluminium housings had been reduced, as they appeared to be thicker than the design pressure required. What was not carried out was a pressure test to check that decision. In the ICBM, the stronger housings remained, and the problem was not forthcoming; but in Gemini, in which weight was crucial, they had been replaced. On 9 December, Gemini 2 suffered for that 'improvement'.

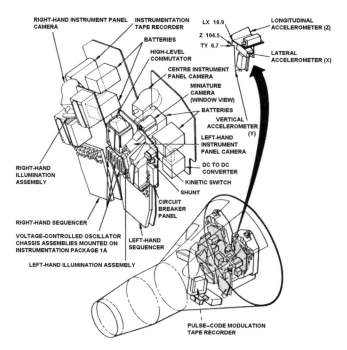

The instrumentation pallet, used for recording flight data, installed in the crew positions inside Gemini 2.

Preparation for the new launch in January ran smoothly, except for the set of six older model fuel cells onboard the spacecraft. These were of the pre-1964 design that was not to be flown on later spacecraft, but they were left in Gemini 2 to test the operation of the system in flight, as long as they did not impact upon flight objectives. These six stacks were known to be a problem design, and those on GT-2 had proved erratic since they were first installed. They also caused trouble during the 9 December aborted launch, and were scrubbed from use. When the mission did not fly, the stacks were powered up and tested, and were then shut down until the next launch attempt. Just 90 minutes after the resumption of the count for GT-2's second launch, attempt the fuel cells indicated further problems, and instead of holding the count once again it was decided that the fuel cell stack would be deactivated and not tested in flight.

The flight of Gemini 2

At 09.03.59 am EST, 19 January 1965, Gemini 2 was launched on a sub-orbital trajectory 1,848 nautical miles down-range into the Atlantic, reaching an apogee of 99.4 miles in an 18-minute flight. It was recovered by the aircraft carrier *Lake Champlain*. The mission achieved the primary objectives of testing the structural integrity of the spacecraft and verifying the integrity of the re-entry heat shield. and was also a successful demonstration of the flight control, life support, retrograde rocket, recovery and parachute landing systems.

At 45 seconds after staging, the spacecraft nose radar and sensor fairing were jettisoned. Separation of the spacecraft was accomplished as the second-stage engine tailed off at the end of boosted flight, and was achieved by firing the spacecraft's two aft-facing OAMS engines at 20.32 seconds after second-stage cut off. This increased the velocity by 15 feet per second. Just two seconds later, the spacecraft completed a 90 left roll, followed by a 180-degree turnaround some 28 seconds later, and a pitch up manoeuvre fifteen seconds afterwards to align retro engines for firing. At 6 minutes 54 seconds after launch, Gemini 2 was commanded to jettison the equipment section and automatically fire the retro-engines in sequence (engine number 1, 3, 2 and then 4) to begin its descent. The retrograde section was jettisoned 45 seconds later, and blackout began at GET 9 minutes 5 seconds into the flight, ending four minutes later. At GET 9 minutes 20 seconds, at a deceleration of 0.05 g, the Attitude Control Manouver Electronics (ACME) programmed an average roll rate of 13.6 per second to provide zero lift. This was maintained for 149.78 seconds, when GT-2 assumed the maximum lift attitude.

The parachute activation and recovery sequence began at 10,600 feet with the pilot parachute deployed at GET 14 minutes 31 seconds, and the main parachute at GET 14 minutes 35 seconds. The spacecraft splashed down after a flight of 18 minutes 16 seconds. Some small problems were recorded during the flight, but there was nothing that would prevent astronauts from riding the next mission.

The condition of the heat shield

Post-flight examination of the recovered spacecraft allowed a detailed inspection of the heat shield. The peak heating rate of 71.8 Btu/ft/sec was reached at GET 10 minutes 43 seconds. The Réné shingles had provided thermal protection for the spacecraft structure under the most critical heating conditions expected to be encountered on any manned mission. In addition, temperatures on the beryllium shingles on the spacecraft were well below both predicted and design levels, and it was not expected that any problems would be encountered during re-entry from any of the planned orbits.

The bond line temperatures of the heat shield were also well below predicted values. In summary, the heat shield maintained its structural integrity, and the amount of lost ablative material was less than expected. This allowed the shield to be fully qualified for maximum lift re-entries in all further planned missions.

Houston monitors the flight

One other aspect of the mission that was directly related to future flights was that Gemini 2 was monitored from the new control centre in Houston by a team headed by Flight Director John Hodge (Blue team). The plan was to monitor both Gemini 2 and Gemini 3, and then Mission Control would handle all MCC activities from Gemini 4, after the vehicle cleared the launch tower at the Cape.

As the clock reached zero, Flight Director Gene Kranz was watching activities at MCC-H as the lights turned on and the room came alive. He was tempted to shout 'Lights! Camera! Action!', but decided against it. Suddenly the room was plunged into blackness, as all power was lost at the moment of lift-off. Extra TV lighting had

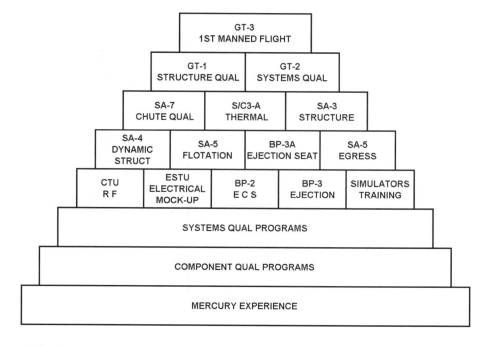

This chart reveals how experience gained during the Mercury programme helped in the test and development programme of Gemini, using mock-ups, structural test units, boilerplates, and flight spacecraft, and culminating in Gemini 3.

caused an overload. Luckily, they were not planning to control the mission, just to observe it, but even that was impossible. All the controllers could do was to listen to reports from the Cape. Despite meticulous planning and preparations, the one event that they had not planned for was a total blackout. In the post-mission debriefing, Lead Flight Director Chris Kraft's directive was to the point: find out what happened, and make sure it does not happen again. As a result, important systems were routed to separate channels of power, with back-up. In future, the media people would have to supply their own power.

GEMINI 3 (GT-3)

Gemini 3 was set to launch on 23 March 1965, and the prime crew of Gus Grissom and John Young had been training for it for almost a year. By the beginning of March, certification pledges from all leading contractors had been received at NASA, clearing everything for manned spaceflight. The media was gearing up to report on the first manned Gemini; the first spacecraft to change orbit; the first two-man American flight; the first Mercury astronaut to return to space; and the first flight of one of the new astronauts. The flight of Molly Brown was to make the headlines, but not before the Soviets declared another scoop for themselves.

Virgil Grissom and John Young – the prime crew of Gemini 3 – discuss the flight plan before entering the spacecraft for a communications test at the Cape.

Leonov steps outside

On 18 March, Voskhod 2 was launched with two cosmonauts on board – Pavel Belyayev and Alexei Leonov. As they completed their first orbit, a news release indicated that they had accomplished yet another space first for the Soviet programme, ahead of the Americans. Leonov had left his spacecraft and had spent ten minutes on EVA before returning inside to rejoin his Commander.

The Americans were planning a stand-up EVA from the open hatch of Gemini 4, but this news, coming as it did just days before Gemini 3, was a further disappointment for the Americans, who were once again playing 'catch-up with the Soviets'. It would be years before the emergence of the truth about the dangers that faced Leonov, but at the time he appeared to have had no problems during his short venture outside, and had made it look so easy.

'You're on your way, Molly Brown'

The astronauts, at the Merritt Island crew quarters, were woken at 5 am on the morning of 23 March, and after a brief medical check-up they took their launch-day breakfast of steak and egg – a tradition that was established during the Mercury era. During their medical, they handed Nurse Dee O'Hara their urine samples as part of the medical experiments. Young was especially careful with his sample, as during an earlier training session where they were to spend a day in the simulator, he had brought his morning sample to O'Hara inside a briefcase full of flight papers. When he opened the briefcase, to his horror the top had not been put on correctly, and all his papers were soaked, Dee O'Hara could not stop laughing at the look on Young's face, although the medics did not find it so funny.

However, on this day there were no spillages, and after the medical the astronauts left for the pre-flight ready room situated at LC 16. Here they were assisted by suit technician Joe Schmitt, who had suited up all the Mercury astronauts, and was now carrying on another tradition that would last into the early Shuttle programme. Grissom and Young squeezed into their pristine new G3C pressure suits and took a short van ride to LC 19 to ascend the gantry, to be assisted into the crew compartment of Gemini 3.

Waiting for Grissom and Young in the White Room were their back-ups, Schirra and Stafford, who had already spent a couple of hours in the capsule, setting up the switches and controls. Schirra – always the practical joker – greeted his Mercury colleague Grissom dressed in an old silver Mercury training suit, its colour tarnished and worn. Slightly tattered, it had been 'improved' with extra holes cut into to it by Schirra, who adorned himself with a couple of dozen security passes to gain access to the pad. With the authority to fly dangling around his neck, he told Grissom once more that if he was not up to making the flight, he was ready to step in for him.

By 7.30, both astronauts were inside the spacecraft with the hatches closed as the count headed toward zero. A forecast of bad weather threatened the launch as a hold was called at T–35 minutes (after they had been inside the spacecraft for an hour) while a leak in an oxidiser line was remedied. When the count resumed 24 minutes later, the clouds had begun to disperse. After the flight, Grissom recalled that the countdown went smoothly, apart from that leak, but although he was aware of the spacecraft's status, he would have liked more information on the Titan on which he was laying. Also, being inserted into the spacecraft 20 minutes early added to the time that they spent on their back. Grissom felt that it would have been better if they had been inserted into spacecraft when the count called for crew ingress, not beforehand, to reduce the time that they were prone. He pointed out that although he was uncomfortable, his condition did not become painful or affect his actions at any time. However, his feet seemed to go to sleep, and he was irritated by something restricting his helmet movement. He suggested that the time in the capsule before launch should be reviewed for future missions.

At 9.24 am, on Tuesday 23 March 1965, launch CapCom Gordon Cooper announced the start of manned Gemini operations with the call, 'You're on your way, Molly Brown.' The initial motion from the pad was so smooth that at first neither astronaut felt anything. Grissom could hear the engine ignite, and he saw the clock running, and then a few seconds after the call he noted vertical motion out of his window as the vehicle rolled. Young felt that the lift-off was much less rough than during training simulations.

With the two men pinned against their seats as the Titan climbed, the time for staging approached, as they were reminded by a call from CapCom. As the first-stage engine shut, down Grissom recalled, 'The sudden drop from approximately 6 g to 1 g or less was, of course, very apparent, but quite smooth.' Young, on the first of his six trips into orbit, was amazed at the sensation and sight of staging. 'I was surprised at the separation noise and debris. The vacuum start of the second-stage engine produced a momentary yellow orange flame around the spacecraft, which

Molly Brown is on her way. The manned flight operation phase of Gemini begins with the launch of Gemini 3 at 9:24 am EST on 23 March 1965.

also surprised me. Consideration of the event, however, made me realise that this was normal for 'fire-in-the-hole' staging.'

Young later called this event 'the great train wreck'. Grissom – a Mercury veteran – was expecting it. As the first stage shut down, they were forced against their seat harnesses, until second stage ignition pushed them back into their couches again. As they pitched over, Young noted that the view of the Earth's curved horizon out of the window was, 'a beautiful sight.' Initial forward velocity was not apparent, but during second-stage flight the motion over the ground became obvious. Grissom reported that no noticeable pogo effects could be detected inside the crew compartment as they headed up and out towards orbital insertion.

Into orbit

Grissom noted that the second-stage cut-off was both clean and sharp. Five minutes 30 seconds into the flight, the second stage shut down, prompting Young to note the sudden drop of acceleration to zero, but with no feeling of vertigo or disorientation as he entered space for the first time.

According to Young, the pyrotechnics that separated Molly Brown from the spent stage barked like howitzers. The thrusters were much quieter, however, and gave little indication of capsule separation. As the spacecraft separated, a shower of white flakes floated in front of the windows

Grissom had forgotten to check the time as he started the separation burn of the OAMS engines, and so continued to thrust until GET 6 minutes 9 seconds had elapsed. This caused a slight over-burn, and resulted in a small over-speed that placed Molly Brown into an orbit of 75.8 x 108.7 miles, although this was very close to the intended 75.8 x 113.0 orbit.

Upon entering orbit, the first task was to run through a post-insertion checklist, which took a little time. Grissom noted how well the training had prepared them for the launch phase, and how the real event was far smoother than any simulator had portrayed. Grissom then had to align the platform for navigation, using the horizon for reference. He found this method worked well, and that using the instruments or the view out the window was satisfactory; but he stated that he would have preferred a window display on the simulator to boost his confidence concerning visual alignment in flight.

While his Commander aligned the spacecraft, Young was preparing to begin work in the right-hand seat. Part of his early orbital activities required him to unstow cameras and film magazines and the blood-pressure bulb for the medical experiment, and to evaluate the environmental control system. He found that his suit temperature was fairly comfortable, but while working, the oxygen flow through the suit (to cool it) was marginal. Young recorded that the suit inlet temperature on the day-side of the orbit was 58°–59° F, while on the night-side it was 54°–55° F. Cabin temperature varied from 90° F to 93° F, which was far too warm, especially when the sunlight entered the crew compartment. Young suggested that the best mode of operation for ECS was with the suit faceplate open and the re-circulating valve also open. Even with the faceplate open and the suit unpressurised, there was very little difference, as the suit ballooned out from the body, and therefore no significant decrease in cooling was noticed in this mode.

Young also noted that a pressure gauge on the ECS had recorded a sudden drop. The pilot initially assumed that something was wrong with the system, but after a quick glance at other instruments that were also recording strange readings, he deduced that the problem was in the power supply to the instruments and not in the systems themselves. Instinctively, he flipped the electrical power converter from primary to secondary mode, and the problem vanished. This prompt action reflected the intense training both men had completed in preparation for the mission, in that, from his first awareness of the problem to the switching of power mode, Young took no more than 45 seconds.

The manoeuvres of Molly Brown
Grissom evaluated the OAMS thruster system on the Gemini, and checked out the pulse mode in making fine adjustments to attitude or rate. Visually tracking a town just north of the Gulf of California, Grissom found it a little difficult to keep the spacecraft in line with his visual target as he orbited. The biggest problem was the lack of good tracking targets because of cloud cover. He tried several ground targets,

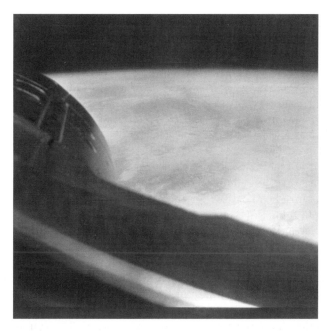

Flying on Molly Brown, John Young took this photograph of eastern Africa with a hand-held modified 70-mm Hasselblad camera. This was one of the first views of Earth taken from a Gemini spacecraft.

with varying success in retaining attitude as he flew over the target, but found it difficult to keep the target in line using the pulse mode of the thrusters, as Gemini was moving too fast. He therefore changed to the direct mode, and was able to hold Molly Brown to within 0.5 of the target although it was sometimes a rush to align the bore sight reticule before the target was obscured, and the 'pipper' tracker was sometimes lost against the Earth, as it was not bright enough. Grissom proposed that the pulse mode would probably work best if the target was acquired early and the bore site reticule pipper was bright enough. When the weather closed in, the use of the direct-control mode would be required.

Grissom also noted a difference using both the thruster rings on the spacecraft. The use of two rings gave the spacecraft a sharp kick, while a single ring, 'felt more like a nice soft push.' He found that there was a tendency to over-control when using two rings, and obviously the fuel consumption was much higher.

During the first two orbits, Molly Brown continually drifted to the left in yaw. Grissom readjusted the position to zero, and then the spacecraft would move back to the left at a rate of 3 per second. On board, the astronauts assumed this was caused by a leaking thruster, but despite several attempts to stop it, nothing worked. However, this yaw-left drift decreased as the flight continued, and by the third orbit there was no apparent drift.

An historic space first was achieved about 90 minutes into the mission when the OAMS engines were fired for 75 seconds, reducing velocity by about 50 fps, and lowering the spacecraft to a near-circular orbit. Grissom noted that he could not hear the forward firing thrusters, but he could hear the attitude thrusters. He also noted a good visual clue that he was still translating, as inside the crew compartment the debris moved towards, and stuck, to the instrument panel! During the second orbit, Grissom tested the translation control to slightly shift the orbital plane of the spacecraft, and then on the third orbit he completed a 2-minute 27-second burn using aft-firing thrusters. The perigee was then lowered the to 44.7 miles, as a precaution to ensure entry should the retro-rockets fail.

A bite to eat

One of Young's tasks on the flight was a brief evaluation of the food and the waste collection system. There was no official 'meal break', but with about a five-hour flight, plus about two or three hours between breakfast and launch and an hour or so for recovery, Young reasoned that he would probably feel hungry during the mission. Schirra had managed to obtain a corned beef sandwich from a nearby restaurant on Cocoa Beach, and had kept it in a refrigerator. When he could, he passed it to Young, who placed it in one of the pockets of his suit.

When food was mentioned during the mission, out came the sandwich, which was offered to a surprised Grissom. He took only a couple of bites, thinking that crumbs floating around the cabin could be a problem. Nothing was mentioned until after the recovery, when the press heard of the incident and of course printed the story. When Congress and the legislators heard about the incident, they were not happy; but contrary to complaints, however, it was not true that this episode had ruined the delicate diet of the astronauts, nor had it disrupted the medical experiments. The food experiment assigned to the astronauts was only for them to evaluate the taste, the ease of use, and how the food was reconstituted in zero g, and there was nothing to link it with any of the biomedical studies. After the flight, Young received a mild rebuke for smuggling the sandwich on board, but the incident was not serious enough to affect his future career as an astronaut. Many at NASA could not understand the reason for the fuss, but to satisfy the uproar, further such incidents had to be prevented.

The resulting memo from Deke Slayton took the official line for future missions: 'The attempt ... to bootleg any item on board not approved by me will result in appropriate disciplinary action. In addition to jeopardising your personal careers, it must be recognised that seemingly insignificant items can and have affected follow-on crews.'

With the sandwich packed away, Young began the 'official' evaluation of the Gemini food samples. He reconstituted some apple sauce and grapefruit juice, and opened a package of chicken bites. The bites were not that tasty, and proved very difficult to remove from the package while wearing the pressure glove. The drinks packaging was satisfactory, although Young noted droplets creeping out of the package, even though it was folded after use. He recommended that tape be added to close used food packages, to prevent unwanted leakage, and also suggested that future crews should 'attempt to eat as much of every meal as possible to prevent possible putrefaction and minimise the stowage problems.'

In evaluating the waste disposal system, Young observed: 'I believe that some of the problems of waste disposal in zero g will have to be endured in Gemini.' He did not have sufficient time to properly evaluate all the food and waste systems on this flight, but (rather accurately) forecast that on future flights, adequate time had to be allowed to prepare a meal, eat it, and then deal with all the matters concerning waste. In all, the drinking water and the reconstituted food was satisfactory, but not really appetising.

The astronauts reported seeing the lights of Perth, Australia, turned on especially to greet the two men, in the tradition begun on John Glenn's flight in 1962, and the view out of the window continued to absorb Young. In the post-flight reports, he noted that observations out of the window included his observation of white flakes and small items of debris outside the cabin. Using the cameras, he recorded the plumes of the thruster firings as, 'a larger yellow glow with lots of sparklers extending for 30 to 40 feet, [and] it is a very spectacular view at night.' He also took photographs of the Earth and the horizon, and although most of the southern part of the US was covered in cloud, they managed to track a town (later identified as Mexicali) and note the street divisions. Young also recalled: 'During the first night pass, the first stars I recognised were the Southern Cross and α and β Centauri. On the last night pass, I picked out the northern constellations very well; Taurus, Auriga, Orion, and the Pleiades.' To Young, they appeared as they would from an aircraft flying at 40,000 feet!

Grissom and Young had very little to say to the scientists or reporters about their three experiments, although Young noted that the clearance between the hatch and the human blood irradiation experiment was much smaller than on the simulator, causing some difficulty with its operation in flight.

The return of Molly Brown
The jettison of the Equipment Module was a jolt clearly felt in the crew compartment. Ground control's count-down to retrofire was an exact match to the event time and the computer onboard the spacecraft – the first retro-rockets fired right on the count of zero. The four retro-rockets were fired in a cross-fire sequence of 1, 3, 2, and 4, and caused Grissom momentary concern as he wondered what happened to retro number 2, forgetting that they did not fire in numerical sequence.

When the four retro-rockets burned themselves out, the retro adapter was jettisoned. The crew noted that this was quite audible and certainly could be felt, again noticing a significant amount of debris as a result. For orientation, Grissom used the rate command and both RCS rings for control, and the eight-ball for attitude, with rates displayed on the flight director indicator (FDI). This was the method he had used during training. After the retros had been jettisoned, he selected pulse mode, and found the attitudes easy to control; but as entry was one of the biggest unknowns, he reverted back to the two-ring system for the remainder of the entry phase.

Grissom found that attitudes were easy to control, as expected. He turned Molly Brown to the 180°-roll, maximum lift attitude two minutes after retro-fire. Prior to retro-fire, the crew had received details of re-entry banking angles and times. These

After three orbits, Molly Brown floats in the Atlantic as pararescue divers assist the crew out of the spacecraft. One of the astronauts is being hoisted up to the helicopter.

were updated just as the plasma sheath surrounding the spacecraft blocked communications. As Grissom rolled the spacecraft, indications showed that they were short and to the north of the landing point, and he tried to adjust this by steering out the error. It was found that data from wind tunnel tests provided a higher angle of attack than was actually being flown.

For Young, the view out of the window was, 'exactly like the technical pictures out the window of Gemini 2 ... same colour gases and the same pattern [of] spiral flow.' Grissom was watching the fireball, and was distracted and the bank angle deviated slightly, giving him some difficulty in holding the correct angle as the g-loads increased. The spacecraft remained very stable during re-entry and although loads did not exceed 4 g, they seemed to remain there some time. Out of the window, Young could see the retro-adapter following them into the atmosphere, and watched it burn up to destruction.

On a real mission, re-entry was completely different from the simulations, and the astronauts suggested that in future it would be beneficial to have a good out-of-the-window display and practise more entry simulations alongside ground support teams. When the parachute deployed, the spacecraft computer indicated a 50–60-mile shortfall in the landing point.

'That was no boat'
The RCS was turned off and the drogue parachute released at about 50,000 feet. Initially, this stabilised the spacecraft, but then oscillations began, and Young turned

on the RCS again, allowing Grissom to dampen them out. The propellant lines were turned off at 30,000, feet and rate command was left on to deplete any fuel and oxidiser still left in the supply lines. At 10,600 feet, the main parachute was deployed, and to the crew's relief there were no holes or torn panels. Descending at a rate of 30 feet per second, Grissom then activated the two-point suspension, 'and got the biggest surprise of the entire flight, [as] the spacecraft literally dropped from a vertical position to a 35 position! My head snapped forward with sufficient force to break my faceplate on the [windshield] mounting bracket.' At first, Grissom thought they had lost the main parachute. His parachute harness was not locked, and as they recovered themselves, the astronauts could not determine just when or how they would hit the water. So they braced themselves for landing as best they could. Young's harness was tightened, but he too still hit the faceplate of his suit, causing it to be scratched.

When Gemini 3 hit the water it was a milder impact than expected. Initially dragged by the parachute, it seemed as if Molly Brown wanted to nose into the water; and with both windows covered by water, Grissom released the parachute. For some time the thrusters spluttered and vented yellow smoke, and so they closed the snorkel valve for a while. They also sealed their face plates and breathed from the spacecraft systems when they noticed some greyish smoke and fumes inside the crew compartment, which they assumed was from the hot shingles after re-entry.

Having taken off their spacesuits, Young and Grissom wear towelling robes over their underwear as they walk across the deck of the prime recovery vessel *Intrepid*. The space mission is over, but the celebrity 'mission' is just beginning.

With memories of his escapade on Liberty Bell 7 (which took on water after landing, during which he nearly drowned), Grissom was not going to open the hatch until the recovery ship was alongside. Upon checking how long they would have to wait, they were first informed they were only five miles from *Intrepid*. After twenty minutes, this was revised to 55 miles (which involved another two hours wait). Grissom asked for helicopter pick-up, but still refused to crack the hatch until the pararescue men had secured a flotation collar to his spacecraft.

It would still be thirty minutes before they could exit the spacecraft climb into the life raft for helicopter pick-up. Gemini 3 proved to be watertight, but it was becoming hot and extremely uncomfortable in their suits, and they elected to take them off. In the confines of the closed cabin, this proved very difficult, especially as the spacecraft pitched and tossed in the ocean swell. The heat and effort took its toll on Grissom, and although Young managed to keep his breakfast down, he later observed: 'It is essential to get out of the pressure suits as soon as you land if there is to be any delay in recovery. Further, it is important to position all spacecraft controls, switches and safety pins as soon as possible, because I noted that seasickness reduces crew efficiency.' The former navy pilot also made a slightly less technical observation on Molly Brown's seaworthiness: 'That was no boat.'

Grissom noticed the first pararescue man drop into the water 20 yards in front of them, and the second only after they had attached the flotation collar and when the diver 'walked' past Grissom's window. The two astronauts wasted no time in ascending into the helicopter, and immediately the post-flight medical and debriefings began.

Tests concluded
On the carrier, the crew received a telephone call from President Johnson, who congratulated them on 'a very thrilling and wonderful flight.' He also joked that, 'Apparently, Molly Brown was as unsinkable as her namesake.' The President had sat unmoveable in front of a television in the White House as he watched a 'live' transmission of the launch, and smiled as the spacecraft was confirmed as entering orbit. In Houston, at the new Mission Control that was monitoring the flight ready to take full control from Gemini 4, Chris Kraft commented: 'If there had been a target rocket up there today – as there will be on future Gemini missions – the Molly Brown could have pulled alongside it.'

Gemini 3 and the three orbits of Molly Brown were a complete success, and ended the testing phase of Gemini. Now it was time to reach for each of the major objectives of the programme, the first of which would be to extend the duration from Gemini 3's 41 hours in orbit, up to 14 days.

Another key development from the following mission would be the activation of Mission Control to full control status from tower clear to landing. After Gemini 3, the flight controllers who monitored the flight were assembled for a debriefing, and after an open meeting of staff officials, engineers and astronauts, Gene Kranz reviewed the activities and events leading up to and during the mission. After highlighting both the highs and lows presented by the experience, he reminded them of the code of the flight controllers. Like the aura of the astronauts' 'Right Stuff',

Gemini 3 homecoming ceremonies in front of the Project Management Building at MSC in Houston. (*From left*) Paul Haney, Public Affairs Officer; William Schneider, Deputy Director, Gemini Program, Office of Manned Spaceflight, NASA HQ; Dr Charles Berry, Chief of Medical Programs; Chris Kraft, Assistant Director for Flight Operations; Dr Robert Gilruth, Director MSC (at podium); Virgil Grissom; John Young; Max Faget, Assistant Director for Engineering and Development; Deke Slayton, Assistant Director for Flight Crew Operations; and Wesley Hjornevik, Assistant Director for Administration.

this would become part of NASA legend: 'Our mission will always come first. Nothing must get between our mission and us … discipline is the mark of a great controller.'

Endurance

Extended-duration spaceflight had been in the minds of planners since they had first envisaged a programme to follow Project Mercury. By 1963 the first US manned space project had established that humans could survive in space for a day; and the Soviets had clearly demonstrated that this could be extended to as much as five days, and that a woman could also endure the rigours of spaceflight. However, alongside the highly promoted achievements of the cosmonauts, there were also reports that their condition during flight was not always as good as they implied: and for some, adjustment after a flight was as uncomfortable as their experience during the mission.

Although it was known that a man (or woman) could endure a relatively short and uncomplicated spaceflight for a few days, doubts remained about just how long someone could remain in space. It had been determined that a flight to the Moon and back would take at least a week; and with several proposals for rendezvous and docking, exiting and working outside the spacecraft, performing scientific experiments and observations from orbit, and exploring the surface of the Moon, a period of up to two weeks would be more suitable.

THE 14-DAY TARGET

By the summer of 1961, only four men had flown into space, on missions of 108 minutes (Gagarin), 15 minutes (Shepard and then Grissom), and 25 hours (Titov), and there were still many doubts whether 14 days was feasible in the near term. However, from the very beginning of Gemini, an objective of two weeks in space was listed as a first priority. After evaluating supporting one man for 14 days in an improved Mercury spacecraft, the advent of a two-man spacecraft design meant that the long-duration objective would be the first to be attempted in the flight programme – initially with men up to seven days, and animals up to 14 days. The two manned missions would be attempted on missions 3 and 4, while the animal flights would be manifested for missions 6 and 8. The idea of initially flying the animals on duration missions reflected uncertainty whether the spacecraft's systems (especially the retro-fire system) could support a long mission. While astronauts were on board, a manual back-up was available if the primary system failed. Once proven, then the animals could be targeted for the longer duration. There was also the effect of prolonged radiation exposure on the crew and concerns about their effectiveness and responses at the end of what was expected to be a demanding mission.

By the time the revised project development plan was issued in October 1961, the animal flight and high-altitude radiation research studies were deleted, to be replaced by an 18-orbit manned flight to test crew performance over a day in space and qualify the spacecraft for longer missions. The two long-duration flights of up to 14 days each had the sole objective to gather biomedical data on the psychological and physiological effects on a crew of lengthy periods in the space environment.

On 4 June 1962, a 14-day simulated long-duration Gemini 'mission' began at the Air Force School of Aviation Medicine, at Brooks AFB, Texas. In a representation of the Gemini environment, two volunteers lived in a 100% oxygen atmosphere maintained at 5 psi. Two months later, on 28 August, the Gemini Project Office (GPO) began a co-ordinated programme to develop long-duration (and rendezvous) missions, which would be completed by a mission planning and guidance analysis group with assistance from three working panels. Then, in October 1962 a joint McDonnell and Lockheed report on the hazards and constraints for Gemini missions found only that the 14-day mission would need to be limited to no higher than 115 nautical miles.

By April 1963, programme planning indicated that the third flight would become the first manned orbital flight, with probably no more than three orbits instead of the eighteen original planned. The fourth mission would be a seven-day flight with the use of a radar evaluation pod as a preliminary exercise to Agena docking missions. Flight 6 would attempt the 14-day mission.

The fuel cell issue
To support longer flights, the development of the fuel cell was crucial to mission planning, as chemical batteries would not be able to support a flight longer than four days. As early as 1959, plans for a 14-day mission had proposed that the spacecraft's power should be supplied by fuel cells, replacing the chemical batteries. But the development of the fuels cells had taken longer than anticipated and had forced NASA to include batteries on spacecraft 3 and 4, as the fuel cell would not be ready in time. Early in 1964, the radar pod was taken off Gemini 3 and Gemini 4 and was instead placed on Gemini 5, which also indicated a slip for the first the first Agena docking mission.

The original plans included the three-orbit flight of Gemini 3 followed by a seven-day flight of Gemini 4. This concerned Charles Berry, the medical director for Gemini, in that there would be such a large increase from a 4½-hour mission to a 168-hour mission on the following flight, and he proposed at least a 50% cut. At the same time, the installation of batteries on Gemini 4 also limited meant the mission to four days.

Plans were therefore revised for the Gemini 4 mission to be a battery-powered long-duration flight (four days), and Gemini 5 would double this duration (eight

Gemini 7 fuel cells, located in the adapter section of the spacecraft, are checked prior to flight.

days). Although an improved fuel-cell design was integrated into spacecraft 5 production, the issue of batteries or fuel cells for Gemini 5 continued into September 1964. A study suggested a possible combination, using batteries for leading load requirements, and fuel cells to supply the mission's remaining power requirements. Spacecraft 5 eventually flew with fuel cells as the primary fuel source and batteries as a back-up and re-entry mode.

With Gemini 6 targeted for the first Agena docking, Gemini 7 was to become the target for the 14-day space marathon. The crewing for these missions has been discussed (on p. 101), but there remained several issues to resolve before flying the longer missions.

Biomedical instrumentation
It was obvious that during the long-duration flight, most of the research would be biomedical in nature, and research areas were evaluated to meet this objective. In May 1962, the MSC Life Systems Division had proposed measuring seven parameters to determine the crew condition on *all* flights: blood pressure, with electrocardiogram and

phonocardiogram as first and second back-up; electroencephalogram; respiration; galvanic skin responses; and body temperature. It was determined that the mass of the biomedical equipment would equate to 31 lbs per man, and would require two watt-hours and the shared use of a six-channel telemetry system. From this study, a series of measurements were approved: electrocardiogram, respiration rate and depth, oral temperature, blood pressure, phonocardiogram, and nuclear radiation dosage. The biomedical instrumentation was still to be designed, developed, qualified and purchased – a process that would take almost three years.

In October 1964 a developmental test of the instrumentation was required, and astronaut Rusty Schweickart was assigned to spend eight days in a Gemini pressure suit. During the simulation he would be evaluating the biomedical instrument recording capabilities, including the completion of runs in the centrifuge, flying several parabolic curves on the KC-135, and simulating a four-day Gemini mission. By 15 January 1965 the Crew Systems Division at MSC had qualified the bioinstrumentation equipment for flight.

The conditions that the astronauts might face on the 4-day mission – let alone the 14-day flight – was cause for concern. Cardiovascular problems had surfaced during Mercury 8 (eight hours) and Mercury 9 (34 hours). There were so many unknowns in the early days of spaceflight that no-one could be sure that the astronaut would not pass out or even die at the end of the mission due to the stress of longer flights. High g-forces could induce fainting, and no-one knew what to expect if an astronaut was forced to eject due to a failed parachute after re-entry. Could he survive the stress, or would he even be conscious to effect the ejection?

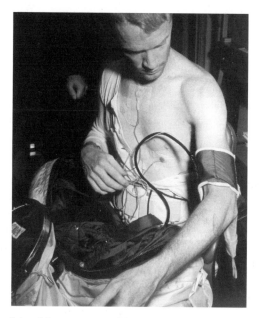

Astronaut Rusty Schweickart removes his Gemini pressure suit after a seven-day bio-instrumentation test.

As the cramped confines of Gemini would allow very little room for conditioning exercises, the crews were given a bungee cord devised from a nylon strap rubber extension cord and a spear-gun handle. This would extend to twelve inches if a force of 70 lbs was applied. Their diet and calorie intake could be controlled, medical monitoring could be studied both in-flight and on the ground, and physical training and conditioning could be applied pre-flight, to a certain level in-flight, and post-flight. But no-one really knew what would happen until the men flew the missions. Even studies of Russian data proved inconclusive about the effects of space motion sickness and disorientation. The longer missions would certainly separate the 'straw men' from those with an iron nerve.

The long-duration missions

The flight of Gemini 3 had established the integrity of the Gemini mission profile, but had only lasted three orbits and 4½ hours. Gemini 4's duration would be four days (96 hours) – well beyond the longest American flight of Mercury 9 (34 hours) in 1963, but short of the 119-hour Vostok 5 mission the same year.

FOUR DAYS OF GEMINI 4

The manoeuvring fuel used early in the mission, to remain in close proximity to the Titan stage, had been more than planned. Gemini 4 carried smaller fuel tanks than did later spacecraft, and in order to conserve battery power and fuel, McDivitt powered down the spacecraft shortly after completion of White's EVA, allowing Gemini 4 to drift during the rest of the mission. The objectives of the mission and its increased length required both a well-defined co-ordinated crew performance and a rigorous housekeeping regime.

Gemini 4 habitability

Eating The crew reported that eating became very important during the mission, with both astronauts feeling hungry every four or five hours, and experiencing a 'run-down' feeling. As soon as they had eaten they felt an energy surge, so they felt they had to eat regularly to function efficiently; and as a result, they consumed 31 of the 32 meals available on the flight. For the menu, the crew had a choice of beef pot roast, banana pudding and fruitcake. As a catholic, McDivitt was catered for by having fish dishes (tuna, salmon and shrimp) for Friday. White hated fish, and had chicken and beef. However, as all the food was either freeze-dried or dehydrated, they had to inject water into the bags, knead it to a mulch, and then squeeze it through a feed tube. Despite having the consistency of baby food, the astronauts reported that it was palatable and provided a varied diet. They commented that the dry bite-sized toast slices and peanut cubes tended to crumble, and that the toast supplied by the vendor was from the wrong type of bread! They also enjoyed the bacon in the menu – so much, in fact, that they recommended that more emphasis should be placed on providing smoked meats on future flights. One other problem was that eleven out of the 70 dehydrated food packages leaked around the valves.

Jim McDivitt (*left*) and Ed White leave for the pad to begin their four-day Gemini 4 on 3 June 1965.

Drinking Juice bags tended to leak the most. The astronauts found this most annoying, although they were able to consume the liquid both from the bags and from mid-air wherever possible. There were also several problems associated with the drinking water supply. It was found to be too easy to crimp the hose where it was attached to the trigger gun and shut off the water supply, and the hose also appeared old and cracked. The gun also stuck in the open position and had to be pushed into the closed position to shut off the water supply. There was also an excessive amount of air mixed in with the water, so that when the food bag was filled, one part air and three parts water was mixed with the food. Dehydration was a concern, and both men were constantly reminded to drink up to two quarts a day – double their Earthly requirements. This was to replace evaporated perspiration in the dry spacecraft environment. The crew also suggested that a meter be added to record the amount of water they had consumed.

Sleeping Sleeping was a problem on Gemini 4. The only way the crew could completely turn off the noise of the radio was to remove the helmet quick-disconnect, which was not wise. The crew had intended to sleep alternately four hours each; but this proved

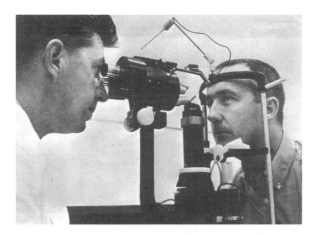

Jim McDivitt receives a pre-flight eye examination from Air Force Lieutenant Colonel James F. Culver as part of a comprehensive physical examination which both astronauts undertook two days prior to launch.

difficult with the noise of the radio, the occasional thruster firing, and frequent contact with the other astronaut moving about the confines of the cabin. The disconnects were eventually removed so that the crew could sleep, and they tried a six-hour sleep period and a later two-hour nap system. The changing of command from one crew-member to another during the alternate sleep–work cycles required considerable briefing and an exchange of information to ensure continuity of the mission, spacecraft safety and status. But this only added to the workload, and delayed sleep times. The crew reported that changes to the flight plan produced an unsatisfactory sleep period, and that this 'must be given more attention in future duration mission planning.' When White tried to sleep with the faceplate closed and his gloves on, he was so uncomfortably warm after 90 minutes that he woke up, although the situation was improved by turning on both suit fans on and having the faceplate open.

Waste management The crew found the use of the waste management system quite satisfactory, although the plastic bag containing the disinfectant was too difficult to break. The one-sided paper was also not sufficiently absorbent, and hygiene tissue was used instead of the paper. The urine collection device was only marginally successful, as a good seal restricted the flow of urine and a loose seal allowed a better flow but also leaked excessively.

Stowage There was not enough Velcro around the spacecraft, and pouches to store books, maps and other items were used continuously, as were the large dry waste bags mounted in each footwell and the centre and side boxes. The crew found that the left and right food boxes were accessible, but the removal and replacement of items proved difficult. The most useful storage area was under the pilot seat, where all the EVA equipment was secured after the spacewalk.

Personal hygiene Wet packs with the food were very useful for other hygiene uses and the larger towels were employed for mopping up urine. Although the tissues were useful, the dispenser was not as well designed, as the zipper failed early in flight. Oral hygiene gum was used only infrequently, and one toothbrush was lost on the very first day and the other not used at all – although the crew was not concerned. They did not shave for four days, but the growth caused no discomfort, and they expressed no particular need for a shaver.

Flight planning GT-4 was the first experience with real-time flight planning and there were many changes to the flight plan which needed co-operation between MCC and the flight crew. One of the problems encountered was the use of two time systems (GET and GMT), which complicated and confused the transmission of data to the crew.

Flight control Gemini 4 was the first operational use of MCC-Houston and the first three-shift operation. Kraft acted as Mission Director for the flight and Lead Flight Director for the first shift, with Kranz taking the second and Hodge the third. After the EVA, Kraft's team supported the crew in completing their flight plan, while Kranz's team was assigned to system performance and consumables. Hodge's shift conducted real-time flight planning on what had occurred, and updated the flight plan to relay to the crew the next 'morning'. This became known as the 'planning' or 'graveyard' shift, as it normally operated through the crew sleep period. As Kranz took over MCC Flight Director position for the first time, he wore a bright white waistcoat to denote his White Shift. The controllers commented, 'Nice vest, Flight,' and the next day Kranz's photograph, with him wearing what would become a tradition of MCC, appeared in all of the newspapers.

Chris Kraft, Assistant Director for Flight Operations, sits at his console in MOCR in his role as Mission Director during preparations for Gemini 5 in August 1965.

Medical The crew wore identical biomedical instrumentation packages, with data transmitted in real time to MCC or recorded on tape-recorders that were recovered after the flight. Exercise came from the bungee cord, where they placed their foot in the loop, grabbed the handle, and pulled upwards thirty times in 30 seconds. For comparison, White was to exercise four times a day, and McDivitt just once a day. By Day 2, McDivitt requested more exercise, as he had not moved much. Both men felt that their capacity and desire to exercise decreased to a certain point and remained there throughout the flight. They completed a programme of leg and back stretching – probably more so than using the exerciser.

Environment The astronauts both wore the G4C suit for the duration of the mission, with McDivitt wearing the standard cover layer. He was comfortable throughout the flight, and found mobility to be adequate. White wore the same suit, but with added EVA layers (see EVA operations, p. 269). He was warm throughout the flight and a little restricted in mobility, but was generally comfortable. After about six hours into the mission, White noted a smell that he could not identify, which also tended to keep him awake. Both men then felt a burning sensation in their eyes, which worsened throughout the following 24 hours, making their eyes very red. This irritation then decreased for the rest of the flight, and it was thought that this had been caused by out-gassing from the water-absorbent material or its additives. Relative humidity on a hand-held electronic sensor was recorded as 62% during the flight, and temperature measurements taken every four orbits recorded around 70–79° F. The crew also recommended better lighting in the cabin, and thought that although communications were satisfactory, the lightweight headsets were a nuisance as they repeatedly floated off their heads.

Are you being good?
During the mission the crew were able to talk to their wives, which provided them with the chance to catch up with news from home. News of their sons' efforts in baseball cheered them, and McDivitt's wife asked 'Are you being good?,' to which he replied, 'Don't have much choice. All I can do is look out the window.' Later, Pat White reminded her husband to take a drink of water, and said that during the EVA he seemed to be having, 'a wonderful time ... I can't wait to talk to you about it.' He agreed that he had had 'Quite a time, quite a time.' Pat McDivitt offered advice that was more like that of a flight controller than a wife: 'Disconnect your headset from now on at the start of the sleep period. Did you get that? Did you get that message?' 'I sure did,' McDivitt dutifully replied.

The day the straw men fell down
During their 22nd orbit, McDivitt and White surpassed the 34-hour 20-minute record of Gordon Cooper on Mercury 9, to the congratulations of the flight control team. White thanked them, but pointed out that they still had a few more hours to fly.

On 7 June the astronauts returned home, completing a mission of 97 hours 56 minutes. During the helicopter ride back to the recovery vessel, the medics reported that both men appeared to be in good shape, although a more intense post-flight medical examination awaited them. Both astronauts alleviated fears concerning their physical condition by walking across a red carpet on the deck of the carrier, smiling and waving. Over a period of 66 hours after landing, a post-flight medical test and evaluation programme revealed no serious problems. The day after landing, White demonstrated his readjustment to gravity by taking part in a 15-minute game of tug-of war with the ship's crew, between the marines and midshipmen. After his midshipmen team lost, he then went to continue the medical tests.

Both astronauts were fatigued at the end of their mission, although neither of them fainted. The loss of bone mass in their heels and fingers was not a surprise, but the loss of plasma volume in the blood was unexpected. Both lost weight (as had all previous astronauts) – McDivitt 4.5 lbs, and White 8.5 lbs – but the success of Gemini 4 boosted confidence for even longer missions. When the astronauts walked across the deck of the carrier, any feelings of apprehension about long missions in space began to quickly diminish. The flight of Gemini 4 was a valuable stepping stone towards the 14-day goal, which only five days previously had seemed to be a huge hurdle. In dedicating photographs of themselves at the end of the mission, the astronauts inscribed the phrase: 'The day the straw men fell down'.

EIGHT DAYS OR BUST!

The intended duration of Gemini 5 meant that management of supplies of oxygen and hydrogen was of concern, as no manned spacecraft had flown for that long. To cope with this, Cooper planned to fly with the cells operating at the lowest possible pressure. Early in the flight, Conrad pointed out that the pressure had dropped too low, and while out of communications with the ground, Cooper elected to power down the vehicle in case the fuel cell ceased to function. Without electrical power, the completion of all the rendezvous exercises with the released pod was not possible (see p. 239), and even the duration attempt might have been 'busted' almost immediately. After the ground had worked on the problem and realised that the fuel cells could indeed operate at lower pressures, flight controllers advised the crew to again turn on the fuel cells. The batteries could support the flight for a day while the cells were powered up. When the pressure remained stable, it was greeted with relief, on the ground and in space, as a good sign for completing their planned mission, although they still endured long periods of free drift to save power. On the fifth day the OAMS seemed sluggish, and one thruster failed. Shutting off more systems to save fuel and power involved longer periods of drifting flight that affected the experiment programme planned, although only one of seventeen was scrubbed (see p. 331).

Gemini 5 habitability
Eating During pre-flight training, a spaceflight crew has the chance to taste the food and practice with the facilities that they will use in space. When the Gemini 5 crew tried the bite-sized food during the flight, they found that it was not as appetising as it had been during training. As a result, from the third day neither astronaut ate bite-

Gemini 5 astronauts arrive at LC 19 for a simulated flight as part of their preparations for their 8-day mission. (Courtesy NASA via British Interplanetary Society)

sized food, and used only the dehydrated food supplies for the rest of the mission. Of these, four bags failed because they were unable to extract the food past folds in the packaging.

Drinking Acting upon the advice of the Gemini 4 crew, the flight included a method of recording water intake. A careful log was kept to ensure that each had sufficient water during the flight, and the astronauts found the water both good and cold. It still retained a quantity of air, but this had no adverse effect on the two astronauts. They also found that the gun worked well, and did not leak or suffer any operational problems.

Sleep As with Gemini 4, the crew experienced difficulty in sleeping. As the vehicle was powered down it became very quiet in the crew compartment, and when one of the crew communicated with the ground, operated a system or experiment, or moved something, he disturbed his sleeping colleague. At times the crew slept simultaneously to alleviate the disturbances. They also found that the Polaroid window shades were useful in restricting the distraction of sunlight. The problem of disturbed sleep was highlighted in this exchange:

Ground: 'You have a busy flight plan ahead. We recommend you to sleep during your planned sleep period if you can, so as not to get behind in the fatigue curve.' Conrad: 'We try to, but you guys keep giving us something to do.'

Cooper later asked for some uninterrupted sleep, and finally received it.

Waste management The urine collection device worked well, with the crew operating two new procedures during the mission. They preheated the system for at least four minutes prior to flushing, and once the receiver bag was empty they flushed air through the system for 30 seconds and then cycled the flapper once or twice to help in drying. The new design of rubber receivers worked well, and each lasted a couple of days before requiring a fresh one. However, even when cleaned they became 'gummy and sticky'. The post-flight report indicated that defecation had to be performed carefully and slowly. The astronauts had to ensure that the bag was completely open all the way to the bottom, that it was aligned correctly, and that it was firmly attached. The unit was well known as having the best adhesive in the space programme, and once attached correctly it remained firmly in place – although removing it posed its own, painful problems. The Gemini 5 crew reported: 'The whole procedure was difficult and time-consuming, but possible.' They also evaluated that stowage of used bags required as much volume as one food bag. During a long mission with several crew-members, this would pose a stowage nightmare unless there was, 'a change to the equipment or procedure.'

Stowage Very early in the planning for this mission, it was recognised that one of the largest and most critical problems to deal with was stowage, and during the flight this proved to be the case. The decision to remove all food from the right-aft stowage box involved the relocation of food around the cabin, in stowage bags, while the box itself was re-employed for all wet waste, defecation bags, and general garbage. The flight documentation was stowed at the sides of the seats and other pouches about the vehicle, but the larger stowage pouches were found to be unsatisfactory in-flight, as the lids, with the bungee cord on top, were too difficult to open. If a large item was stowed in the pouch, the lid remained open, allowing small items to float out, and as the flight progressed the pouches were soon worn out.

Helmets and gloves were removed early in the orbital phase of the mission and only put back on prior to re-entry. They therefore had to be stowed in the cabin for almost eight days – which was not found to be a problem, as they were stored in bags in the footwell in front of the ejection seat, attached by Velcro. This actually created additional stowage on each side of the helmet, which the crew found useful.

Trash management One problem with stowage that soon became evident as the flight progressed was that for every food bag removed for consumption, twice as many were required for stowage. When the empty aluminium stowage bag holding the meal was filled with trash, tissues, empty food dispensers and residual food, it took up almost the same volume as when it was filled with food. A defecation bag used and stowed in a similar aluminium bag with tissues and wet wipes also took up the volume of a food container. Every day, the trash and garbage required orchestrated management and careful stowage in order not to reduce the already limited available volume. It was also important to put the trash in the right place, as it could block access to equipment or other items needed later in the day. Care also had to be taken when opening a larger stowage bag to insert new items of trash in case the existing contents floated out into the cabin. Every sheet of paper had to be folded to reduce the size as much as possible, and empty food bags had to be rolled tightly to vacate all the internal air, and then secured with tape or rubber bands.

Environment Generally, the operation of the ECS to support the eight-day mission worked well, although with the spacecraft drifting, the cabin temperature was reduced, and the astronauts continued to shiver, even after turning on the airflow. The pressure suit coolant circuit was also too cold for them (as low as 44°), and they therefore removed the hoses and turned off the system. As Gemini 5 tumbled in orbit, the constant view of stars passing the window disturbed both men, and so window shades were used.

Time management The crew managed to successfully fulfil most of their pre-flight objectives, despite the problems with the fuel cell and thrusters. One of the less highlighted aspects of their management of time was in the careful preparation of the cabin for re-entry. This had to be orchestrated at the very beginning of the mission to ensure all pre-retro-fire stowage was accomplished early enough to allow it to be initiated on time. The spacecraft interior was badly restricted, but the crew's efforts were successful in determining that a rigorous housekeeping and stowage programme was conducted, because they had sufficient free hours available to complete the configuration for re-entry. The crew estimated that re-entry stowage took about twelve hours during the last two days of the mission, but they were able to stow most items in their correct places, allowing Gemini 5 to land in a 'clean, well-stowed configuration'. The astronauts recommended that on any future mission at least four hours should be devoted solely to re-entry stowage issues.

Cabin lighting The crew station design was generally satisfactory for the Gemini 5 crew, except for lighting conditions. Some of the console lights were too bright and could not be dimmed, while others were not bright enough even when turned up. Utility lights were weak, and broke under normal handling, and some of the lights reached high temperatures, adding to the warmth inside the small cabin. Inadequate lighting of critical items on the main instrument panel and water management panel, and low-level lighting in overhead stowage, hampered crew activities in these areas. The stowage in this area was difficult to manage in good lighting, and the lack of it in Gemini 5 compounded the problem.

Observations The increased periods of drift enabled the crew to perform additional visual tests, including observations of a rocket sled test as they flew over Holloman AFB, New Mexico. They later observed the ignition of a Minuteman missile and the contrail of one of the chase planes, and over the Atlantic they observed their prime recovery vessel *Lake Champlain* with an escort destroyer astern. They also observed rectangular marks laid out in fields near Laredo, Texas, although sixteen patches (2,000 feet square) of fine white sea-shells near the tracking station Carnavon, Australia, were not observed, as they were obscured by bad weather and operational difficulties. These observations revealed that the duration of flight had had no adverse affect on their vision, which was important for rendezvous and docking missions, the Apollo lunar missions, and the proposed strategic observations from

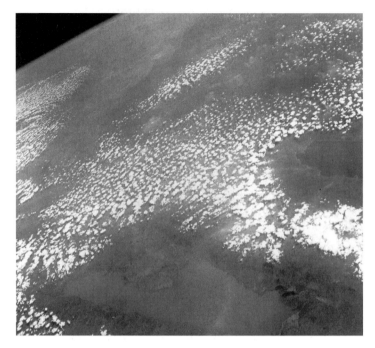

South-east Africa (Tanzania and Zambia), looking south-west from Gemini 5. Lake Tanganyika is below, and Lake Mweru is at centre right.

future military manned spacecraft. Using visual aids and the naked eye provided interesting results and, also confirmed what Cooper had alleged he had seen on Mercury 9. Sceptics had doubted that he could see the contrails of aircraft, the wakes of ships, or streets in towns, with the naked eye from space. But after Gemini 5 confirmed such observations, Conrad – who before the flight was also dubious – now supported his Commander's claims, as he too could see them.

Evaluation of the mission
Gemini 4 had been the first mission to use the Mission Evaluation Team to gather information from the flight as it happened. A 150-person team worked from the moment of lift-off through to post-flight evaluation, to assemble as much information as possible from the flight in progress in order to provide rapid results, and to use this information on the next and subsequent missions. With Gemini, the reduced turnaround meant that information from one mission had to be gathered, analysed and presented very quickly for the next flight to learn from both the successes and the mistakes. Gemini 5 was the first flight to reap the rewards of this team, which afterwards became an important part of every mission – not only for Apollo, but also for the Shuttle and beyond.

From outer space to inner space
While on the long flight, the astronauts had the opportunity to break the routine by

Pete Conrad, seen in his Pilot position during the eight-day mission of Gemini 5.

talking to one of their colleagues, Scott Carpenter, who had flown on Mercury 7 in 1962. During their 117th orbit, they held a radio conversation with Carpenter, who was in the USN Sealab experimental laboratory, 200 feet below the surface of the Pacific off the Californian coast. He was also enduring a 'long-duration mission' by remaining on the ocean floor for at least 15 days. During the four-minute conversation, each passed on their greetings, while Carpenter told his former Mercury colleague Cooper that he was 'doing a great job', and hoped that he would shortly have a pleasant re-entry.

Up to the ears in garbage
During the flight the crew also took the opportunity to exercise their vocal cords by singing, and they also composed poetry, despite being, as Conrad put it, 'up to our ears in garbage.' Apparently, the music was not just a one-way transmission. In his autobiography (published in 2000) Cooper recalled that on their first pass over China, they picked up a radio broadcast dedicating opera music to Gemini 5. The feminine voice of the announcer reminded Cooper of 'Tokyo Rose' from Japanese propaganda broadcasts during World War II – but in this case he thought it more appropriate to call her 'Peking Peggy'. China soon changed its opinion of Gemini 5, and began to brand the astronauts 'spies in the sky' when it was found that it would orbit over their borders – especially since they had cameras with telephoto lenses on board.

'Busting' the record
At just short of 98 hours into the mission, the crew surpassed the recently set Gemini 4 endurance record. But Cooper had already become the most experienced astronaut (at 84 hours 47 minutes into the mission) when he surpassed the duration time set by Bykovsky in 1963. As the mission also flew past this duration and set a new record, they entered uncharted territory for manned spaceflight. Mission Control suggested that a 'victory roll' might be in order, but Conrad replied that they did not have the fuel to spare.

Pete Conrad tugs playfully at his Commander's beard at the end of the record-breaking mission on 29 August 1965. The flight lasted just under eight days.

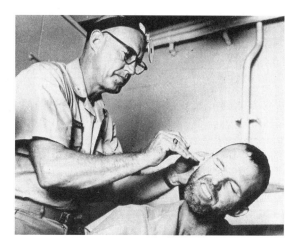

After the spacecraft has landed, the experiments continue. Here Gordon Cooper endures an ear examination as part of post-flight medical tests which both astronauts endured on the recovery ship.

After 3.3 million miles in flight, an elapsed time of 190 hours 27 minutes, and the completion of 120 orbits, Gemini 5 was floating in the ocean at the end of a highly successful mission. In a post-landing statement, Hugh Dryden reported that Gemini 5 had, 'demonstrated the capability of man to withstand prolonged periods of weightless ... this has assured us of man's capability to travel to the Moon and return.'

Medically, both men had lost more calcium and plasma volume than had the Gemini 4 crew, and their heart rates had gradually dropped below pre-flight levels until they adapted to their flight on the fourth day, when they levelled out. Upon their return, the rates increased as expected, but by the second day back on Earth they had returned to pre-flight levels. They had not exercised as much as they should have done, but they remained fairly healthy. They were fatigued, and as a result of their report, the crew of Gemini 7 would be allowed to sleep simultaneously during their 14-day mission. Gemini 7 would also include a review of the stowage issues and food complement as a result of experience gained from the previous two long-duration missions.

IN THE FRONT SEAT OF A VOLKSWAGEN

The success of Gemini 4 and Gemini 5 provided great confidence for the prospects of Gemini 7 completing its two-week flight. The media caught the mood, and described life inside the Gemini spacecraft as being like trying to spend two weeks cooped up in the front seats of a Volkswagen Beetle. It was that small, even without gravity.

The crew of Gemini 7 had to be carefully chosen for compatibility for just this reason (see p. 105), and Borman was forever grateful to Slayton for assigning Lovell as his Pilot. The crew also took great care in ensuring that they talked to the Gemini 4 and Gemini 5 crews, and it was an advantage that they were the back-up crew for McDivitt and White. On Gemini 4 the crew had become fatigued with the attempted rendezvous and then the EVA, and on Gemini 5 the crew was tired and bored. There was no EVA planned for Gemini 7, and the crew had already taken the advice about sleeping simultaneously and working hard at cabin stowage and housekeeping. With twenty experiments also assigned, they did not think boredom would be an issue, although they had received official permission to carry a couple of paperbacks to read during the flight. Deciding on subjects that had nothing to do with the space programme, Borman took Mark Twain's *Roughing It* – a story about a nineteenth century journey – while Lovell selected Walter Edmond's novel *Drums Along the Mohawk* – a story of the eighteenth-century Indian wars. They had little time to read, however, and neither of them finished their book during the flight.

'We're on our way, Frank'

The longest spaceflight in the American space programme (prior to the Skylab missions of 1973) proceeded from Earth on 4 December 1965, with Lovell informing his Commander: 'We're on our way, Frank.' But this was obvious from the vibrations and the noise of the engines as the heaviest Gemini began its historic mission.

The medical experiments began just thirty minutes into the flight, when the cardiovascular conditioning cuffs on Lovell's legs were activated. The medical tests lasted throughout the entire 14 days (see p. 331). The flight progressed well, apart from concerns over a fuel cell failure and a thruster problem during the latter part of the mission. Towards the end of their mission the crew was briefly joined in orbit by the Gemini 6 astronauts for the world's first space rendezvous between two manned vehicles (see p. 241). The sight of visitors was one of the welcome breaks in their long flight, which the crew reported as being boring and tedious.

A suitcase for a fortnight. Some of the equipment stowed on Gemini 7 for the two-week mission.

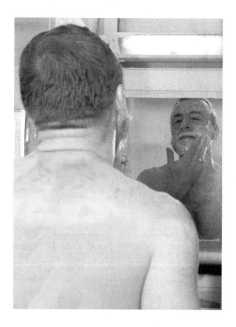

Jim Lovell takes a shave prior to suiting-up and boarding Gemini 7 for what was to be a record-duration mission.

The grass needs mowing

To help keep the astronauts informed of life back on Earth, regular news and sports reports were read up to them. In one news report early in the mission, they were told of an air accident between two commercial jets. Borman commented that it appeared to be safer up in orbit than down on the ground, at which point Lovell looked across and observed, 'We're not down yet.'

News from home was always emotive – especially with the festive season approaching. Lovell's daughter requested CapCom to play a special song for him so that he might return home in a hurry, and Borman's sons worried that he might not be able to take them fishing after the mission because he might be too tired. But he was not tired, and they went fishing.

During one communications pass, Lovell was in the middle of medical data readings, when – probably hoping to save himself a more earthly chore after the mission – he passed word down to CapCom to tell his family that the grass still needed mowing. They also regularly received good wishes on the progress of their flight and at one point received a message that the postal service had said that the public were being kept waiting too long for their Christmas gift. Once again, Lovell took the bait: 'I have a stack of stuff up here, but I can't find a post office.'

The G5C spacesuit

The success of the Gemini 7 mission was due in part to the use of the new lightweight pressure garment, designated G5C. Designed especially for the 14-day mission, it was developed at MSC and took advantage of G4C (the EVA suit) material and technology, but eliminated the restraint layer and the inner liner. The helmet of the suit was, in effect, a soft hood with an integrated visor and no neck ring and when not required it could be unzipped at the neck and folded behind the head. For head protection during launch, an aircraft crash helmet ('bone dome') was worn. The communications systems remained the same as on the other Gemini flights. The crew was also provided with tailored HT-1 nylon two-piece orbital flight suits, which could provide additional warmth when not wearing the G5C suits.

Confinement in the Gemini spacecraft for 14 days was recognised as a hurdle for crew comfort, as well as a medical issue. Development of the softer suit was in progress after tests revealed that the astronauts could remain almost suitless in the crew compartment. Astronauts Cooper and See had evaluated the use of Air Force flight suits (supported by medical monitoring devices and emergency oxygen masks) in an altitude chamber test in June 1965. Both astronauts were enthusiastic about the results – especially Cooper, who was in training for the long-duration Gemini 5 mission – but officials were not enthusiastic about allowing astronauts ride into space without a pressure suit for two weeks. If there were to be a loss of cabin pressure, the two astronauts would be lost; or, in the event of a cabin fire, the quickest way to extinguish it would be to vent the atmosphere, which the astronauts could not do without suits.

Crew Systems Division at MSC produced the soft suit concept, and by removing as much of the stiffness and bulk as possible, reduced the weight from 23.5 lbs to 16 lbs. The suit was not ready in time for Gemini 5; but in training, Borman and Lovell found it much more comfortable.

A technician models the G5C suit used on the Gemini 7 mission. This photograph depicts the suit in fully pressurised configuration, featuring the 'bone dome' helmet worn beneath the larger soft pressurised helmet of the suit.

On the mission it had been planned for both astronauts to take off the suits once the ECS had been confirmed as operating normally, about 48 hours after lift-off. Management soon reversed that decision, as they did not want both men to be unsuited at the same time, and they allowed only one of them to remain unsuited, alternating between them. During launch, re-entry and critical phases of the mission, both would wear suits.

According to the pre-flight test, taking off the suit should have taken only minutes, but when Lovell began to remove his suit 45 hours into the flight it took him more than an hour, because of the tight confines of the compartment. The cabin had been warm with both men suited, and although this was alleviated a little when one of them was unsuited, it was not of much help to Borman as he continued to perspire. Lovell was the larger of the two men, and he found it difficult to don and doff the garment, so with Borman's agreement he remained unsuited while Borman continued to literally sweat it out. After 100 hours, Borman asked for permission to remove his suit. Meanwhile, Lovell had briefly worn the flight overalls, but again found them too hot, and so remained in his underwear.

The Gemini 7 astronauts, wearing pressurised G5C suits, leave the suit trailer for LC 19. Lovell leads the way, and Borman follows.

Both astronauts wanted to swap configurations of the suit after the rendezvous with Gemini 6, but were told that they needed to exchange configurations much later into the flight to provide comparative sets of medical data on men suited and unsuited. So, 148 hours into the flight, Borman took his suit off and Lovell put his back on. In his first sleep period without the suit, Borman slept continuously for six hours, and later reported that he felt 'like a million dollars.' On the ground, mission management was polled as to the best configuration – suited or unsuited. Most agreed that the reasons for taking off the suits far outweighed those for leaving them on. Even medical data indicated that blood pressure, pulse rate and sleep quality were closer to normal when unsuited, and the crew were therefore told that they could take off their suits for the rest of the flight, up to preparations for re-entry.

Gemini 7 habitability
Eating Many of the astronauts considered that Army C rations were a gourmet meal compared with what they had to eat on Gemini. In effect, the astronaut guide for eating in space was 'add water; ignore taste.' However, with each flight the dieticians, like the astronauts, gained experience, and food systems improved – eventually. Gemini 7 carried 86 man-meals (14.2 days for two men), with a 'typical' meal consisting of one dehydrated food selection, one dehydrated drink and two bite-sized food packets. The crew followed a three-meal-a-day system, and of the 43 individual meals available to each man, 41 were consumed, providing an approximate intake of 2,200 calories per man per day. The crew found that the food was adequate and

palatable for the 14-day flight, but that there should have been little more variety. The astronauts' favourites were the juices and puddings, and once again the dehydrated food far outweighed the bite-sized selection in popularity. Although one bag burst at the seam during use, the packaging was generally acceptable. The astronauts discovered that the feeding spouts for semi-fluid meals, such as shrimp, sauce and tuna salad, were far too small to extract the food, although they were ideal for purely liquid refreshment and the puddings. As discovered on earlier flights, the beef bites crumbled too easily, and contaminated the cabin rather than being eaten. The crew also found that both the beef and egg bites were also too dry, and left a bad taste in the mouth and a coating on the tongue. The desserts were too sweet and, there were not enough fruit juice drinks. They recommended that provision and stowage of food should reflect a normal meal for the day. They would have preferred eggs and bacon for breakfast instead of the shrimp and bacon with which they were provided. They also suggested that chunks of food should be extractable without breaking the bag, and the bite-sized chunks of food should be manufactured so that they did not crumble so easily.

Drinking Forcing themselves to drink more water than they really wanted was the best way to replace lost fluid as result of the dry cabin atmosphere (which, according to Borman, resulted in flaking skin and two cases of terminal dandruff). The relocation of body fluids in the early days of adjustment to zero g was a discomfort for a few days as pressure built up in the head, giving both men stuffy noses and sore eyes until they adjusted to the condition. A new water-metering device (replacing the dispenser used on Gemini 5) supplied the drinking water. A measuring device in the pistol measured and dispensed ½-ounce increments. Trigger-operated bellows measured and dispensed the water, and a mechanical counter was used to calculate the number of cycles on the trigger.

Sleeping Both men slept simultaneously during the mission. A ground rule adopted for this mission was to follow the schedule based on local time at Mission Control (Houston Time). The crew followed the work/rest cycle with which they had become accustomed on the ground during training, and a ten-hour 'night' rest cycle proved substantially better than on Gemini 4 or Gemini 5. Mission Control had been instructed not to disturb the crew unless there was an emergency situation. Even so, Borman found it difficult to sleep soundly due to not being able to stretch out to his full length. They also found that the Polaroid window filters failed to sufficiently block the Sun's heat and light, but the placing of aluminium sheets from the food packages between the window and the Polaroid filter rendered the crew compartment completely dark and allowed the crew to follow the accustomed 'day–night' cycle without any regard to the changing orbital lighting conditions.

Waste management As with Gemini 5, the crew took care of their accumulated waste at the end of every day. They adopted the experiences of Gemini 5 by packaging and stowing all used food packages and trash as compactly as possible. Both astronauts experienced problems with leaking urine receivers, and one of the urine bags leaked. They realised the importance of accurately measuring and recording urine output for

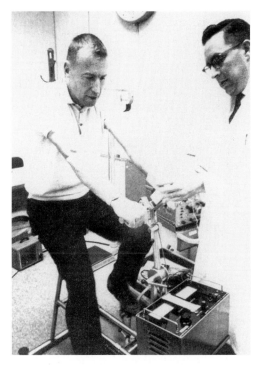

'It's exercise Jim, but not as we know it.' Lovell tries out a bicycle ergometer as part of the pre-flight conditioning and medical tests prior to Gemini 7. Dr Charles Berry is at right.

the medical experiments, but found that the hardware procedures and the sample-making system was not satisfactory. The receiver was large enough, but they found that back pressure developed when urinating in the bag, and repeated use of the receiver (for which it was designed) only loosened its grip and increased the leakage problem. They suggested one new receiver per man per flight day would be more suitable. The defecation bags were used several times with no major problems, and the crew discovered that the system was much easier to use with the suits removed, although they had to make a small cut in the disinfectant packet in order to extract the liquid into the defecation bag. Storage was not a problem on the flight, although they experienced problems removing food from the aft storage box location – especially the first unit from each box – as it was so tightly packed.

Flight plan updates Lessons learned from earlier missions had provided both the astronauts and the flight planners with experience in planning activities during long flights. The flight plan for Gemini 7 was essentially only a guide, to be updated real-time each day and to have experiments and activities inserted as and when the controllers and the crew could complete the task. The only 'scripted' flight plan was included in the first four hours of the mission and for re-entry at the end of the flight. Updates to the flight log were hand-written in the experiment log sheets of the

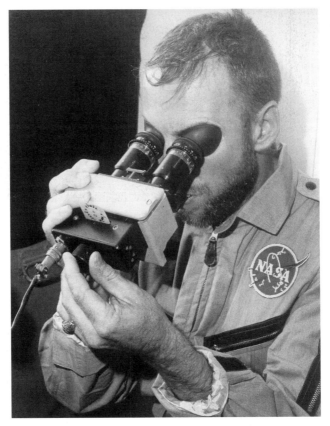

Jim Lovell undergoes an eye examination during post-flight examinations onboard the ship *Wasp*, to provide post-flight data, additional to data obtained pre-flight and during the mission, for the series of visual tests and experiments carried out on the long mission.

procedure book, and the crew then transferred these notes and placed them in chronological order in the flight plan book. At the end of the day the experiment log sheet was rechecked to ensure that all tasks had been completed and logged correctly. The crew also suggested that future flight plans should include – during retro-fire, re-entry and landing – only those items that were necessary to accomplish these tasks, and should not include experiments, operational tests, medical tests or scientific observations.

Housekeeping One of the bonuses of flying with the lightweight suits was that the crew could remove them and use the suit bypass hose as a vacuum cleaner to remove small crumbs and debris from the cabin. The screen on the end of the hose prevented contaminants entering the ECS system. The crew reported that the disinfectant pill tended to crumble when being placed in used food bags, and minute pieces floated about the cabin. When they entered the eye, it caused burning.

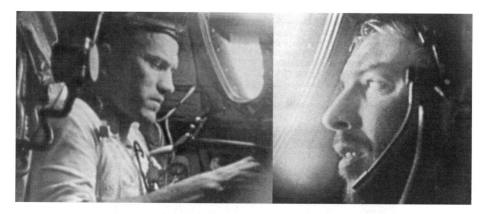

During the Gemini 7 mission, Frank Borman, without his G5C suit, sits in the Command Pilot seat and performs one of the onboard experiments. Lines of strain are etched on Jim Lovell's face as sunlight shines through the window.

Habitability Overall the crew reported that the crew station was reasonably satisfactory, although both astronauts experienced some discomfort in the cramped compartment. Frequent use of the exerciser and the removal of suits was particularly helpful in alleviating the discomfort. Cabin temperature was around 80° F throughout the flight, although at 158 hours into the flight this was reported to be down to 66° F. The chance to take off the suits – or parts of them – helped alleviate the discomfort, and the amount of liquid which the astronauts drank varied according to whether or not they were wearing the suits. This was picked up in the reports on daily water intake. CapCom asked why one of the crew took more water than the other. Were they thirsty? Were they perspiring? Were they unsuited, partially suited or fully suited? The cause for concern was not just that they kept up the water intake, but also that the medical data being collected related to the way they were dressed. The two weeks inside the cramped spacecraft, coupled with the heat and the lack of adequate personal hygiene, resulted in a distinctly odorous pair of astronauts by the end of the flight.

'Going back to Houston'
Taking the advice of the Gemini 5 crew, Borman and Lovell took time to pack up for the trip home. The last three days after Gemini 6 had departed were the most difficult of the entire flight, with fuel cell and thruster problems still plaguing them. The fuel cell had shown a warning light on the very first orbit two weeks earlier, but had continued to supply electrical power for 126 hours. As the astronauts prepared to begin their final orbit (number 206), Mission Control played the tune 'Going back to Houston' up to them.

After 330 hours 35 minutes second, the spacecraft, containing two tired but happy astronauts, was floating in the ocean waiting for the recovery team. Lovell had worn the leg cuffs, pumped up to prevent pooling of the blood; but Borman, acting as a

The coast of British Guyana, as seen from Gemini 7 during the 133rd orbit.

control subject, had not worn them. Although he felt a little dizzy and was expecting to pass out, he did not do so. He asked if Lovell felt as if he might pass out, and Lovell replied he did not. 'Neither do I,' replied Borman, which made him wonder if the leg cuffs were, after all, of any use.

It was not long before they were each hoisted up into the helicopter, and as they received the greeting from the aircrew, neither of them felt any queasiness or dizziness, although they felt as though they had rubber legs. Post-flight medical results yet again indicated weight loss – Borman 10 lbs, and Lovell 6.5 lbs. Despite Borman's doubts, the leg cuffs that Lovell wore had created less pooling in his legs than in Borman's legs, but it took only a night's sleep on the carrier for both men to report that they felt much better.

What pleased the medics was the astronauts' ability to walk off the helicopter and across the carrier deck. Pictures relayed to MCC Houston showed two tired-looking, bearded, but happy astronauts, walking slightly stooped and a little wobbly-legged (due in part to the roll of the carrier), but unaided. As Charles Berry reported on the post flight medical tests: 'The most miraculous thing was that they [did] not flop on their faces. They were in better physical shape than the '5' crew. Initially, their tilt table responses were not as bad and did not last as long. It looked more like four-day responses, by far, than eight-day. The calcium loss was the same way. Amazingly, they maintained their blood volume. They did not get any decrease, but they did in a peculiar way. They lost the red cell mass still, but they replaced the plasma – they put more fluid in. Apparently, there had been enough time for adaptive phenomena to take place.'

Full biomedical studies would take some time, but confidence was boosted by the

The Gemini 7 astronauts arrive onboard the prime recovery vessel *Wasp* on 18 December 1965. The mission set a new world endurance record of 330 hours.

result of Gemini 7 (and that of Gemini 5 and Gemini 4 before it) that the time it would take for Apollo to reach the Moon and back was survivable. 1965 was a milestone year for NASA: five manned spaceflights accomplished in eight months, the first demonstrations of rendezvous and spacewalking and, above all, the achievement of one of the objectives of the December 1961 development plan of Gemini – regular spaceflight operations.

It was also, of course, the year in which the Americans finally caught up with – and significantly surpassed – previous Soviet achievements. The prospects for 1966 and the first missions to attempt docking with the Agena seemed promising; and with the impending naming of the first Apollo crews to test the new spacecraft in Earth orbit, at the beginning of 1966, the Moon appeared to be much closer than it had only twelve months earlier.

Soviet and American mission duration records, April 1961–December 1965

Launch date	USSR Mission	Crew	h:m	Orbits	USA Mission	Crew	h:m	Orbits
1961 Apr 12	Vostok	1	01:48	1				
May 5					MR-3	1	00:15	Sub-orbital
Aug 6	Vostok 2	1	25:18	17				
1962 Feb 20					MA-6	1	04:55	3
Aug 11	Vostok 3	1	94:22	64				
Oct 3					MA-8	1	09:13	6
1963 May 15					MA-9	1	34:19	22
Jun 14	Vostok 5	1	119:06	81				
1965 Jun 3					GT-4	2	97:56	66
Aug 21					GT-5	2	190:55	127
Dec 4					GT-7	2	330:35	206

MR = Mercury Redstone; MA = Mercury Atlas; GT = Gemini Titan

Rendezvous and docking

The series of rendezvous and docking operations carried out during Gemini were perhaps the most significant contribution of the programme to human spaceflight exploration. Extending the duration of human endurance to four, eight, and finally 14 days, was equally as important for the longer-term programme, and the lessons learned on EVA were valuable for future activities outside the spacecraft. But it was the achievement of joining two spacecraft together that was a significant leap forward in technology and operations.

In 1962, the selection of the Lunar Orbital Rendezvous method for the Apollo programme prompted NASA to develop new techniques before a lunar flight was attempted. If President Kennedy's goal was to be met, NASA had just over seven years to learn how to join spacecraft in orbit, and to practice it enough times to be confident of sending three men to carry out a similar job a quarter of a million miles away around the Moon.

Gemini – conceived just prior to the decision to use the LOR technique – was NASA's testing ground on a series of missions prior to the development of operational procedures on Apollo. The lessons and experiences learned from Gemini during 1965 and 1966 (after three years of research) laid the basis for rendezvous and docking operations that continued beyond the Apollo programme into the Skylab and Apollo-Soyuz programmes, the Shuttle docking to Mir, and more recently to the ISS. In the mid-1960s the Soviets were also developing procedures to join one or more spacecraft together, and this led to the highly successful Soyuz dockings with Salyut and Mir over a period of thirty years. Soviet unmanned docking operations with Progress have, for the most part, also been outstanding achievements – not only in the actual operations, but also in their consistency.

THEORY AND PRACTISE

The idea of rendezvous and docking had been put forward several decades before anything flew in space, in the works of notable space theorists, artists and planners in Russia (Konstantin Tsiolkovsky and Sergei Korolyov), Germany (Herman Oberth and Wernher von Braun), America (artist Chesley Bonestell and author Willy Ley, in the series of *Collier's* magazine articles), and Great Britain (Arthur C. Clarke, R.A. Smith, and the British Interplanetary Society).

Although plans to dock variants of the Vostok spacecraft had been on the drawing board in Soviet design bureaux for some time, it was the creation of Gemini – for which rendezvous and docking was listed as a primary objective – that spurred Korolyov, particularly after the success of John Glenn's first American orbital flight in February 1962. If the Soviets could launch two manned spacecraft and then bring them together to meet in orbit, it would hint at docking capability.

Space brothers

This is exactly what the Soviets achieved in August 1962, with Vostok 3 launched on 11 August and then Vostok 4 launched a day later. The Soviets indicated that the 'Nik and Pop' space duo were in similar orbits and close to each other, and intended to provide data on the possibility of obtaining direct contact with each other; and Western observers initially interpreted this as a practice docking. Vostok did not have the capability of manoeuvring, however, and the launches were timed so that as Vostok 3 flew over the cosmodrome, Vostok 4 was launched into a matching orbit, but only for a short time. The closest proximity was about four miles, and the cosmonauts could see each only other as small pinpricks of light. A day later they were 528 miles apart, and at the end of the flight they were 1,770 miles apart.

A year later the feat was repeated, with the flight of Bykovsky on Vostok 5 (also setting the world endurance record of 119 hours, subsequently surpassed by Gemini 5) and the first woman in space, Valentina Tereshkova, on Vostok 6. At the time, Western journalists misinterpreted the achievements and capabilities of Vostok, and hailed the missions as a milestone on the Soviet road to the Moon. With the secrecy of the programme, however, the Soviets did not reveal the whole truth (the first Soviet docking hardware was still some years in the future), and had no intention of shattering the illusions created for them by the Western media.

Heavenly twins

With the development of Mercury Mark II, NASA was seeking to improve the Mercury design and to include the capability to move orbit. Although the emerging Apollo programme was to be at the forefront of any effort following Mercury, the need to define clear roles for Mercury Mark II was imperative. When Apollo was committed to the Moon in the summer of 1961, the thought behind Mercury Mark II would be quickly evolved. One question that remained unanswered for Apollo was whether rendezvous and docking was a feasible option for the missions to the Moon. Clearly, Mercury Mark II could address this question.

In the August 1961 development plan, the fourth objective listed was rendezvous and docking, which was manifested for flights 5, 7, 9 and 10 in the programme. The spacecraft would launch towards a target vehicle in space in a given period of time (launch window), which would place it a reasonable distance from the target in orbit so that it could catch up with it and then dock to it. One element would be the manoeuvring vehicle (active), and the other would be the stationary target (passive).

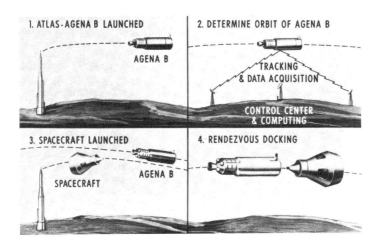

This artist's impression, dating from 1962, depicts the four stages of a Gemini rendezvous mission.

The primary spacecraft contractor, McDonnell, had been evaluating rendezvous for months, and initially there were two different ideas about what it should include.

McDonnell supported the idea of using their spacecraft as the 'active' element, and the non-manoeuvrable target as the 'passive'. The extent of operations depended upon the amount of fuel the spacecraft and target could carry, with the crew in control of the target from the spacecraft. MSC, on the other hand, supported only semi-automatic control of the target from the spacecraft, preferring it to be manoeuvred by ground commands. The result was a compromise between both ideas, by adapting the spacecraft to carry only the equipment necessary and allowing the target to be controlled either from the ground or, during docking operations, from the spacecraft.

With the approval of Mercury Mark II on 7 December 1961, and the renaming of the project as Gemini early in the new year, the programme could progress from theory to flight. When LOR was selected for Apollo in the summer of 1962, it moved rendezvous higher in the priority of objectives to be achieved. Now that rendezvous and docking was a clear objective for Gemini, decisions rested on how to achieve it and what hardware to employ.

THE AGENA TARGET VEHICLE

The Agena vehicle was an early proposal for an upper stage to the Atlas SLV to launch Mercury Mark II, until the Titan II was selected as Gemini's launch vehicle. However, the Agena offered the programme a suitable target vehicle with which Gemini could dock.

Agena B

During August 1961, contact was made for the first time with the Lockheed Missile and Space Company, of Sunnyvale, California, manufacturer of the Agena stage. At that time the current version was Agena B, which had been in use for about a year. Like Titan, the Agena used storable hypergolic propellants (unsymmetrical dimethyl hydrazine as a fuel, and inhibited red fuming nitric acid as an oxidiser). One attractive feature of its pump-fed rocket engine was that it was capable of restart. Termed 'dual burn', the engine could be fired, shut off and then re-ignited. This was of especially value for the Gemini missions, during which large-scale manoeuvres – and perhaps even high altitude orbits – could be attempted.

Agena was the second stage of the Thor launch vehicles used to lift the USAF Discovery satellites. It was modified for use as the second stage of the Atlas SLV in the USAF Midas and Samos military satellite programmes, and the NASA Ranger and Mariner programmes. The Atlas–Agena B combination was capable of placing a 5,000-lb payload in a 300-mile circular orbit.

As both Atlas and Agena were continuing programmes, little modification was required to adapt them to Gemini's needs. NASA was already using the combination in its own unmanned programmes, and there was an existing Agena Project Office located at the Marshall Space Flight Center in Huntsville, Alabama. It was a simple matter to let Marshall also handle the Atlas–Agena requirements. In December 1961, discussions on the specific role and requirements for Agena in Gemini were conducted. There was very little change required with the Atlas, as it was already being used to launch Agenas, but there were significant alterations needed to change Agena from an upper stage to a docking target vehicle. These included the installation of radar, the restartable engine, more stability control, and the docking unit. As with other elements, one of the first impacts on the programme was the rising cost.

The change to Agena D
While NASA was evaluating the use of Agena B for Gemini, the USAF was pursuing the development of the more advanced Agena D. The concept for Agena B was to configure the stage to the mission, whereas the newer version would be standardised, with mission modules incorporated for the specific requirements of a given launch. By 11 June 1962, Agena D was assigned to be the target for the Gemini missions.

As the docking target programme evolved, so the costs escalated – as reflected in many areas of the programme – and eventually the Agena testing programme, like the Titan programme, was curtailed to alleviate growing costs and perhaps fend off the possible cancellation of the stage in Gemini. The GPO had informed Marshall that budget limitations had forced a 'reshape and reschedule' of the Gemini–Agena programme to fit the budget – which was due to be cut from $27 million to $10.3 million for 1963. With a one-third reduction, development would be a difficult task, and with fewer test firings of the engine it was expected that system reliability would be unachievable. With the first launch of Agena planned for October 1964, Bell Aerospace Systems, prime contractor for the Agena engines, stated that it would be unable to produce engines to meet that date if funds were not increased. The initial response was that there was no choice but to remain within the limitations. The programme was again reviewed, but even with an increase to $12.7 million, there would still be an estimated 14-month delay in Agena development.

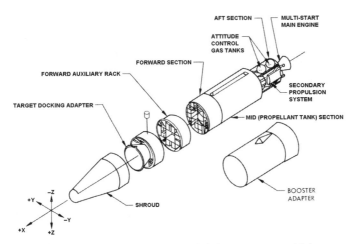

The configuration of the Gemini Agena target vehicle.

Management meetings in November produced a stop-gap solution which proposed significant admendments to development, qualification testing, test firings, production lead times and man-power management. The development would run at a reduced level for the rest of 1962, and then on full man-power from 1 January 1963. Agena would be four months late, but its costs would be as low as $44.1 million – a massive reduction of $32.7 million.

Throughout 1963, development of the Agena progressed slowly. In addition to a lack of programme funds, the reduced testing programme also encountered hardware problems, the remedy of which cost more money, added to the delays, and also slipped the delivery dates back further. On 3 June the USAF, NASA, and Lockheed began the Gemini Extra Care Program in an attempt to reduce the cases of equipment failure and careless workmanship on the Agena target.

Problems continued through 1964, with further budget and development hurdles to overcome as the first launches approached. Doubts as to whether Agena would achieve its assigned mission were not alleviated by the slow development of the propulsion systems, and a non-rendezvous Agena mission was suggested to boost confidence. This was not accepted, but it was decided to assign the first Gemini Agena Target Vehicle (GATV-5001) as a ground test vehicle. With the probability that this stage would not fly, the Atlas to launch it was also terminated, which saved another much-needed $2.15 million which could be used elsewhere.

All these delays posed difficulties for the GPO in assigning missions to the objectives. For the rendezvous, it was planned to fly GT-6 to the first Agena and then the 14-day mission on GT-7, so that if problems occurred on GT-6 there would be sufficient time to solve them before flying the next docking mission on GT-8.

Alternative concepts
Ironically, it was the choice of LOR for Apollo, in the summer of 1962, which raised doubts about using the Agena in the Gemini programme. The development of

techniques for the Apollo parent spacecraft to rendezvous and dock with the landing module had helped promote Gemini as an important step to the Moon. In a meeting of the Manned Spaceflight Management Council on 30 October, Joseph Shea, Deputy Director of Systems Engineering, questioned whether the rendezvous and docking phase of the Gemini programme actually needed the Agena at all, as all of the other objectives were achievable without the use of the target vehicle.

There was a group at NASA Headquarters who considered that the Gemini rendezvous objectives could be achieved by flying a small 'piggy-back' package in the Adapter Module of the Gemini. This could solve the problem of dual launches with the Titan and the Atlas, and the increased costs of developing the Agena. It would provide a stable but non-manoeuvring target for the Gemini to develop the techniques required for Apollo. However, it was severely limited in its capabilities, and its expected weight of nearly 400 lbs was a threat to the weight envelope for the proposed paraglider system. It would not therefore provide docking experience, but with the Earth orbit Apollo Block I series of spacecraft planned from late 1966, docking techniques could be practised on those missions instead.

At the same time, MSC was considering a Radar Evaluation Pod as an experiment for an early mission. This was a small deployable package that had a battery, radar transponder, beacon and flashing lights. Weighing just 66 lbs, it would be deployed from the Adapter Module to allow the astronauts to practice terminal rendezvous manoeuvres before later missions approached an Agena. Although the 'piggy-back' proposal did not materialise, the REP remained, although its allocation changed from the first manned mission (third mission) to the first long-duration flight (fourth mission) and then to the fifth mission.

PLANS FOR GEMINI–AGENA

The methods of rendezvous and docking assigned to Gemini had been devised over a period of three years since the programme was authorised. From a variety of profiles, the Gemini rendezvous method resulted in three main approaches: tangential, co-elliptic, and first apogee. Each one used rendezvous from different orbital altitudes, and ensured that the target moved at different speeds. The Hohmann transfer orbit principle of minimum energy (as little engine burn as possible) to move from one orbit to another was adopted, so that the Gemini in a lower orbit would gradually gain on the Agena in the higher orbit. When the Gemini performed a manoeuvre at the low point (perigee) of its orbit, it would affect the high point (apogee) around the other side of Earth. Equally, a manoeuvre at apogee would affect the perigee of that orbit. Using engine burns to change the orbital parameters provided the Gemini with the capability to change its orbit to intercept the Agena.

Tangential approach With Agena in a circular orbit, Gemini would be placed in an elliptical orbit profile, but with its apogee at the same altitude as the target. Gemini would gradually gain speed in its approach to the target, and to ensure that the orbit intercepted at the right time a sequence of smaller course corrections would be required on the final (terminal) ascending part of the ellipse.

Co-elliptic approach Similar to the tangential method, but the apogee of the Gemini would be lower than the orbit of the Agena. The crew would have the opportunity to make a series of trim burns before making the final burn to circularise the orbit inside that of the target. From this slightly lower circular orbit, the crew could then fire the thrusters to raise Gemini's orbit towards the target, arriving from below and positioning in front of it. Another engine burn in the direction of flight would slightly raise Gemini's orbit, as well as allow the Agena to match its position in the orbit.

First apogee approach This method would allow for very little margin of error, and would enable the Gemini reach the Agena on the very first apogee of its initial elliptical orbit. A series of rapid mid-course manoeuvres would have to be conducted, but when the spacecraft arrived at the target there would hopefully still be a significant quantity of propellant left on board to achieve the final rendezvous and docking. This was considered to be a useful method for a rapid rendezvous and docking profile.

After an analysis of each of these concepts it was decided to prepare the first docking mission, Gemini 6, for the co-elliptical method on the fourth orbit (also called M = 4 and conducted on Gemini 8 and Gemini 10). Gemini's first orbit would then be free of rendezvous manoeuvres to verify the correct operation of the spacecraft, with rendezvous occurring during the fourth orbit near the end of the fourth darkness period.

A second primary technique was adapted for an M = 3 mode, during which Gemini would rendezvous at the end of the third darkness period. This was performed during Gemini 9 and Gemini 12. The first orbit (M = 1) method was completed on Gemini 11.

Further types of rendezvous conducted during Gemini included optical rendezvous, rendezvous from above, stable orbit rendezvous, and optical dual rendezvous. In all, ten rendezvous operations (excluding those with the Titan stage) were achieved:

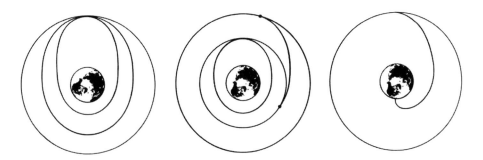

The three basic rendezvous methods employed during Gemini rendezvous and docking missions: (*left–right*) tangential, co-elliptical and first apogee.

GT   *Type of rendezvous*

6   Fourth orbit (M = 4)
8   Fourth orbit (M = 4)
9   Third orbit (M = 3)
       Optical re-rendezvous
       Re-rendezvous from above
10   Fourth orbit (M = 4)
       Optical dual
11   First orbit (M = 1)
       Stable orbit
12   Third orbit (M = 3)

Rendezvous considerations

In preparing any of the rendezvous profiles for a mission, several factors had to be considered during mission planning:

Launch procedures Agena was launched first in the sequence, and once the target was securely in orbit the Titan launched the Gemini astronauts to begin the chase towards the target. The Agena was normally placed in a 185-mile circular orbit, from which precise orbital velocity and trajectory data was analysed and computed into the launch azimuth (direction) trajectory planning for the Titan, which could launch to the Agena in a two-hour launch window each day for the five days that the Agena could wait for rendezvous and docking. Launch considerations originally envisaged a large launch window, but in the event this was significantly reduced, and most Gemini missions were launched on time.

Early in the programme, the development of on-time launches was critical to achieving rendezvous and docking. If a launch was scrubbed, then it was recycled by

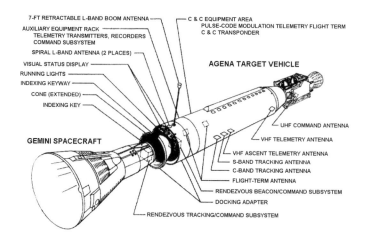

The Gemini Agena docking configuration and primary components.

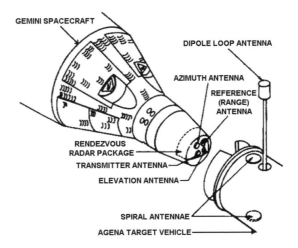

The Gemini Agena rendezvous system.

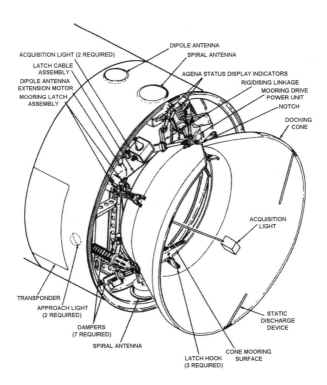

Major features of the Agena docking cone.

Launch time performance

GT	Launch	Launch date deviation	Launch time deviation	Rendezvous target
1	1	1964 Apr 8	On time	–
2	2	1965 Jan 19	+ 4 minutes	–
3	1	1965 Mar 23	+ 24 minutes	–
4	1	1965 Jun 3	+ 16 minutes	Titan second stage
5	2	1965 Aug 21	On time	Rendezvous evaluation pod
6	1	(1965 Oct 25 – target vehicle launch failure)		Agena 5002
	2	1965 Dec 15	On time	Gemini 7
7	1	1965 Dec 4	On time	Titan second stage
8	2	1966 Mar 16	On time	Agena 5003
9	1	(1966 May 17 – target vehicle launch failure)		Agena 5004
	2	1966 Jun 3	On time	ATDA
10	1	1966 Jul 16	On time	Agena 5005
11	3	1966 Sep 12	On time	Agena 5006
12	3	1966 Nov 11	On time	Agena 5001R

at least 24 hours. Experience in count-down and launch preparations allowed an absorption of small launch delays into the countdown without affecting the final launch.

Systems requirements During the development of rendezvous, the systems designed to achieve it (radar, computers and so on) were also developed, and in each mission the capability of these systems to support the desired rendezvous profiles was evaluated. On Gemini 12, it was planned to attempt an $M = 2$ second orbit rendezvous. Unfortunately, limits to the systems (in this case, radar) placed the target out of range at $M = 2$.

Crew procedures There were crucial questions to be answered for the crew to be able to operate reasonable working procedures. Would there be sufficient time for the crew to complete the activities? Would the approach trajectories allow for small errors during insertion of manoeuvres? Would the back-up procedures also allow for adequate lighting conditions? Would their be sufficient margin for low terminal approach velocities and line-of-sight angle rates? And would there be a back-up procedure for failure of guidance facilities?

DOCKING WITH THE AGENA

With the selection of the Agena as the target vehicle, there had to be some consideration for the system with which Gemini was to dock. This involved considering the strength of the structure to support manoeuvres of the combined vehicle, and the 'bending' of the combination due to stresses on the structure.

A range of docking systems was evaluated for use on Gemini–Agena missions, including a selection of 'original' concepts (shown in the illustration opposite). The

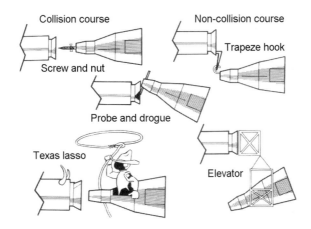

Various docking concepts evaluated for Gemini.

evaluation included decision on whether the Gemini should physically link with the target, how the astronauts could control the target, how to avoid damaging the thermal protection system on the nose of the Gemini, and how to guarantee separation from the target.

It was determined that the best formation for Gemini would be a docking at the front of the target, with the crew looking down the length of the vehicle. This was not the best formation for manoeuvring burns, as it involved backward acceleration for the crew (eyeballs out), and would cause stress loads on the centre of the vehicle. It did, however, offer the best display of the target, its systems, and engine performance.

Collision course docking

The method chosen for joining the two vehicles was the 'crash-course' approach – similar to the technique for refuelling aircraft in flight. This was suitable for safety and control, and also proved to be the simplest configuration. The nose of the spacecraft (with an indexing bar for alignment) would be the probe, and the docking collar on the Agena would be the drogue. All electrical and primary mechanical devices were installed on the target, which saved weight on Gemini. McDonnell constructed the docking hardware that was installed on the Agena stage. All systems and equipment were self-contained on Gemini, with only electrical, control and telemetry crossing to the Agena vehicle.

The final docking approach was visual, with cues that could be viewed on either day- or night-time passes of the orbit. Docking was achieved by connecting three latches on the target with three corresponding fittings on the spacecraft nose – which automatically rigidified the connection between the two vehicles – and seven dampers at three locations were used to soften the docking 'impact'. They could also absorb motions in all three axes (to two of freedom) and return to the central location. The cone was then retracted by using a small electric motor and a cable drive linkage,

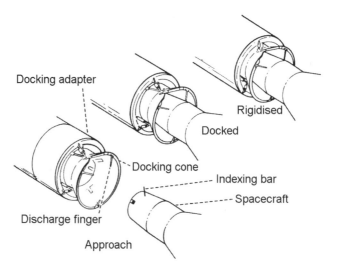

Docking adapter

Rigidised

Docked

Docking cone

Indexing bar

Spacecraft

Discharge finger

Approach

The docking and rigidising sequence of Gemini docking to an Agena.

The Agena target vehicle display panel. The displays located above the docking cone include green DOCK and PWR lights that indicated when it was safe to dock. The target-vehicle systems were then verified by a green MAIN light for pressure levels, green SEC HI and SEC LO for secondary propulsion condition, and a green ATT light indicating that the target-vehicle cold-gas attitude system was activated. Upon docking, the green DOCK light was turned off, and when rigidised, the green RIGID light was observed.

which closed down the latches and completed the rigid linkage. Discharge probes handled the static electric charge that was transmitted across the vehicle. Undocking was a reversal of this process. Tests on the configuration evaluated the structural strength of the combination to ensure its integrity.

Rendezvous and docking activites
Physical docking between a Gemini spacecraft and an Agena target occurred on four of the ten manned missions, but all of the missions contributed to the success of the rendezvous and docking objectives. Gemini 3 evaluated the Gemini propulsion and guidance system, while on Gemini 4 an attempt was made to rehearse a rendezvous with the spent second stage of the Titan. The REP was used on Gemini 5, and a phantom rendezvous was performed that duplicated the docking profile planned for Gemini 6. When the Gemini 6 target was lost in a launch accident, plans were amended to allow a rendezvous with the previously launched Gemini 7. It was during Gemini 8 that the first docking in space was achieved, and four subsequent Gemini missions expanded the experience of rendezvous and docking operations. Gemini 11 and Gemini 12 also took the opportunity to evaluate tether dynamics between the Gemini and the Agena, and by using the Agena's propulsion system, Gemini 10 and Gemini 11 set new apogee altitude records.

THE LEARNING CURVE

Gemini 4 and the other first step
The first opportunity for a Gemini rendezvous came on 3 June 1965, shortly after the launch of Gemini 4. The idea was for McDivitt to separate Gemini 4, and to then manoeuvre as close as ten feet from the spent Titan stage. If all went well and the stage was not spinning, Gemini 4 would fly in formation at 100 feet for the next orbit (90 minutes), allowing White to approach the stage and photograph it during his second-orbit EVA.

Just six minutes after leaving LC 19, Gemini was separated from the second stage of the Titan. McDivitt initiated a small separation burn as he used the spacecraft's thrusters to turn Gemini around in order to view the stage; and as the vehicle turned, both astronauts reported a great deal of debris floating around on all sides of Gemini.

After the turnaround, the Titan stage was in view at about 200–500 feet behind them and to the left. Apparently, a 3-fps retrograde velocity had been imparted on the stage as Gemini separated, effectively pushing it further than predicted, and therefore affecting the rendezvous plans. The stage was clearly visible to the naked eye, against the black of space, with the flashing lights also recognisable. McDivitt pointed the nose of Gemini at the stage and applied forward thrust for about six seconds, but he did not have enough time to place the computer in the catch-up mode. When it appeared that the stage and the Gemini were still separating, he applied thrust for a further four or five seconds, after which it appeared that they seemed to have closed in slightly. The guidance platform was realigned as the stage dropped below the spacecraft and out of sight of the windows. Gemini then thrust downward using the top thruster, and was pitched down so that the crew could see the stage. When they did so, they found that it had dropped much further below them than they had expected and it was difficult to distinguish against the backdrop of the Earth. They therefore realigned the platform again, and placed it in the orbit mode.

The 'brute force' method
Once again the crew tried to catch up with the stage, which by then was at least 1,000 feet below them. McDivitt was faced with two options: perform a retro-thrust to a different orbit and attempt rendezvous, or use the OAMS in a posigrade firing to force a direct route towards the stage. As time was against them, the crew decided on the 'brute force' method. The Titan stage was 1,200 feet away and as soon as it entered the darkness it disappeared, although the flashing lights (which could be seen from as far as 300 miles) remained visible.

The crew continued thrusting in an attempt to close the gap, and by the time the Carnarvon ground station came into range, McDivitt had been able to manoeuvre to the approximate orbit of the stage. They were at last closing in – only to discover that the stage was tumbling at about 40–50 feet per second. The lights began to disappear from view as it tumbled against the backdrop of the Earth. With only a single visible light with which to judge both range and rate, they could not accurately estimate their distance until they re-entered sunlight. As they did so, they found that the stage was below them, and about two miles away, but with the Earth's surface passing beneath them, the water, clouds and landmasses made it difficult to distinguish ranges greater than a mile.

Rendezvous or EVA?
McDivitt tried thrusting again, but found that without burning more fuel the distances would not be reduced significantly. The astronauts were mindful that this was a brand new technique, and as pioneers they could very well be completing the wrong manoeuvres. The flight plan called for the rendezvous on the first orbit and, on the second, Ed White's EVA, for which he needed time to prepare. McDivitt called MCC in Houston, having previously discussed the importance of the rendezvous against the spacewalk, and had the EVA confirmed as top priority: 'We just can't close on it … I have been struggling here not to let it get too far from me … It's out further than we expected … I don't want to get down where it's going. Do you want a major effort to close with this thing, or save the fuel?'

Flying with smaller fuel tanks than those on later Geminis, they had used 212 lbs of the 410 lbs on board during the first orbit of a four-day mission. With the need to conserve fuel and prepare for the EVA, and with the booster now three miles way and increasing its distance, Flight Director Chris Kraft ordered termination of all further efforts to rendezvous with the stage: 'We're more concerned about the lifetime [of the mission] than making the close [rendezvous].' McDivitt agreed: 'I guess we're just going to have to watch [the stage] go away.'

Orbital 'catch up'
The crew of Gemini 4 had discovered that using brute force to chase a target does not work in orbit. When they fired the engine posigrade (in the direction of flight) it raised the orbit instead of decreasing the distance. The manoeuvre slowed the Gemini relative to the stage, and placed it further away. What the crew should have done was to fire the engines retrograde (against the direction of flight), effectively applying Gemini's brakes. This would have dropped Gemini into a lower orbit than that of the stage – a shorter orbit which would have allowed it to 'catch up'. Once within range, a short burn in the posigrade direction (Gemini's accelerator) would raise the orbit to match that of the target, which could then be approached directly. The learning curve of rendezvous and docking in space had begun.

Gemini 5 and Conrad's 'little rascal'
The Radar Evaluation Pod, deployed from Gemini 5 shortly after entering orbit, was to have been used by the crew to practise rendezvous. After the experience with Gemini 4, this became even more important, as Gemini 6 was to attempt the first Agena docking. The Gemini 5 crew training reflected this both in the simulators and in the many hours spent with Gemini 5 on LC 19 and an Atlas–Agena on LC 14 on 22 July, in a simulated double launch.

An artist's concept of the Radar Evaluation Pod during approach by Gemini 5 – achieved by superimposing the REP on a photograph of an Earth–Sky panorama taken from Gemini 4.

Activities with the REP began on 21 August 1965, at 2 hours 7 minutes into the mission. The unit was ejected from the Adapter Module, and Cooper turned Gemini around to achieve radar lock-on. With the REP moving out directly from them, the radar tracked its progress, providing both range and rate of the separation. For 40 minutes, out to 7,721 feet maximum range, the REP (which Conrad had called the 'little rascal') was moving behind and above the Gemini. The crew then noted a sharp fall in the fuel cell cryogenic oxygen supply pressure. The decision was made immediately to power down to conserve what battery power they had if the fuel cell shut off completely, which terminated any further manoeuvres with the REP.

The REP remained in sight of the crew for some time. Conrad reported that it was about 2,000 feet from them at 2 hours 42 minutes into the mission, while Cooper reported at 9 hours 30 minutes into the mission that it was still in sight and had approached as close as 1,000 feet. With no power from the fuel cell, the planned tracking and rendezvous on the fourth and fifth orbit was impossible, and the REP drifted out of range.

Phantom Agena

While the problems with the fuel cell were handled by one flight control team, the Planning and Analysis Division team worked with astronaut Buzz Aldrin to develop a procedure to allow the crew to perform some type of rendezvous, if electrical power could be restored to support it. The subsequent restoration of the fuel cell to support the full duration of the mission also allowed the crew to complete this rendezvous with the 'phantom Agena' on Day 3.

On the second day in orbit, the Gemini 5 crew locked their radar on to an L-band transponder at the Cape, and completed a test of the accuracy of the radar by measuring the distance from Gemini 5 to the transponder. It was measured at 167 miles, while the transponder recorded the distance to Gemini as 170 miles. This result was termed as the closest approach that Gemini 5 had made to the Cape since leaving there two days earlier. There were some reservations that the Gemini radar could track a ground target – but it had done so, and the system had been successfully qualified for use on a docking mission.

The next day, the flight control team created a theoretical target vehicle in a 141 x 210-mile orbit, and then relayed this information to Cooper and Conrad on Gemini 5. They were to perform four manoeuvres over two orbits. The plan was to achieve a 124.2 x 193.2-mile orbit, and at the end of the manoeuvres Gemini 5 actually arrived in a 124.0 x 192.6-mile orbit. The crew successfully demonstrated the ability to use both spacecraft and ground computations to arrive at an orbit where a target vehicle was located. Ground tracking determined that had an Agena actually been in orbit, Gemini 5 would have been within 0.3 miles of it, which would have allowed visual acquisition. It was an important step towards the physical docking of Gemini and Agena in orbit.

SEVEN ... SIX ... COUNT-DOWN TO RENDEZVOUS

Astronauts Walter Schirra and Tom Stafford had been preparing for Gemini 6 for about a year when they settled themselves into the spacecraft on 25 October. The plan was to launch the Agena target into a 185-mile altitude orbit, and then, 1 hour 41 minutes later, launch Gemini 6 into an orbit some 1,200 miles behind the target. Gemini 6 was the final battery-powered spacecraft of the series, and was scheduled to fly a 46-hour 47-minute mission. If the docking was achieved on time during the first day, the option to land might be taken after just 24 hours in space.

Using the M = 4 co-elliptical rendezvous mode, Gemini 6 would gradually close in on the Agena in a 100 x 168-mile orbit for three orbits. Then, prior to the fourth orbit, Gemini would be placed in a 168-mile circular orbit. On the following orbit, Schirra would nudge it into an orbit that matched the Agena's orbit for the final docking approach at about 5 hours 30 minutes into the mission. The flight plan

At LC 14. the Agena stage is hoisted to be mated to the Atlas for launch as a Gemini docking target. (Courtesy British Interplanetary Society/NASA.)

Ready, Steady, Go.! The Atlas Agena seems to be held at the traffic lights and awaiting the green light for launch. The Agena stage is the thinner structure at the top of the vehicle above the chequered banding. (Courtesy British Interplanetary Society/NASA.)

called for two hours of docked tests, undocking and a fly-around photo-inspection, and then three further dockings (two during night-side passes of the orbit). After undocking for the final time, Schirra would separate to a safe distance while the Agena engine was commanded to fire, and the crew would photograph the event as the Agena pulled away from them. After ten hours, the docking phase would be completed. At the pre-flight press conference, Schirra expected the link-up to be 'a piece of cake' after all their training, and Stafford pointed out that if they could not rendezvous and dock, 'We are stalled on the Moon trip.' The launch of the Agena was all that was required to begin this sequence.

A dramatic loss of telemetry
At 10.00 am EST on 25 October, Atlas was launched from LC 14 to begin the important mission sequence. Just minutes after launch, the Agena separated from the Atlas, but it appeared to be wobbling as it ascended. Tracking stations at Bermuda and Antigua received erroneous signals, and Canary was unable to make

any contact with the Agena at the time that it should have flown over the islands. At 6 minutes 20 seconds, MCC Houston reported 'a dramatic loss of telemetry'; and from the Carnarvon tracking station came the dreaded result of trying to find the Agena as 'No joy … No joy …' At 10.54 am, the launch of Gemini 6 was officially scrubbed.

Stalled on the pad

The Agena had separated from the Atlas, and the main engine on the target stage had ignited as planned; but shortly after that it had apparently exploded, as USAF radar tracked at least five pieces of debris where the target vehicle should have been.

After 1 hour 27 minutes lying in the Gemini, Schirra and Stafford were helped out. It appeared as though it would be early 1966 before another Agena would be ready. The disappointment was obvious – especially to Schirra's eight-year-old daughter Suzanne, who reportedly commented: 'Daddy will be upset. I thought he looked mad on television.'

The investigation into the loss began only an hour after the failure. The Agena programme was barely managing to cope with the developments of the flight programme in 1965, consideration was given to bringing the 5001 target out of the test programme for use with Gemini 8; but when 5002 blew up, the question was not whether 5001 should rejoin the flight programme, but how quickly it could be readied to support Gemini 6.

By 12 November, the most probable cause of the explosion had been identified as a 'hard-fire back-start' (equivalent to a car back-firing), or an electrical short circuit, creating oscillations and mechanical damage and leading to a premature shutdown. Indications of a fuel spill and an electrical failure caused engine operations to cease. However, a fuel tank pressure valve remained open and as pressure built up with no fuel to pump, it ruptured, and destroyed the Agena. Recommendations for preventing such a failure led to Lockheed's formation of the Project Surefire Engine Development Task Force to complete the alterations and qualifications, as recommendations and test reports continued to appear until early March 1966.

A rapid-fire launch

The Agena launch attempt at the Cape was watched by the Gemini 7 crew, and standing nearby were McDonnell's Gemini Spacecraft Chief, Walter Burke, and his Deputy, John Yardley, who were discussing the possibility of using Gemini 7 instead of an Agena as a target. This was an idea that the Martin company had first proposed several months earlier in which one Gemini would be launched on a duration mission, then a few days later a second Gemini would be launched to rendezvous with the first (a rapid-fire or salvo launch demonstration).

The astronauts liked the plan – apart from the idea of an inflated collar on the back of the Gemini 7 to allow Gemini 6 to dock with it. Borman would not permit anything to come that close to the back of his spacecraft, as it was just too dangerous. However, a rendezvous to a few feet seemed more promising. Gemini 7 was set for a December launch on a 14-day mission, and in order to meet it in space, Gemini 6 had to be launched only a few days afterwards, and from the same pad.

Launch-day breakfast for the crew of Gemini 6 in October 1965. Stafford and Schirra are in the centre of the frame facing the camera. Behind Schirra (in the dark jacket) is Deke Slayton, Director of Flight Crew Operations, while Gemini 7 Pilot Jim Lovell sits next to Stafford. On the right (facing away from the camera) are Al Bean and Pete Conrad (and two unidentified astronauts), while on the other table, behind Lovell and Stafford, are *(left–right)* Al Shepard, Chief Astronaut, an unidentified astronaut, and the Gemini 8 crew, Armstrong and Scott.

Studies had evaluated the activation of a second Gemini pad to shorten the time between launches, provide an unmanned Gemini for a possible rescue, or provide a dual manned launch capability. But with the budget restrictions affecting the basic programme, a second pad – although an attractive idea – was not worth the cost.

While the idea of the dual launch was circulated, preparations continued for the launch of Gemini 7. There was thought about saving time by placing Gemini 7 on the already installed Gemini 6 launch vehicle, but the 14-day spacecraft was much heavier than the two-day version, and the GLV-6 did not have the power to lift the heavier spacecraft. Work therefore began to destack Gemini 6 and install Gemini 7 for an early December launch. Gemini 6 would fly in late February or early March, and for a second time would attempt to dock with the Agena.

'You're out of your minds'

The day after the aborted launch of Gemini 6, the results of the discussions about the possibility of launching Gemini 6 to meet in orbit with Gemini 7 had reached Houston, where no-one could produce a reasonable opposing argument. With all the hardware at the Cape and most of the pre-launch checking completed, the turnaround to launch Gemini 6 would still be a race against time, but it would be possible. But could two manned Gemini craft be simultaneously tracked in the same orbit? Chris Kraft's first thought was that it could not be done. 'You're out of your minds,' he said. But support for the project was growing. The four astronauts were in favour of it, and flight controllers soon realised that the tracking system could indeed be adjusted to handle two manned craft if the older Mercury-type teletype tracking network was used for the passive spacecraft (Gemini 7) and the more sophisticated computer system used in Gemini handled the active (Gemini 6) spacecraft. Flight Director John Hodge called it 'A hell of a great challenge.'

As long as there was no serious pad damage from the launch of Gemini 7, all indications pointed to a challenge that could be met. On 28 October, just three days after the aborted launch, NASA issued a statement that Gemini 7 would be launched on its 14-day mission on 4 December, and that Gemini 6 would launch from the same pad on 12 December. Gemini 6 would be launched to rendezvous with Gemini 7 in orbit, and then land after one day, allowing Gemini 7 to complete its two-week mission.

Gemini 7 and a tumbling Titan

On 4 December, Gemini 7 was launched from LC 19. Six minutes after lift-off the spacecraft was separated from the Titan, and Borman fired the OAMS to turn the spacecraft around to perform a rendezvous with the Titan – a repeat of the Gemini 4 experiment. A 2-second separation burn and a 5-second rotational burn positioned Gemini to keep station with the Titan second stage. Within 30 seconds Borman saw the stage through the windows, although debris from staging had clouded the window pane. The second stage was spewing fuel from a broken fuel line, and small globules crystallised and shone in the Sun as the venting caused the stage to tumble at about 2 rpm.

For the next 15 minutes Borman and Lovell took turns in controlling Gemini 7 at distances of 60–150 feet from the tumbling stage while they photographed it. They reported no difficulty in station keeping, but with such an extended mission ahead of them they did not wish to use too much fuel in maintaining position near the booster. After 15 minutes, a separation burn was performed to separate from the stage.

'We saw it ignite – we saw it shut down'

Almost as soon as Gemini 7 had entered orbit, work began on LC 19 to prepare for the launch of Gemini 6. On 12 December, Schirra and Stafford once again slipped into their seats on Gemini 6 at LC 19 for a planned launch at 09:54:06 local time. Just 26 seconds prior to that time, Gemini 7 flew overhead. As the count-down reached zero, the crew felt the whining roar of ignition as launch control went through the sequence: '… three, two, one, LIFT-OFF!' Then, 1.2 seconds later, both Titan engines shut down automatically. 'We have a shut down on Gemini 6.'

Schirra knew from past experience that there had been no 'launch'. The Titan was still sitting on the pad. Electing in a split-second decision not to eject, and trusting that the Titan would not blow up or topple over, both astronauts remained inside the Gemini capsule as the vehicle was safed. Each astronaut had his hand on the D-ring ejector handle, but if Schirra was not about to eject, then neither was Stafford. However, Stafford's original comment was 'bleeped' over by NASA, in the short time delay to 'live' radio and TV, as 'Oh shucks'.

Up in Gemini 7 the astronauts reported, 'We saw it ignite – we saw it shut down.' Schirra's decision not to eject saved the mission for another attempt, although this was not his initial intention. He had immediately realised that the booster had not lifted off and would not fall over, that there was no need to eject, and that it was safer to remain where they were. With six days left on Gemini 7, the race to attempt a third launch in three days' time began as soon as the vehicle was safed. Once again, two tired and frustrated astronauts walked from the pad, after 99 minutes inside the spacecraft.

It was found that an umbilical connector had prematurely separated from the base of the Titan. Sensing no upward movement, the Malfunction Detection System shut down the supply of fuel to the engines and cancelled the lift-off. Later examination revealed that a plastic dust-cap had been inadvertently left in the engine oxidiser inlet port. It could have blocked the port, and would have shut down the engine down one second after the computer did so.

Third time lucky
At 08.37 am on 15 December, to Schirra's call of 'For the third time, GO!', Gemini 6 at last left LC 19 and headed for orbit. As cloud covered the Cape area, the Gemini 7 astronauts could not see the launch, but they were informed of the success. The chase was on.

'A lot of traffic'
Gemini 6 was placed in orbit 1,238 miles behind Gemini 7, which entailed a 3-hour 15-minute chase until the radar could pick up the Gemini 7 spacecraft as a flicker on the radar screen. Schirra and Stafford followed the M = 4 flight profile using gradual increases in velocity and altitude over three elliptical orbits to manoeuvre the spacecraft. At 3 hours 47 minutes, a 54-second burn of the OAMS placed the spacecraft in an almost circular orbit of 167.7 x 170.2 miles at a range of 198.2 miles from Gemini 7.

The astronauts placed the rendezvous mode into automatic at 3 hours 51 minutes, which continued the catch-up with Gemini 7 for the next 1 hour 25 minutes, until the spacecraft was at a point where Schirra could resume manual control to complete the final approach. At 5 hours 16 minutes into the flight of Gemini 6, Schirra was close enough to thrust directly towards Gemini 7, and after one or two small manoeuvres, he began slowing his spacecraft as the Gemini constellation stars of Castor and Pollux became aligned with Gemini 7.

Onboard Gemini 7, Borman was reading out figures to Houston while Lovell was concentrating and looking out the window at the approaching Gemini 6. Mission Control asked if Lovell was out to lunch. Borman replied, 'Not exactly, but he is busy.'

Schirra performed a braking manoeuvre when Gemini 7 was just 130 feet in front. With no relative motion, the spacecraft had achieved the world's first true space

Gemini 6 astronauts walk dejected from the launch pad after ignition and shut-down of their Titan II vehicle on 12 December 1966.

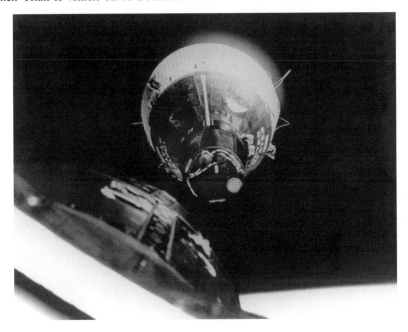

On 15 December 1965, Gemini 6 approaches to within a few feet of the nose of Gemini 7, marking the world's first rendezvous of two manned spacecraft.

A view of Gemini 7 as seen from Gemini 6 during a flyaround of the spacecraft during the 14-day space marathon.

The rear view of Gemini 7 as seen from Gemini 6. Note the trailing streamers from the separation of the spacecraft from the Titan. These flapped against the spacecraft, and the noise they made was a mystery until the Gemini 6 astronauts' observations revealed the cause.

Celebrating success. (*Left*) Chris Kraft, Red Team Flight Director, (*centre*) astronaut Gordon Cooper, and (*right*) Dr Robert Gilruth, MSC Director, celebrate the rendezvous of Gemini 6 with Gemini 7 on 15 December 1965.

rendezvous, and down in Houston controllers celebrated with cheers, while onlookers waved US flags and lit up cigars. Schirra later commented that despite Soviet claims that Vostok 3 and Vostok 4 had 'rendezvoused' at three miles, this was in reality only the beginning. Stopping at 120 feet was the point at which station keeping began.

A comment from Gemini 6 that 'There just seems to be a lot of traffic up here' was answered with the suggestion of calling a policeman! From Gemini 7, Borman was amazed by the 40-foot length of flame and fireworks from the Gemini 6 thrusters as it approached, and the amount of stringers, cords and tapes seen flowing out behind the other spacecraft at it pirouetted in front of him. 'You've got a lot of stuff all around the back end of you,' he told Gemini 6. 'So do you,' replied Schirra – apparently revealing the source of the noise which the Gemini 7 crew had heard early in the flight, caused by their streamers flapping against the hull as the spacecraft moved.

'We'll see you on the beach'

Schirra had been 'light on the pedal' because, arriving at Gemini 7, he had 62% fuel reserves left for additional fly-around operations. Approaching the end of their flight, Borman and Lovell, in Gemini 7, were not so fortunate, with less than 11% remaining. Over three orbits (41 hours) Gemini 6 flew around Gemini 7 at distances between 300 feet and 1 foot, and moved around it as slowly as 0.1 fps while travelling at 17,500 mph.

It was not all work for the four astronauts. Three of them (Schirra, Stafford and Lovell) were former graduates of the USN Academy, while Borman was a West Point graduate. With the 1966 American football season underway, Schirra, taunting his colleague, placed the sign in his window 'Beat Army'. But Borman, not to be defeated, informed Mission Control that the sign said 'Beat Navy'. Then, with the rendezvous completed, the two spacecraft prepared to separate. Schirra told Borman and Lovell that they were doing a good job, and 'We'll see you on the beach.' Schirra and Stafford moved away from Gemini 7, having created a piece of space history, and confident that Gemini 8 would take the final step of docking with the Agena in March.

AGENA, ATDA AND AN ALLIGATOR

The success of Gemini 6 and Gemini 7 hit the headlines just before Christmas 1965, and ended a highly successful year for Gemini. Less satisfying was that the Agena testing under Project Sunfire had revealed a series of problems. These still threatened the continued use of the Agena in the programme, as an alternative design was made ready. The Augmented Target Docking Adapter was a spacecraft rendezvous and recovery section bolted to a docking adapter. Despite being lighter, it did not possess the engine to place it in orbit, although the Atlas could do so. It became known as the 'poor man's docking target'; but it threatened to replace the Agena, and so Lockheed redoubled their efforts to finish the Agena test programme and overcome its problems. Gemini 8 was set for 16 March, and with ATDA in storage as a stand-by, the Agena was requalified on 4 March, just two weeks before its next mission.

'We'll take that one'

The Atlas–Agena for Gemini 8 left the pad at 10:00:03 on 16 March 1966. There were a few tense moments, as the October loss of the Gemini 6 target had been the previous launch of the series; but just over five minutes later the stages separated and the Agena ignited, with engineers, controllers, managers and astronauts holding their breath and crossing their fingers to help it on its way. This time everything went flawlessly, as the Agena placed the docking target into an 185-mile circular orbit.

The launch of the Titan, carrying astronauts Armstrong and Scott, was planned for 1 hour 40 minutes 59 seconds later. There was a six-minute launch window to achieve M = 4 rendezvous on the fourth orbit, although a fifth and sixth orbit were also possible, depending on the time that the Titan left the pad in that window. Waiting inside Gemini, Armstrong and Scott were told that their target vehicle was safely on orbit, 'Beautiful. We'll take that one,' Armstrong replied.

At 11:41:02 (three seconds late), the Titan, carrying Gemini 8 towards orbit, roared off Pad 19. The initial elliptical orbit was 99 x 169 miles, and upon entering orbit Gemini 8 was 1,050 miles behind the Agena. The first manoeuvre to adjust the apogee of the spacecraft was completed 94 minutes after lift-off, followed by a second adjustment to the plane of the Agena 2 hours 45 minutes into the mission. A circularisation burn 3 hours 47 minutes into the mission placed Gemini 8 in a circular 169-mile orbit, 170 miles behind their target, and two further manoeuvres brought it into a matching orbit with the Agena. At 206 miles, the crew reported a solid radar lock-on, with a visual sighting later, at 86 miles. During the manoeuvres,

Armstrong had noticed a problem with a slight residual thrust when he cut off the spacecraft thrusters.

'It's a real smoothy'

The terminal approach phase began after a burn at 5 hours 45 minutes GET. Six hours into the mission, Armstrong reported through the Hawaii tracking system that they were just 150 feet from Agena and station keeping, with the target appearing to be in good condition. Given the 'go' to approach, Armstrong gently nudged the nose of the Gemini towards the docking cone at 8 fps, with both vehicles matched in velocity at 17,500 mph. Just two feet from the target he held his position, and awaited being in range of the tracking ship *Rose Knot* in the south Atlantic. After a final check via the RNV to Houston, the 'go' to dock was given, and at 6 hours 33 minutes 22 seconds into the mission, Gemini 8 docked with the Agena and completed another milestone in space history. Armstrong: 'Flight, we are docked! It's a real smoothy. No noticeable oscillation at all.' MCC: 'Hey, congratulations. Real good.'

For the second time in less than three months, Mission Control erupted in loud cheers, hand-shaking and back-slapping, and once again the flags were waved and the cigars lit. Armstrong: 'You couldn't have the thrill down there we have up here. We've got a real winner here!'

The crew yawed the combination around 90°, which took about 55 seconds, and initiated a sequence of sixteen commands to complete and stop the manoeuvre. They then began their checklist. Their major mission objective achieved, they could now concentrate on several hours of docked activities and prepare for Scott's EVA, planned for Day 3 – or so they thought.

Gemini 8's view of the Agena docking target during approach on 16 March 1966. The Agena is about 190 feet away, and reveals the engine bell to the astronauts. The docking collar is at the other end (at the left) of the spacecraft.

The final moments prior to docking with the Agena. The view from Gemini 8 Pilot Dave Scott's window.

Tumbling end over end

During the yaw operation, the crew noted that the combination was in a 30-degree roll when in fact they should have been in a level attitude. Armstrong fired the thrusters to compensate, but after the deviation was corrected it started again. Both astronauts assumed that the Agena was being non co-operative, and so Scott turned off the Agena attitude control system. As the vehicle stabilised, the problem appeared to be solved, and he recycled the switches.

Out of radio contact with the ground between tracking stations, Armstrong was using the Gemini thrusters to attain the correct horizontal orientation, when the roll suddenly started again – but this time, much faster. As they considered the problem, they thought that the planned test objective to determine stress tolerance of the docked vehicle was rather academic. While Armstrong struggled with the controls, Scott quickly photographed the interaction and display readings of the Agena, to capture all possible data. With OAMS propellant reading as low as 30%, the astronauts assumed that they had a stuck thruster.

Scott cycled the Agena switches off, on, and off again, as Armstrong tried to determine if cycling the Gemini thruster switches could isolate the problem. But nothing seemed to solve the situation and they began spinning in a second axis (yaw) at the same time as the roll. They realised that the only way to stop the wild ride they were experiencing was to undock, and quick thinking by Scott enabled him switch control of the Agena from the spacecraft to the ground, which protected its command control for future use. Scott then hit the undocking control, and with a long burst of the forward facing thrusters, Gemini 8 pulled away from the Agena. Almost at once, the suspicion of a stuck thruster was confirmed, as instead of decreasing the rotation rates they increased alarmingly, just as communication with the tracking ship *Coastal Sentry Quebec* (*CSQ*) was established:

MCC: 'How does it look?'
CSQ: 'We're indicating spacecraft free ... indicating they are not docked ... let me check with the crew... Gemini 8; CSQ CapCom; how do you read?'
GT-8: 'We consider this problem serious. We're tumbling end over end but we are disengaged from the Agena.'
CSQ: 'What seems to be the problem?'
GT-8: 'It's a roll, and we can't turn anything off.'
MCC-H (to CSQ): 'Did he say he could not turn the Agena off?'
CSQ: 'No, he says he has separated from the Agena and he is in a roll and can't stop it.'

A violent roll
Onboard Gemini 8, the rotation had increased to one complete revolution each second. With fuel close to depletion and the crew approaching dizziness and then black-out (with its fatal consequences), the astronauts responded with their only option. They disengaged the OAMS system and initiated the RCS in the nose of the Gemini. Checking that the hand-controller still worked, one RCS ring was then turned off to preserve it as they tried each OAMS thruster in turn in an attempt to identify the source of the problem. They determined that thruster number 8 had failed 'on', and was stuck in the open position.

Armstrong reported that they were in a violent roll with the RCS armed. CSQ confirmed that they seemed to be in a violent tumble and could not release the stuck thruster, and that the Agena was also tumbling violently as a result of the situation when docked to Gemini.

Slowly, Armstrong gained control of the spacecraft using the RCS system, but precious fuel was used in stabilising the spacecraft. Asked if they had a visual sighting of the Agena, the astronauts replied that they had not seen it since undocking. CSQ asked how they were handling the situation.

GT-8: 'OK. We're working on it.'
CSQ: 'OK, relax. Everything is OK.'

As they gradually brought Gemini back under control using the RCS, mission rules stated that any emergency use of the RCS system necessitated a return to Earth as soon as possible, so that any further leak or use would not jeopardise re-entry fuel levels. The recovery of Gemini 8 would be achieved at a contingency site in the Atlantic, just ten hours after launch.

The Agena stabilised quite soon after the incident. A total of ten major manoeuvres were performed over the next few days to deplete the heavy fuel load, raising and lowering the orbit and evaluating its performance with the view of revisiting the target on Gemini 9 or Gemini 10. By the time a second Gemini visited the Agena, the electrical system would be long dead, and with no way to control the target, a heavy fuel load would be a potential hazard for any rendezvous. There was some difficulty in learning to control the vehicle in orbit from the ground, and a centre-of-gravity miscalculation in conjunction with a sluggish control system would later be solved by adding ballast on future vehicles. By 9 March, the Agena was in a 191 x 193-mile orbit, and had an estimated orbital life of at least 134 days.

Post-flight analysis of the problem was hampered because the Adapter Module containing the stuck thruster had been jettisoned prior to re-entry, and had burned up. But NASA evaluated the recovered Re-entry Module, reviewed flight data and the astronauts' reports, conducted ground tests, and concluded that the stuck valve on number 8 thruster was probably caused by an electrical short circuit. The system was changed so that in the 'off' position no power could be fed to the thrusters.

New approach, new target, old problem

By 10.00 am on 17 May 1966, Gemini 9 astronauts Stafford and Cernan were inside their spacecraft awaiting the final moments of count-down for the launch of their Atlas–Agena. They were to follow it 99 minutes later. Their mission was ambitious, with three types of rendezvous assigned to it.

The first rendezvous would be an $M = 3$ profile on the third orbit, which would be of use to Apollo as a close approximation to a lunar orbit rendezvous between the returning lunar lander and the orbiting Command Module. The crew would later attempt a second rendezvous and docking – from above the target vehicle instead of from below it – and finally a third profile would include a high-altitude burn using the Agena propulsion system, to attempt a phantom rendezvous with a second 'target'. Although this third approach was deleted due to doubts about the Agena engine's reliability, and the risk of radiation to the crew, it was soon replaced with an approach that Gemini 8 could not achieve due to its shortened mission. During this equiperiod orbit, the Gemini thrusters would be fired to pull the spacecraft from the Agena to a slightly higher elliptical orbit, which theoretically would see the two spacecraft arrive at a point where the two different orbits intersected and only the braking manoeuvre to re-rendezvous would be required.

Achieving all this with the small fuel tanks which the Gemini 9 carried (330 lbs), as opposed to the later larger tanks (460 lbs for Gemini 10), plus a complicated EVA using the Astronaut Manoeuvring Unit, presented the crew with a great deal of work during their training programme – and they had only recently replaced the original prime crew of See and Bassett, who had been killed in February.

At 10.15 EST, the Atlas left the pad to begin the complex mission. After two minutes of powered flight, one of the Atlas outboard engines did not cut off, and instead gimballed over and locked, causing the whole stack to pitch over. Rather than head for orbit, it nose-dived back to the ocean!

Telemetry indicated engine cut-off on the Atlas and separation of the Agena, which continued transmitting for a few seconds until all contact was lost. Once again, problems plagued the Gemini Atlas–Agena programme. The booster fell into the Atlantic 123 miles from the Cape – another Agena lost. But this time, there was a back-up plan. The 'insurance' was the ATDA at the Cape, and if that also failed, then Gemini 9 would attempt to rendezvous with Gemini 8's target. However, trying to reach that target at 250 miles might use too much fuel, as there would be no Agena engine to make the transfer orbit burn.

Mission Director William C Schneider announced: 'We have lost our bird. The mission is scrubbed. Gemini will not fly today.' As Stafford clambered out of the spacecraft half an hour later, he noted that this was not a new situation for him: 'You can't get your hopes up until that Agena comes across the States [on its first orbit]. I've been ... here a number of times before.'

If at first...

On 1 June, all was ready for another attempt, and at 10.00 am EST, the Atlas–ATDA was launched into a near-circular orbit of 185 miles altitude. Following cut-off of the launch vehicle's engines, and separation, the target launch shroud was supposed to separate to reveal the docking cone of the ATDA. But ground telemetry did not confirm this, and indicated that it had failed to jettison, or had only partially opened.

Stafford and Cernan's 'angry alligator', as seen from their vantage point onboard Gemini 9 on 3 June 1966, high above the Caribbean island of Los Roques (at left), with the coast of Caracas, Venezuela, hidden under clouds (at right). The retraining straps on the fibreglass launch shroud failed to release and presented problems during docking manoeuvres.

Inside Gemini, Stafford and Cernan awaited their own launch 98 minutes later, with Mission Control remaining optimistic about the orbiting of the target at the second attempt: MCC: 'We knew that if you stuck there long enough we would get you a good one.' With only a 40-second launch window, a subsequent hold in the count again led to the postponement of the launch of Gemini.

'Oh shucks,' commented Stafford once again, while Cernan could not believe that it had happened again. A 48-hour delay involved another trip down the launch tower for Stafford and Cernan. Stafford later commented that at 14 days, Borman and Lovell might be more experienced in space, 'but nobody had more pad time in Gemini than I did!' On Gemini 6 and Gemini 9, he would proceed through launch preparations six times.

At 08:39:33 am EST, 3 June, Gemini 9 finally left the pad to the call of 'For the third time, GO!' from Stafford, echoing Schirra's comments on the Gemini 6 launch. Six minutes later they were in orbit and 658 miles behind the target. To perform the $M = 3$ rendezvous they had to perform seven manoeuvres over the following four hours or so, to both catch up with the target and to change the orbital plane to match that of the ATDA.

At 2 hours 24 minutes the orbit was circularised to close within 109 miles behind and 12 miles below the ADTA, at a rate of 126 fps. At 57 miles the crew gained a visual on the optical sight, and it appeared that the shroud had ejected; but as they approached it was clear that this was not the case. By 4 hours 11 minutes they were just one mile from their objective.

'An angry alligator'
Closer inspection confirmed that the shroud had not ejected after all. 'The shroud is half open on that thing,' exclaimed Stafford. Cernan added that it looked as though it could be knocked off, although Mission Control informed them not to try to nudge it loose. While Mission Control discussed what to do next, Stafford described his view out the window as he moved Gemini from 100 feet out to as close as 30 feet from the target.

'We have a weird looking machine here. Both the clamshells of the nose are still on, but they are open wide. The front release has let go, and the back explosive bolts attached to the ATDA have both fired.' He then added the comments that became the trademark of the mission: 'The jaws are like an alligators' jaws open about 20–30°. It looks like an angry alligator out here, rotating around.'

Revised plans
Over the next orbit, several proposals were put forward to release the shroud. These included an EVA to perhaps cut the straps free, and recycling the jaws to make them spring free; but each idea was soon rejected, and a revised flight plan was sent to the crew to complete a rendezvous from 21 miles above and 11 miles behind the target, in the equiperiod profile that had been planned for the 28th hour of the mission. Using the computer and the hand-held sextant on board for guidance, Stafford complied, and found the manoeuvre relatively easy.

The next rendezvous began an hour after arriving at the ATDA for the second time, but would not be completed until the next day, after a sleep period. This third rendezvous was to simulate a lunar module rendezvous (from above). With Gemini about 70 miles in front of the target, the first manoeuvre would be to lower the spacecraft's orbit to allow the ATDA to close on the spacecraft, and then the crew would fire the engines to again raise the orbit of Gemini 9 to approach the target from 'above'. The velocity of the target would reduce Gemini's lead in orbit, and all that the crew needed to do was to carry out small adjustments to place their spacecraft above and ahead of the ATDA.

The crew found that it was visually difficult to track the target against the backdrop of the Sahara desert. Stafford called it, 'looking for a pencil dot on a sheet of paper,' and if they had not had the radar they would have failed in the attempt. They could not visually acquire the ATDA until they were three miles from it, and even then they occasionally lost sight of it as the terrain behind the target changed. This problem almost had more disastrous consequences thirty years later, in 1997, when a cosmonaut tried to automatically dock a Progress resupply craft to the Mir space station.

Gemini 9 approached to within 3 inches of the ATDA, to take close-up pictures of the shroud wires for post-flight examination. As Stafford later stated: 'We rolled Gemini on its side and right up to where the Gemini 'X' axis was 90° to the ATDA 'X' axis, and snapped the pictures, making sure the alligator wouldn't bite us that way.'

At 21 hours 42 minutes into the mission, a tired but happy crew had successfully completed three different types of rendezvous. This completed their work at the ATDA, and Stafford thrust away to begin the rest of their mission and the preparations for Cernan's EVA.

AGENAS, ALTITUDES AND TETHERS

For the final three Gemini missions, rendezvous and docking featured as primary objectives, to expand upon the experience and to perform manoeuvres to assist in docking operations on Apollo or follow-on programmes. The Agena engine would be used to increase the altitude and to evaluate large propulsion systems in space on Gemini 10 and Gemini 11, and tether dynamics would be explored on the final two flights, Gemini 11 and Gemini 12.

Gemini 10's ambitious flight plan

The primary objective for the tenth mission was to rendezvous and dock with a new Agena (designated Agena 10). The secondary objectives were to achieve an $M = 4$ docking, to use the Agena propulsion system to attempt a dual rendezvous with the Agena from Gemini 8 (termed Agena 8), to conduct docking practise, to evaluate the bending of the combination while docked, and to park the new Agena for future use. As Commander John Young observed, with fourteen experiments, and also EVA, it was an ambitious flight plan.

This time all went well with the Atlas–Agena launch, which took place at 15:39:46 pm on 18 July. It entered its prescribed orbit, and was followed 100 minutes later by John Young and Michael Collins in Gemini 10, to begin the chase. The launch times had been determined by the requirement to dock to Agena 10 and then use the target propulsion system boost up to the Agena 8. Several difficulties affected the plans when this was proposed. Firstly, no one had ridden Agena in the docked configuration before. Could the astronauts stop in time, and not hit the Agena 8?; and could they find the old target by using only optical methods? All electrical power on the older Agena was depleted, and this also took out the radar transponder. As these questions were addressed, the planning of the rendezvous was being tackled – Agena 10 docking on Day 1, dual rendezvous with Agena 8 on Day 2, and chasing the passive target on Day 3. Tracking Agena 8 was the responsibility of NORAD, and from the orbits, trajectories, objectives and capabilities, they worked back to the launch pad and the window in which to launch the two vehicles.

Riding a rascal

The fourth orbit rendezvous was hindered by the difficulty of using the sextant against the star-field, and ground computations were instead utilised, but two large course corrections were required to reach the target after a misalignment during docking approach. Fuel consumption amounted to 400 lbs – almost three times that on any previous mission – which caused the docking practise to be cancelled. Hard docking took place 5 hours and 52 minutes after launch, and the astronauts were to remain docked to Agena 10 for the next 39 hours, to take full advantage of the target's fuel capacity.

A side view of the Gemini 10 Agena prior to docking. For the first time on a Gemini docking mission, the astronauts later ignited the propulsion system of the Agena to raise their altitude.

Six manoeuvres were made using the Agena – three with primary systems, and three with secondary systems. The 'go' was given to fire the large engine for the first time some 7 hours 30 minutes into the flight. The burn lasted 80 seconds, adding 423 fps to the velocity and resulting in an orbit of 98.4 x 256.1 miles. During the event the crew did not say much; but this was not surprising, as they were flying backwards (eyeballs out) with g-forces against their fronts instead of their backs (as on launch), and were looking back at the firing Agena,. Young later described the experience: 'Mike threw the switch, and 24 seconds later ... it was really something. We had negative 1 g, and were driven forward in the cockpit. We had a shoulder harness fastened. The first sensation was that there was a pop, then a big explosion, and a clang [in front of their eyes]. Fire and sparks started coming out of the back end of that rascal ... the light was something fierce, and the acceleration was pretty good. We got a tremendous thrill ... on our way out to apogee and a new world record for altitude. The shut-down was just unbelievable. It was a quick jolt. And the tail-off ... I never saw anything like that before – sparks and fire and smoke and lights.'

Young and Collins were looking at Earth from a vantage point that no other human had ever seen, but in keeping track of the spacecraft, systems and radiation dose (which remained at a tolerable level), they did not have much time to look out the window. Besides, the Agena blocked most of the view anyway, and they had been so impressed by the burn that the vista in front of them did not really register. Young later stated that they took some pictures at apogee, but he was not sure of the location, although it showed the curvature of the Earth. 'We took some pictures coming down the hill ... I think it was the Red Sea.'

After a sleep period, the orbit was adjusted to lower the apogee. The Agena was turned to the direction of flight and fired for 78 seconds, reducing velocity by 344 fps and lowering the apogee to 237°.1. The force was 1 g, but as Young observed, 'It may be only 1 g, but it's the biggest 1 g we ever saw! That thing really lights into you.' Subsequent smaller manoeuvres adjusted the orbit to circularise it, by raising the low point to 234.6 miles – just 10.5 miles below Agena 8.

From Agena to Agena
After a second night's sleep, the crew prepared to undock from Agena 10 and complete the final manoeuvres to reach Agena 8. Young remarked that after a day and a half docked to Agena, the view was, 'just like backing down the railroad in a diesel engine, looking at a big box car in front of you.' At GET 44 hours 40 minutes, Young backed away from 'the boxcar', and after a small manoeuvre Gemini then closed in on Agena 8 for the next three hours. The crew reported first sighting at 45:38 GET. After calculation of the distance, Houston informed Young that they were still 109 miles from their target. It was pointed out that it was their own Agena 10, 3.4 miles away, at which Young was looking, but added, '109 miles is a pretty long range.' Young agreed. 'You have to have real good eyesight for that.'

NORAD data proved accurate, and Young found Agena 8 in the predicted position. Closest approach to Agena 8 was about ten feet above it at 48.00 hours into the flight, and during the three hours that Young performed the station keeping, Collins conducted his tethered EVA (see p. 290) to retrieve a sample package from the side of the Agena. Following completion of the EVA, with Collins back inside, a separation manoeuvre was performed to lower the orbit to a safe altitude for re-entry.

Following the departure of the Gemini, the Agena completed three manoeuvres over a period of 12 hours under ground command. This resulted in a final circular orbit of 163.2 x 218.7 miles, and after the final burn was completed, control was handed over to the Hawaii tracking station, which monitored the target for the next week until its power depleted. Agena 10 was to remain in this orbit, and it was hoped that it might serve as a passive target on a future mission, although this opportunity never arose.

To the high frontier

The penultimate Gemini – 11 – was assigned a rendezvous in the very first orbit in support of Apollo rendezvous development. A second task was to take Gemini to a high-apogee orbit, using the Agena engine. There had been several proposals to take Gemini to high-altitude orbits, and even towards the Moon (see p. 285), but these were soon dismissed. However, the proposal for a high-apogee Gemini mission appealed to Gemini 11's Commander, Pete Conrad, who pursued the idea on the grounds that it would provide some interesting photographs to evaluate the use of colour imaging on weather satellites. When the fear of high radiation doses threatened the idea, Conrad sought the help of fellow astronaut and nuclear engineer Bill Anders to help evaluate the problem.

A third task – discussed for years, but never attempted before Gemini – was to attach a 165-foot dacron or nylon tether to the Agena and Gemini by EVA. Inducing a spin rate of several degrees per second on the undocked combination could then produce enough tension to practise formation flying, or create a small amount of artificial gravity. NASA decided to investigate both. Gemini 11 would feature the M = 1 profile and stable orbit profile, high altitude, tether exercise, and EVA – another packed mission.

FROM PAD TO AGENA IN ONE ORBIT

Agena launched at 08.05.02 am EST on 12 September, into an orbit of 133.1 x 141.8 miles. In order to achieve a first-orbit rendezvous, Gemini 11 had to follow it 97 minutes later, but, in order to arrive at the correct point in orbit, Gemini 11 also had the shortest launch window in the whole programme. Gemini 10 was tight at 35 seconds, and Gemini 12 would have 30 seconds, but Conrad and Gordon on Gemini 11 had only a *two*-second window to leave the pad; and at precisely 09:42:26.5 – half a second into that two seconds – Gemini 11 was launched.

Upon reaching their initial orbit of 75.6 x 131.3 miles, the crew performed a manoeuvre that corrected their flight path, which refined their position by either moving the Gemini up, down, left or right, and speeding it up or slowing it down. This Insertion Velocity Adjust Routine (IVAR) was a difficult and delicate operation to carry out so close to the spent second stage of the Titan, as there was a

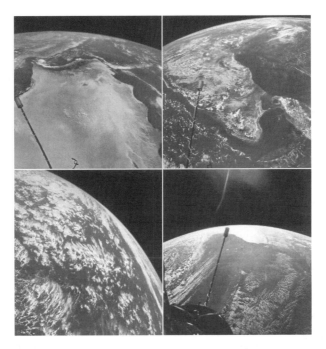

Photographs of the Earth obtained while docked to the Agena target vehicle. The altitudes were 250 nautical miles, 410 nautical miles, 670 nautical miles, and a record 740 nautical miles.

possibility of manoeuvring into it. But a mission rule was that the booster had to be in sight of the astronauts before they committed to this manoeuvre. Radar contact on the target was established with the Agena at a distance of 267 miles, and out of direct radio communications. The crew completed their calculation and adjustments, so that when contact was re-established they reported being within 58 miles of the Agena. When their calculations were checked against those from the ground, the difference was so minor that either would have achieved the rendezvous.

During the post-flight press conference Conrad recalled: 'We came upon the Agena in daylight a bit sooner than we expected. I think I shot two or three percent of fuel fussing around getting used to seeing the bright Agena when I couldn't see the instruments. I fumbled for my sunglasses and didn't get them.' Gordon described the first-time space-flyers' difficult choice of either working or looking out of the window: 'When Pete separated from the booster ... I couldn't help myself. I really had to look outside and I got the biggest shock of my life, because floating all around the spacecraft was all kind of garbage. I said 'Hey, look at all that junk out there.' The next word I got from him was 'Hey, get to work,' but I just couldn't resist that first quick look.'

Just 85 minutes after leaving the pad, Gemini 11 was positioning in front of the Agena with 56% of its fuel remaining. The crew happily informed the Flight Director, 'Mr Kraft, would [you] believe M equals 1?' Seconds later, after permission to dock, the two spacecraft were joined. Each astronaut then performed a task that NASA had been waiting to do for some time – docking practise – by pulling away and then redocking, which was accomplished in daylight and then in darkness. This was followed by a short burn of the Agena engine 90° to the flight path, to 'shift' across to a new orbit. The crew was also impressed by the power of the Agena, as had been the Gemini 10 crew.

'The world is round'

After a night's sleep, the crew prepared for Gordon's first EVA. It was during this excursion outside (see p. 293) that he attached the tether stored in the Agena target docking adapter to the docking bar on the nose of Gemini. The crew then took a second night's sleep after a very exhausting EVA.

Day 3 was to include the trip to the higher orbit, and the crew skipped breakfast in order to prepare everything inside the cabin on time. During pre-firing checks they noted a problem in the Agena displays. The ground assured them that it was not a serious problem, but Conrad observed, 'It's a heck of a time to have a glitch like that.' They were given a 'go' for the ignition at 40 hours 36 minutes into the mission, during the 26th orbit.

'Whoop-de-doo!', yelled a surprised Conrad as the large engine ignited in front of them. 'That's the biggest thrill of my life.' He later added that on passing over India it became apparent to the both of them that they were climbing at a fantastic rate. 'We just had the impression that we were looking down at the ground [and] going straight up,' and Gordon added, 'We were wondering if we were going to stop.' The engine burned for 26 seconds, taking them to 850 miles above the Earth in two highly elliptical orbits.

From their vantage point they took more than over 300 photographs of the weather patterns and terrain below them. Radiation doses showed 0.2 rads per hour, which during their two high-altitude orbits amounted to less than the total to which the Gemini 10 crew had been exposed for the entrie time that they were at 515 miles. Choosing the region over Australia where the radiation belts were thinner helped keep the doses low. CapCom Al Bean observed: 'Sounds like it's safer up there than a chest X-ray.'

Conrad: 'I tell you it's 'GO' up here ... and the world is round ... you can't believe it. I can see all the way from one end around the top, about 150°.' Asked to expand upon his observations of Earth, the astronaut continued: 'It really is blue. That water really stands out and everything looks blue ... the curvature of the Earth stands out a lot. [There are] a lot of clouds over the ocean [but] Africa, India and Australia [are] clear.' Conrad later explained the experience in the post-flight press conference, 'We saw some of the most amazing sights that man has ever seen. The photographs, I think, do some justice to what we saw, but you can't do justice to what you actually see with your eyes.'

Over the United States, on the 28th orbit, the Agena was fired for 23 seconds to decrease the orbital velocity and bring the orbit closer to Earth, lowering its apogee to 189 miles.

All tied up

After a short rest, the stand-up EVA was accomplished before attempting the first tethered activities in this very busy day in orbit. There were two modes of tethered operation. Gravity gradient had the Agena engine pointed to Earth, while the spacecraft backed out to the length of the 100-foot tether. Once taut, a slight thrust of 0.1 fps would keep it taut, and the two vehicles would remain in the same relative position and attitude. In the other mode (Rotating Tethered Vehicles), thrusters would induce a rotation of 1 fps, with the centre of gravity on a point on the tether where both spacecraft would rotate. Centrifugal force would keep the tether taut and the two spacecraft apart, while centripetal force would continue the movement around the centre.

Gemini backed away from the Agena at 49 hours 55 minutes into the flight, but the tether was found to be stuck during the unreeling. It required a couple of short bursts of thruster to pull it free, but this disturbed the Agena, and the astronauts had to compensate. Just short of full extension, they discovered that it was taking longer for the Agena to settle down than anticipated, and so the gravity gradient move was abandoned.

About 17 minutes into the experiment, the rotational rate was introduced, initially at 38 per minute. Conrad reported difficulty in keeping the tether taut, and described it as skipping like a rope between two people. 'Man! Have we got weird phenomena going on here. This will take somebody a little time to figure out,' he commented. But although it was curved, it was also taut. Conrad tried to straighten the arc by using the thrusters, and then they achieved a taut, straight tether, although neither astronaut could recall exactly what they did to achieve this result.

As the combination slowly turned, the astronauts became so accustomed to the effect that they were able to eat a meal. They then increased the rate to 55° per second, but this also returned the oscillations to a point at which Gemini was being see-sawed in pitch to 60 in what Conrad termed, 'a big sling-shot effect.' When the motions again settled down, the astronauts held a camera against the instrument panel, and then released it. It moved in a straight line to the back of the cabin, demonstrating a small amount of artificial gravity (although the astronauts never felt any effects).

After three hours they separated from the tether by jettisoning the spacecraft docking bar. The post-flight report indicated that this had been an economical and feasible method of unattended station keeping. The official history of Gemini – *On the Shoulders of Titans* – recorded it as an interesting and puzzling experiment. Indeed, the economics of the experiment had shown that less fuel was used than expected, and instead of immediately pulling away from Agena, the controllers added some real-time flight planning to the mission in asking the crew to attempt a co-incident orbit (stable-orbit) rendezvous. They would add speed and height to Gemini, so that the Agena would pass beneath them. They then returned Gemini to the same orbital path as the Agena, but 18 miles behind it, and kept at that distance in a long-distance station keeping exercise.

Second rendezvous

The following day the crew initiated a series of manoeuvres, beginning at 65 hours 27 minutes into the flight, to re-rendezvous with the Agena, and after a slight lowering of the orbit, the spacecraft closed in. Just 1 hour 13 minutes later, within one orbit, Conrad had the Agena just above the spacecraft, with the tether pointing straight up and after twelve minutes they made the separation burn to begin the journey home. Conrad suggested that perhaps a refuelling tanker might be sent up in order to extend the mission so that they could carry on with their work – an option not available in 1966, but a forecast of what would follow in space station operations from 1978. The crew left the 'best friend we ever had' in orbit, and were sorry to see it go, as it had been so kind to them on their highly successful mission.

The last mission

The final mission had in its objectives rendezvous and docking, docking practise, tethered vehicle station keeping by gravity gradient, and Agena propulsion manoeuvres – all of which were achieved, marking a successful end to the Gemini rendezvous programme.

The refurbished Agena 5001 was launched as 5001R at 14.08 on 11 November 1966, followed by the final Gemini launch, just 98 minutes – twelve successes in twelve launches. Gemini 12 was to conduct an $M = 3$ profile rendezvous. Radar lock-on was reported by the crew at 270 miles, but the radar system failed shortly afterwards. It would be for the crew to provide the onboard computation to arrive at the Agena. Fortunately, the pilot on the mission was Buzz Aldrin, who had worked on the charts and manual orbital rendezvous procedures, and had gained a PhD in orbital rendezvous from MIT before joining NASA. Now was his chance to put theory into practise and live up to his 'Dr Rendezvous' label, using sextant, charts, slide rule and the onboard computer.

Lovell flew to these co-ordinates, and after 3 hours 45 minutes of flight (and using just 280 lbs of fuel), they docked to the Agena. Aldrin's work had paid off, and had also demonstrated that if radar failed, an astronaut could perform the rendezvous and docking with manual techniques. As Aldrin later wrote in his 1973 autobiography *Return to Earth*, 'The rendezvous and docking went off perfectly, and I had proven a major point. Man, working from the computer on his spacecraft with a series of cryptic charts, could do as effective a job as the hundreds of men and hundreds of computers back on Earth.'

Both astronauts practised undocking and redocking, although Lovell misaligned one approach. He fired the forward and aft thrusters in a move similar to an attempt to rock a car wheel free from the mud. He finally shook Gemini free of the Agena without any damage to either craft.

Unfortunately, they lost their trip to the high-altitude orbit due to a chamber pressure drop on the target vehicle shortly after launch. However, there was a small compensation for the experimenters. When the launch had slipped two days, the planned photography of the solar eclipse had been lost. Now, using the secondary propulsion system, the astronauts performed two docked manoeuvres to adjust the

Gemini 12 photographs of the Agena XII target and its tether as the combination passed over the Texas coast of the Gulf of Mexico during the 47th hour of the four-day final Gemini mission.

Astronauts Buzz Aldrin (*left*) and Jim Lovell (*right*) onboard Gemini 12. During this mission, Aldrin was able to put into practise his manual techniques for rendezvous with a target in space – the basis of his PhD thesis prior to joining NASA.

phase of their orbit, allowing for an 8-second photographic pass of the eclipse on the tenth orbit.

A second tether exercise

During the first EVA on Day 2, Aldrin had installed a tether between the two spacecraft. Then, following the second EVA, the Agena target was used to pitch the combination to a gravity gradient. Lovell undocked Gemini 12 at 47 hours 23 minutes GET, and backed away, thereby extending the tether, which this time hung up only once. Like Conrad before him, Lovell experienced difficulty in keeping the tether taut, but despite this and a one-time 'sling-shot' of 300, they achieved the gravity gradient stabilisation. The tether operations lasted more than four hours, after which Gemini ejected the docking bar at GET 51 hours 51 minutes. The crew performed a separation manoeuvre, and pulled away some 23 minutes later.

The last rendezvous manoeuvres of the programme were successfully concluded, although the plan to park the target after Gemini 12 landed could not be achieved, due to the depletion of attitude control gas onboard the target vehicle.

Summary

The rendezvous operations conducted during Gemini revealed that it was possible to rendezvous with an active or a passive target, and that it could be achieved using only onboard guidance, ground data, or a combination of both. And if the automatic system failed, then the crew could – depending upon the severity of the failure – perform manual rendezvous operations.

It was also determined that docking with a large vehicle in orbit was both practical and, with appropriate conditions, relatively easy. While launch payload constraints were a burden, Gemini proved that a combination spacecraft with propulsion units could manoeuvre to higher orbits, and that the structure of such a combination was sound in design. Tethered operations also offered a potential for gravity gradient and centrifugal stabilisation of larger vehicles.

As well as the crews in orbit, ground operations (especially in the on-time launch of target and spacecraft), orbital tracking, communications and control and personnel training, all contributed to one of the major success of the Gemini programme.

Rendezvous propellant usage

GT	Type	Actual (lbs)	Minimum (lbs)	Ratio
6	M = 4	130	81	1.60
8	M = 4	160	79	2.02
9	M = 3	113	68	1.66
	Optical	61	20	3.05
	From above	137	39	3.51
10	M = 4	360	84	4.28
	Optical dual	180	73	2.46
11	M = 1	290	191	1.52
	Stable orbit	87	31	2.81
12	M = 3	112	55	2.04

Agena launches

Agena	Int. design	Launch	Decay	Lifetime (days)
6 (5002)	n/a	1965 Oct 25	Launch failure	
8 (5003)	1966-18A	1966 Mar 16	1967 Sep 15	548
9 i (5004)		1966 May 17	Launch failure	
9 ii (ATDA)	1966-46A	1966 Jun 1	1966 Jul 11	40
10 (5005)	1966-65A	1966 Jul 18	1966 Dec 29	163
11 (5006)	1966-80A	1966 Sep 12	1966 Dec 30	108
12 (5001R)	1966-103A	1966 Nov 11	1966 Dec 23	41

EVA operations

On 18 March 1965, cosmonaut Alexei Leonov became the first man to work outside his vehicle in orbit. Officially termed Extravehicular Activity (as opposed to pressure-suited work *inside* the spacecraft – Intravehicular Activity) it is more commonly known as spacewalking – although 'walking' is something of a misnomer as the event takes place at 17,500 mph. Although the first EVA was performed by a Soviet, the skills and difficulties of working outside the spacecraft were first investigated during the Gemini programme. The early success and apparent ease of Ed White's first American EVA in June 1965 led to an underestimation of the techniques of EVA work, and as a result, later Gemini EVAs were planned as ambitious activities. When the next astronauts attempted EVA, they found that the task was not as straightforward as the exploits of Leonov and White had indicated. These EVAs were invaluable in determining the parameters of man's capabilities in spacewalking, and allowed three final Gemini EVAs to demonstrate that with adequate training and facilities, EVAs were a valuable resource.

EVA FROM GEMINI

During March 1961, interest was growing within NASA for EVA to be included in future advanced programme planning, although no one was quite sure how it could be done. Leading Mercury spacecraft designer Max Faget, of the STG, had been intrigued with the idea of astronauts leaving their capsule in orbit, and had asked John Yardley, of McDonnell, to investigate the possibility of a two-man version of Mercury that could also support EVA. There was recognition that the two-man spacecraft design being evaluated at that time could offer the best solution, with an element of safety and shared responsibility of work for such experiments. These suggestions for EVA prompted the first drawings of a two-man Mercury from McDonnell.

The inclusion of a large mechanical hatch for the crew station ejection seats on Mercury Mark II offered the chance to seriously consider EVA for the first time. Although initially the large hatches in the side of the spacecraft raised questions concerning structural strength and the integrity of heat protection, the design offered the best prospect for EVAs directly from the crew station without the need for airlocks and complicated hatch designs. In August 1961, EVA was featured in Mark II plans, in which an astronaut would transfer between two spacecraft in orbit as part of the development of the lunar landing programme. Although EVA was not part of the stated December 1961 programme objectives, the prospect of performing EVA from Gemini spacecraft was being evaluated.

Gemini EVA guidelines

During May 1962, the GPO began to look at the requirements for EVA by asking McDonnell to investigate the operation of opening and closing the hatches in the space environment, and the MSC Life Sciences Division was tasked to determine what the additional requirements would be for a 15-minute EVA capability. By 19 September, the Crew Systems Division had reported that a super-insulation overall worn over the pressure garments would be required for thermal protection. In addition, amendments to the life support, ventilation and cooling systems would have to be developed, and would use connections to the main spacecraft (umbilical) or a separate unit carried by the astronaut (portable life support system). There would also need to be a means of manoeuvring close to the spacecraft.

McDonnell reported that hatches could be opened and closed in the vacuum of space, but there would be limitations on the number of times that the crew compartment could be vacated of its atmosphere and then repressurised. For safety, it was also decided to open only one hatch at a time, and for only one astronaut to leave the spacecraft while the other monitored the spacecraft and his colleague and controlled the vehicle in a stable attitude. From these studies, it was decided that the Pilot would leave the spacecraft and the Commander would remain inside – at least initially. This was based on the layout of the displays and controls, and the roles and responsibilities of the two astronauts, evaluated by McDonnell earlier that year.

By 3 February 1963, the Crew Systems Division had established a set of guidelines for proposed EVA from Gemini, and had instructed McDonnell to study these guidelines and report back with a list of requirements to support them. The study area looked at the extent of manoeuvrability inside the spacecraft with the hatches closed, and in unpressurised and pressurised suits, to establish the ability to prepare for EVA. They also investigated the changes that would be necessary to assist in these preparations, to allow the crew to stand in the open hatchway (on the seat, but not to actually leave the vehicle – called stand-up EVA), and the requirements to fully leave the spacecraft and inspect the exterior of the vehicle (EVA). McDonnell were directed to provide capability for SUEVA from spacecraft number 2 onward.

At a meeting at MSC on 21 March, discussions on EVA operations included decisions on the design of the basic pressure suit with additional thermal protection. It was also decided that the tether used to secure the astronaut to the spacecraft should be used only for that purpose, but it should include twelve nylon encapsulated communication wires, and be long enough to allow the astronaut to reach the Adapter Module. Manoeuvring capabilities would be handled by other means, and the supply of oxygen would allow up to 30 minutes of activity outside the capsule, in addition to time allocated to prepare for the spacewalk and to conclude it. The meeting also established that the capability to support EVA could be incorporated on every production capsule from spacecraft 4.

The David Clark Company had been awarded the contract to produce the Gemini (internal) pressure suit, designated G3C, on 13 June 1963, after completing negotiations with NASA on 28 May. In October, requests for proposals for an EVA life support package were issued by MSC, and by 25 November, proposals had been received from prospective contractors. These included a high-pressure gaseous oxygen bottle supply, with regulators and valves for the control of oxygen flow in an open loop design, in which there was no self-correction action (unlike the closed-loop system, where output is used to control input). It was designed to support initial EVA for up to 10–15 minutes using a hard-line tether. The proposals were evaluated by December, and in January 1964 the Garrett Corporation was contracted to supply the system.

The EVA programme
During January 1964 the first Gemini EVA plan was published. The objectives were:

- To evaluate man's capability to perform useful tasks in a space environment.
- To employ EVA operations to augment the basic capability of the spacecraft.
- To evaluate advanced EVA equipment in support of manned spaceflight and 'other national' programmes.

The Flight Crew Operations Directorate (FCOD) had been working on flight planning, and included EVA in this scheduling.

Gemini 4 would be the first flight with full EVA capability. The Pilot would depressurise the cabin, open his hatch, and stand up for a short period, tethered to the vehicle, during the mission in February 1965. After a short evaluation of his mobility, he would possibly take some photographs, and would then sit down, close the hatch, and repressurise the cabin.

The Gemini 5 mission would include the first full exit of the Pilot into space, practising egress and ingress. On the Gemini 6 mission, the astronaut would make his way along the Adapter Module to the rear of the vehicle, and then retrieve data packages from the inside of the Equipment Module. Hand-holds would probably be used during operation, although this was not made clear.

The Gemini 7 and Gemini 8 programme plan indicated that hand-holds, tethers and manoevrability would be evaluated by the EVA astronaut.

An 'astronaut manoeuvring unit' would be tested on Gemini 9, while on Gemini 10 to Gemini 12 the astronauts would 'evaluate other advanced EVA equipment and procedures'.

The plan also indicated that Crew Systems had the responsibility for ground testing of EVA hardware, and had begun egress and ingress exercises in a simulated space environment inside altitude chambers and on the vomit Comet. The training programme was also being established, allowing the astronauts to prepare for EVA experiments on their assigned missions. At the time of the announcement of this plan, there was opposition to any EVA from Gemini until realistic simulations could be performed.

Plans the first stand-up EVA

According to the plan for Gemini 4, the Pilot would do no more than stand with his head and shoulders above the hatch line. But when the crew (Commander Jim McDivitt and Pilot Ed White) were named on 27 July, there was a hint that the plan could involve more than simply standing up on the seat, and might actually include 'stepping into space'. If this did take place, then Ed White would become the first man to perform an EVA, if not actually leaving the vehicle.

The Gemini 4 crew and the David Clark Company representative discussed improvements to the basic G3C pressure suit in order to support EVA, and this evolved into the authorisation for the G4C EVA suit. In July and August, the first design reviews of the chest-mounted life support system highlighted further problems due to insufficient room inside the spacecraft. When stowed, it would take up space allocated for experiments, and it was suggested that some of them could be moved outside – possibly to the Adapter Module. This also required EVA to activate or retrieve some of these experiments. Another development, begun in 1964, was that of incorporating the oxygen supply through the tether. This evolved into the umbilical design that carried communication lines, and oxygen from the spacecraft system for breathing and ventilation, and acted as the restraint for the astronaut performing EVA.

In October 1964, Jim McDivitt evaluated the prototype EVA suit in the Gemini mission simulator. It appeared bulky and immobile while unpressurised, but when pressurised it did not apparently hinder mobility. Another suit had also been sent to Ling–Temco–Vought, with added thermal and micrometeoroid protection for thermal testing in a simulation chamber. In the same month, re-entering the spacecraft was evaluated in the mock-up inside the KC-135. With practise, the astronauts, wearing the suit and chest-pack, could accomplish this and shut the hatch in about 50 seconds.

The following month, the Gemini 3 crew of Grissom and Young offered to use one of the scheduled vacuum tests on their spacecraft number 3 to prove that an astronaut could open a hatch and stand up in a 'space environment'. McDonnell was opposed to the idea because, as John Young later pointed out, 'They did not want to take the chance of bagging a couple of astronauts in the altitude chamber.' With nothing between the astronauts and the vacuum except the skills of the seamstress at David Clark who joined the parts of the suit together (or, as has often been quoted, 'that little old lady from Worcester, Massachusetts, and her glue pot'), even NASA management had reservations. But, as Kleinknecht clearly pointed out, if the exercise could not be carried out in the altitude chamber, NASA had no business in attempting EVA at 100 miles altitude! With that, the test was authorised.

With both men wearing full G3C pressure suits, and the chamber at a simulated altitude of 150,000 feet, Young opened the hatch, stood up on his seat, and then sat down again. He had some initial difficulty in closing the hatch, but the two astronauts proved that the concept proposed for Gemini 4 could work, which allowed more serious planning for a stand-up EVA on Gemini 4.

At the end of December, the first prototype G4C EVA suit had been received by CSD for testing. During KC-135 tests in January 1965, the suit was generally satisfactory, but the heavy cover layer impeded movement. It was therefore redesigned to remove excess bulk, and was satisfactorily re-tested in February. Also during January, both the Gemini 3 and Gemini 4 prime crews practised zero-g training with a mock-up Gemini spacecraft in a KC-135. The Gemini 3 crew first tried stand-up EVA procedures, and the next day these were duplicated by the Gemini 4 crew.

Stand-up becomes full exit

Six days after Leonov completed the world's first EVA, White opened the hatch of spacecraft 4 to complete a stand-up EVA at a simulated 150,000 feet in the altitude chamber at McDonnell Douglas. The Soviet EVA and the success of Gemini 3 on 23 March, plus this recent altitude test, prompted an informal meeting in the office of MSC Director Robert Gilruth on 29 March, to discuss the possibility of moving more than just head and shoulders out of the hatch of Gemini 4.

Joining Gilruth in the meeting was the centre's Deputy Director, George Low, Richard Johnson from Crew Systems, and Warren North from Flight Crew Operations. During the discussions a mock-up of the chest-pack was viewed, and North expressed the confidence of the Crew Systems people that an umbilical EVA could be achieved during Gemini 4. On 3 April, the plan was presented to George Mueller who, though not totally convinced, in turn briefed the relevant directorates at Headquarters, who tentatively agreed that Gemini 4 could attempt a full EVA, based on further engineering evaluations and qualification tests. Upon this authorisation, MSC began preparations and training to achieve the EVA over the next few weeks. After Associate Administrator Robert Seamans reported on the briefing, he received at MSC, on 14 May, the news that all was ready for attempting the experiment. This news was greeted with enthusiasm, and confidence that full authorisation would soon be forthcoming. There remained medical concerns over the stresses placed upon White and the possibility of disorientation, but the astronauts reviewed Leonov's EVA film, and found that he used the Sun, Earth and spacecraft for reference. White knew he had the answer to knowing where he was at all times in relationship to Gemini 4, and would not become disorientated.

On 19 May all equipment was certified for flight, and six days later Seamans wrote to Jim Webb, stating the reasons for and against EVA and the risks involved. The letter ended with the recommendation to proceed with the EVA. The following day, the letter was returned with a hand-written note: 'Approved after discussion with (Hugh) Dryden. Signed: J. E. Webb 25.5.65'.

GEMINI 4 EVA

Once the announcement was made that Ed White would perform an EVA during Gemini 4, the news hit the world headlines the following day. Despite Leonov's historic EVA, the fact that White would do the same for America was too good a

The Gemini 4 EVA spacesuit and life support system.

story to miss. It was also stated that due to the expected physical exertion, White's intake for the mission would be 2,230 calories, which was 164 calories per day more than McDivitt's intake.

Gemini 4 EVA hardware

The EVA suit (shown in the illustration above) consisted of (from the outside inwards) an outer protective layer of HT-1 (high-temperature) nylon and a layer of HT-1 nylon felt for protection from micrometeoroids. Then, seven layers of aluminised Mylar and woven Dacron superinsulation were laid over two inner layers of high-temperature nylon for micrometeoroid shock absorption. Beneath these cover layers were the link net restraint layers, pressure bladder, comfort layer and constant wear (underwear) garment of the basic pressure suit.

The EVA cover layer consisted of two parts: the main part that covered the upper torso, and a removable jacket that covered arms and shoulders and could be removed for comfort after the EVA.

For Gemini 4 only, the helmet included a detachable EVA visor assembly that included two over-visors. The outer one was the Sun visor, made of grey-tinted Plexiglas and, on the outside, a thin coating of gold film to reduce light transmission by 88%. The gold film also served to absorb harmful UV light, thus protecting the astronauts' eyes and reflecting the infrared energy. Over the top of this gold layer was a high-eminence coating, which was applied to prevent the flaking of the gold layer and to help reduce the surface temperature of the visor when in full sunlight.

A second, inner visor was made from a polycarbonate material, and provided thermal control by means of a low-emmitence coating over the exterior surface. It also provided visual protection against ultraviolet energy through the use of an ultraviolet inhibitor in the polycarbonate material, and impact protection for the Plexiglas pressure visor of the inner pressure shell of the suit.

For this mission only, a pair of overgloves were worn for thermal protection, and featured a silastic palm insulation that allowed direct palm contact with objects at 250° F or –150° F for a maximum of two minutes.

The chest-mounted Ventilation Control Module (VCM) controlled the pressure in the suit at 202 mm Hg, and provided ventilation within the suit. It was attached to the chest by two restraint straps snapped around the parachute harness and held by Velcro to the front of the unit. There was also a nine-minute emergency supply of oxygen (0.33 lbs) should the main supply of oxygen fail, supplied if required via an outlet fitting to the feed port on the front of the helmet. That main supply was fed via the 25-foot umbilical to the astronaut's suit. The oxygen line was 27 feet long, but the tether designed to take the strain of the astronaut was only 23.7 feet long. This was to prevent undue stress being placed upon the oxygen line. The tether was designed to take a 99.2-lb load, and was covered with a thin layer of gold film to distribute thermal loads from the Sun.

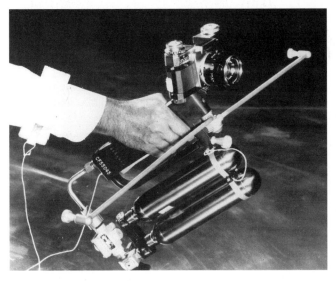

Gemini 4's Hand Held Manoeuvring Unit, showing the attached camera and the trigger system used to initiate the firing of the jets.

White also had the Hand Held Manoeuvring Unit (HHMU – also known as the 'gas' or 'zip' gun) which weighed 7.4 lbs. It consisted of two parts (handle assembly and high-pressure section) for ease of storage in the crew compartment, to be assembled by White in the spacecraft. To do this, he connected a coupling at the regulator and inserted a locking pin to secure the two parts together. Oxygen was supplied from small bottles, for propulsion to a maximum thrust of 2 lbs for 20 seconds, or at lower thrust levels to extend the use. The two 4,000-psi storage bottles were of the same design as the ejection seat emergency bail-out bottles. The regulator was from the Mercury ECS; and as it was stored in the spacecraft, the choice of propellant was important, as it could not be allowed to contaminate the cabin if it leaked. To use the device, White merely depressed the hand trigger to activate the supply of gas through a pressure regulator, which reduced the pressure to 120 psi. From here the gas passed the handle and then through a filter to the two valves. It then passed through the trigger guard to the pusher nozzle, while gas at the other end passed through to a swivel joint, and then through the two arms of the tractor nozzles.

Stepping out

It was already clear that EVA was of more importance than the rendezvous with the spent upper stage of their booster, when Jim McDivitt decided to postpone the EVA from the second to the third orbit. White had been assembling the zip-gun, pulling the umbilical out of its storage bag, and attempting to keep the pull strings clear. He

On 3 June 1965, Ed White became the first American – and second man – to walk in space. The dark land mass to the right of his head is the Texas coast of the Gulf of Mexico. The nose of Gemini 4 is at bottom centre, while the open hatch is at right.

was becoming tired and hot in preparing for his exit into space. The crew had been in space for only about three hours, and had been preparing for the EVA for the last 90 minutes, after abandoning the rendezvous attempts. Although they believed they could begin on schedule, they realised that by rushing, they might skip a procedure of the 54-point check-list.

During the third orbit they went through the check-list one last time. Having started the preparations ten minutes early because of the abandoned rendezvous, McDivitt observed that flight planners had not realised how much time was taken in preparing for EVA. Although they managed to complete the checklist with 15 minutes to spare, it was still too rushed. Finally, with both men suited and the umbilical connected to White's suit, permission to begin the EVA was given as the spacecraft flew over the Carnarvon tracking station. McDivitt began to depressurise the cabin then stopped at 2 psi to check suit integrity before proceeding to a vacuum condition.

The hatch was opened over Hawaii at 4 hours 18 minutes into the mission. White's first task was to install the cameras and umbilical guard on the hatch sill. After checking the camera three times, he stood up on the seat to attach a camera to the HHMU so that he could take photographs of Gemini 4 in orbit. In Mission Control, Chris Kraft informed the astronauts that they were ready for White to go outside.

'Absolutely no sensation of falling'
White tried to use the gun to propel himself out of the hatch, with no push-off from the spacecraft, and left Gemini after 12 minutes of open hatch time. As he exited, he noted that the gun enabled him to control where he wanted to go. The gun was, however, exhausted after only four minutes. In this time, he had managed to travel 15 or 16 feet from Gemini, used the unit to stop his motion, and then completed the trip back to the spacecraft and out into space twice more before the fuel was depleted. This disappointed the astronaut, who wished for more fuel. Even though he had used the supply sparingly, it was exhausted by the time they approached the west coast of America.

'I am looking right down, and it looks like we're coming on the coast of California,' White observed. He later said that 'there was absolutely no sensation of falling. There was very little sensation of speed, other than the same type of sensation that we had in the capsule ... you can't actually see the Earth moving beneath you ... the Sun in space is not blinding but it's quite nice.' The EVA would last for 6,000 miles, and as he flew over MSC he took the opportunity to admire the spectacular view below him. 'Hey, Gus [Grissom–CapCom]. We're looking right down over Houston ... that's Galveston Bay right there.' As White later recalled, 'I was taking some big steps ... the first on Hawaii, then California, Texas, Florida, and the last on the Bahamas and Bermuda.'

White had elected not to use the right-hand over-glove, so that he had an improved hold on the zip gun. Early in the EVA, he saw it float out of the open hatchway and into orbit (and after a few months its orbit decayed and it burned up in the atmosphere).

'You dirty dog'

With the fuel of the HHMU exhausted, all that White could do was use the tether to move about; but he found this was more difficult, and never really reached where he wanted to go. 'The tether was quite useful. I was able to go right back to where I started every time, but I wasn't able to manoeuvre to specific points with it.' At one point he used the tether to pull himself down to the surface of Gemini and actually 'walk' across the hull of the spacecraft for a few steps, until his feet lost contact with the spacecraft, which pitched away from him as he tumbled into space again.

White's manoeuvres on the end of the umbilical had an influence on the attitude of the spacecraft. 'One thing about it. When Ed gets out there and starts whipping around, it sure makes the craft tough to control,' McDivitt commented. The Command Pilot was also careful not to fire the attitude control thrusters near his colleague. By using the pulse control mode, McDivitt knew exactly where the thrusters were firing, so that White would be nowhere near the plumes. He allowed the spacecraft to drift most of the time. It soon became clear that the location of the tether in front of the spacecraft was not the best for McDivitt to view White's actions. Each time White tried to place himself in front of the Gemini, the tether carried him in a wide arc over the top of the spacecraft and to the back of the Adapter Module. He found that the thermal tape and Velcro placed on the Adapter Module before launch was still there, and although the adapter separation plane had a good cut, its edges were rough.

On one occasion White came into contact with the Command Pilot window, and caused two smears on it as he brushed the particles which had been deposited during stage separation. 'You smeared up my windshield you dirty dog! You see how it's all smeared up there?' exclaimed McDivitt.

'Get back in'

As the astronauts were using the voice-activated system (VOX) internal communication for most of the EVA, this cut off messages from the controllers. During the EVA the two astronauts kept up a running commentary with each other, which created difficulty in receiving communication from Mission Control.

McDivitt to CapCom: 'Gus, this is Jim. Got any messages for us?'
Grissom: 'Gemini 4. Get back in! We've been trying to talk to you for a while here.'
McDivitt: 'OK.' (To White) 'Come on in then . . .. Let's get back before it gets dark.'
White: 'I hate to come back to you, but I'm coming . . . it's the saddest moment of my life. I'm trying,' as he tried to position himself above the spacecraft
McDivitt: 'OK, don't wear yourself out. Just come on in . . . how you doing there? Whoops, take it easy now.'
White: 'Aren't you going to hold my hand? I'm fixing to come back into the house.'

Twenty-two minutes after opening the hatch, White was standing on the seat and was preparing to move into the spacecraft. The hatch proved difficult to close from the open position, and the locking mechanism refused to work. As he tried to close the hatch, White floated out of the seat, and McDivitt had to hold his legs to keep him inside.

After 36 minutes the hatch was closed and locked, but the planned reopening to jettison unwanted equipment was cancelled, and everything had to be carried in the cabin for the rest of the mission. White had been outside for 21 minutes, and recognised what he had just achieved. 'I'm very thankful in having the experience to be the first [American to perform EVA].' Mission elapsed time was just 4 hours 54 minutes, and the crew settled down to the remainder of their four-day flight.

Physically exhausted
White was one of the fittest astronauts in the team, but by the time he returned to his seat he was physically exhausted, and the sweat was stinging his eyes and fogging his faceplate. He had performed the EVA, but McDivitt had also been exposed to the vacuum inside the crew compartment. The compartment was then repressurised, and they rested, having completed almost one orbit in an unpressurised spacecraft.

White later explained what they were trying to achieve during EVA: 'We were looking to find out, could man control himself in space? ... and the answer was yes, [but] he needs a little more fuel than was provided to me. We also were tying to find out what were the dynamics of the tether. We found out a great deal.'

The success of the EVA prompted Joseph Shea, the Director of the Apollo Project Office at MSC, to state that White had demonstrated that the Lunar Module crew could cross over to the Command Module by EVA after returning from the lunar surface, if the two spacecraft could not dock. The apparent ease of White's activities tempted mission planners to include even more ambitious activities on the next EVAs under consideration.

A DIFFICULT ACT TO FOLLOW

According to the January 1964 EVA plan, the next three missions would all feature EVA to demonstrate astronaut mobility using handholds and tethers around the exterior of the spacecraft. But the EVA on Gemini 4 had been so successful that there seemed little point in repeating it. However, the more advanced environmental systems would not be available until 1966, on mission 8 at the earliest.

The Gemini 4 crew had noted the difficulties of preparing for EVA and the time it took to stow away items, and there was the report of White's fatigue after only 20 minutes. The missions of Gemini 5 and Gemini 7 were planned for long-duration and for more comfort in orbit, EVA would be excluded from those flights. Gemini 6 was being prepared for the first docking with an Agena, and the addition of EVA to the two-day flight would only complicate matters. Therefore, on 12 July 1965, William Schneider sent a memo to Charles Mathews, stating that there would be no EVA on the next three missions.

The short-lived plans for Gemini 6/7 EVA
With the work involved in preparing for the first Agena docking, Walter Schirra had worked hard to remove any EVA from Gemini 6, and by July 1965 this was official.

But the flight was delayed by the loss of the Agena and the flight plan was changed to allow Gemini 6 to rendezvous in space with Gemini 7 during their 14-day space marathon.

Schirra and Stafford again considered the possibility of EVA – and perhaps even allow Stafford to swap seats with Lovell from Gemini 7. But Borman soon rejected that idea. 'Wally could have all the EVA he wanted, but I wasn't going to open the hatch.' Borman's overriding goal was the endurance mission. Essentially, it would be '14 days or bust' – and nothing would threaten that goal.

There were also several problems and hazards in attempting EVA on this mission. During such a transfer, both Stafford and Lovell would have had to disconnect their life support systems and then reconnect them to their new spacecraft, relying totally on the independent suit systems. Besides both spacecraft were configured to the individual astronauts, especially in the seating for protection during re-entry. Stafford was also one of the tallest astronauts and had experienced trouble when shutting the hatch during ground tests, and White's problems when closing the hatch were still fresh in the memory.

The astronauts also trained as teams of two and interchanging them halfway through a flight was, although possible, an event for which they had not trained; and with little time to train, the idea was soon rejected. Lovell was also flying with the long-duration G5C soft-suit, which did not have the protection for EVA. Indeed, when pressurised it resembled the Michelin Man, and he found it almost impossible to bend his arms or legs, rendering it useless for EVA. Despite seeking help from Headquarters to resolve the matter, Schirra did not gather much support for EVA on the dual mission. Having worked hard to remove it from the Gemini 6 schedule, he now found it impossible to have it restored.

Gemini 8 EVA plans

Although EVA was scrubbed from Gemini 5, 6 and 7 in July, the plans for Gemini 8 were announced in early September. On this flight the Pilot Dave Scott, would evaluate the use of an Extravehicular Life Support System (ELSS), the HHMU, and an Extravehicular Support Package (ESP). The chest-mounted life support system fed oxygen to the EVA astronaut from the spacecraft supply, from primary sources in the back-pack, or from its own small emergency supply. The ESP (more commonly called the back-pack) was more advanced. Like the AMU, it was designed to fit in the Adapter Module at the rear of the spacecraft, and carried its own oxygen supply, a radio, and 18 lbs of propellant to feed the HHMU.

The unit was connected to the spacecraft systems via a 26-foot oxygen hose/tether. However, during the EVA, Scott could theoretically switch over to the back-pack system for oxygen, and disconnect from the spacecraft supply. By adding an extra-lightweight 75-foot tether to the shorter hose still connected to the spacecraft, he would be able to move to about 100 feet from Gemini 8.

This was a considerable advance over the short venture outside that White had accomplished just three months earlier, and despite the fact that no other astronaut would venture outside before Scott attempted his EVA, it seemed a logical step in preparing to fly the more advanced USAF-developed Astronaut Manoeuvring Unit on mission 9. When the crew reviewed these plans, they knew that they had a challenging training programme ahead of them to complete the EVA.

Gemini 8 EVA equipment
Scott's suit assembly was similar to that worn by White, although the anti-micrometeoroid protective layers were modified to include two layers of neoprene-coated nylon, instead of the nylon felt and 6-oz HT-1 nylon layers. The gloves were now integrated thermal gloves rather than the pressure gloves and overgloves worn by White.

The ELSS and ESP hardware proved to be quite complex, and featured a significant number of late modifications to the chest-pack in order to accept connections with the 25-foot ELSS umbilical, the 75-foot electrical tether, and the ESP oxygen line.

The ELSS would become the basic EVA system carried on all the later Gemini missions. It had a much larger reserve of oxygen and a greater capacity for removing heat and moisture than had the VCM worn by White. The central component was the chest-pack that supplied oxygen from the spacecraft umbilical, or from the back-pack at normal rates of 5.1 or 7.8 lbs per hour. This supply was mixed with the secondary (recirculated) ventilation gas, and was then supplied to the suit for cooling and CO_2 wash-out. The suit pressure was controlled at 3.7 psig by a differential pressure valve, from which gas was exhausted to space. This allowed wash-out of carbon dioxide at a rate that maintained an acceptable level inside the helmet. The system also provided a heat exchanger for cooling and removing moisture from the secondary gas, a 30-minute emergency oxygen supply, and an emergency audio-visual warning system.

The ESP back-pack assigned to Gemini 8 supplied oxygen to the ELSS chestpack at 97 ± 10 psig, and was used for pressurisation and ventilation, as well as metabolic applications. The oxygen flow rate was 5.1 or 7.8 lbs per hour for normal modes, and could supply an emergency rate of up to 16.2 lbs per hour. In addition to the oxygen supply, the back-pack also contained the supply of Freon 14 for HHMU propulsion at 100 ± 15 psig. This gave 2 ± 0.25 lb thrust over a period of 200 seconds, which could increase velocity to about 72.5 ± 2.5 fps. The back-pack also included its own self-contained battery power supply and two voice communication modes between Scott on EVA and Armstrong in Gemini 8, these being the UHF voice transducer developed for the USAF AMU and the hard-line connection carried through the tether to the spacecraft. Scott could select the desired mode of communication with a switch located on the ESP.

Scott's EVA preparations
With such a complex and involved system, Scott's training reflected the amount of effort which the astronauts exerted in preparing for the expected two-hour EVA. He completed more than 300 parabolas on the KC-135, and 20 hours on the air-bearing table. On the latter he was supported on a structure that was lifted by a cushion of air, hovering 0.001 inch over a 20 x 23-foot 'table'. Using the zip gun, he was able to glide over the polished surface, to mimic lateral movement in space and to practise starting and stopping with the gun.

During the preparations, numerous problems were encountered – including the Freon, which caused blockages in the nozzles or stuck triggers at low temperatures, preventing shut-off of the gas supply to the gun. There was also the problem of frozen oxygen ejectors in the chest-pack, which could block the supply of oxygen to the astronaut. These problems were solved with new seals and shut-off valves to resolve the Freon problem, and a small heater near the ejector to resolve the oxygen problem.

With so many cables and umbilicals, mobility in the crew compartment was becoming extremely restricted, and a considerable amount of planning was involved in having everything connected and prepared before the opening of the hatch. Tests to 150,000 feet in the altitude chamber at McDonnell (and at MSC in the last few weeks prior to launch) allowed Scott to simulate preparations for organising his EVA equipment and for opening the hatch in a vacuum. The KC-135 allowed him to practise translation to the aft of the spacecraft to strap on the back-pack, and the air-breathing table provided translation practise with the zip gun.

After six months of training, Scott was ready to fulfil his EVA objectives. The primary objective of the three-day mission of Gemini 8 was the docking with the Agena. The EVA was planned to take place after the docking; but unfortunately, other events subsequent to docking forced a premature end to the mission, little more than ten hours after launch. Scott would not, after all, perform his Gemini EVA, although he was able to perform EVA on his next two missions, which included three days on the Moon in 1971.

LEARNING THE HARD WAY

The next three Gemini missions (9, 10 and 11) were all to feature EVA as well as docking to a target vehicle. Each was planned to expand upon experience and to develop techniques that could be used on Apollo or on later missions such as the developing Apollo Applications Programme. Each featured modified versions of equipment that had been used on or was planned for earlier missions. Gemini 9 was also to demonstrate the use of the Astronaut Manoeuvring Unit (developed by the USAF) for the first time and required additional amendments to the EVA equipment in order for it to be used safely. If successful, it would be flown again on Gemini 12.

In the current programme, we see the apparent ease with which astronauts and cosmonauts undertake their EVA tasks – retrieving satellites, repairing the Hubble Space Telescope, or constructing the International Space Station. EVA has become as much a part of spaceflight as the launch and landing, and we often forget that when Gemini 9 was being prepared to fly the first AMU, there had been just two men who had performed EVA – for about 55 minutes, and for a total of only 30 minutes outside the spacecraft. Both excursions had indicated that performing EVA was relatively easy, and this led to more adventurous EVA mission planning by both the Americans and the Russians.

Tether or no tether? That is the question

The AMU had been under development within the USAF for some years, and in 1963 was one of the first 'experiments' to be allocated to a possible Gemini mission. The unit was constructed by Chance Vought in co-operation with the Air Propulsion Laboratory at Wright-Patterson AFB, Ohio. It was designed to allow an astronaut to control his stability in space and to manoeuvre around objects independent of the main spacecraft. The potential for such a device was obvious to the military: maintenance, repair, resupply, crew transfer, inspection of satellites (friendly or otherwise), space construction, and so on. The list seemed endless. For NASA, as this was all in the future it was not a priority for Gemini or Apollo, but might be used in Apollo Applications and later projects. The AMU would fly on Gemini only to prove the concept, and as it was a new 'experiment' and not an operational system, the element of astronaut safety and a tether entered into the equation.

NASA and the USAF continued to argue about the length of a tether, and whether or not to have the astronaut tethered at all. There was concern about how to keep a tether taut as the astronaut moved away from the vehicle, and how to stop the tether tangling as he ventured further and moved behind another object. The simple answer was not to have a tether at all – which, it was reasoned, could induce contractors to provide more reliable, redundant, or alternative systems, to ensure the safety of the astronaut so that he would not be lost in space. After all, the AMU was supposed to maintain his correct positioning and stability, and render the tether redundant.

However, the fear of an astronaut being 'lost in space' resulted in a NASA memo indicating that on all flights up to Gemini 12, tethers would be used. As the Chief of Crew Systems at MSC, Warren North pointed out: 'Tethers are a spaceman's best friend – especially if you have oxygen in them.'

The flight of Gemini 8 was to raise another point. The EVA planning originally included the evaluation of a non-tethered test flight of the back-pack, but the subsequent directive determined that a tether would be used. The spinning of the spacecraft early in the mission terminated the flight before any EVA could be attempted, but it raised the question: what would have happened if the stuck thruster had occurred during the tethered EVA? The answer was obvious and unpleasant: Scott could have been lashed to the outside of the spacecraft. To obviate such a problem, the Air Force proposed that a quick-disconnect device could be installed on the tethers to prevent their becoming tangled. But NASA countered: how would he get back inside? The agency reasoned that the best plan once such a problem occurred was to return inside as quickly as possible. But the Gemini 8 incident happened so quickly that some doubted that an astronaut could have managed to do so.

The arguments continued, for and against, but NASA decided that on Gemini 9 the AMU would be tethered, and that on Gemini 12 it might perhaps be untethered.

Gemini 9 EVA equipment

Gemini 9 pilot Gene Cernan would wear the AMU on his back. Tests had revealed that temperatures reaching 1,300° F could be experienced on the legs from the lower

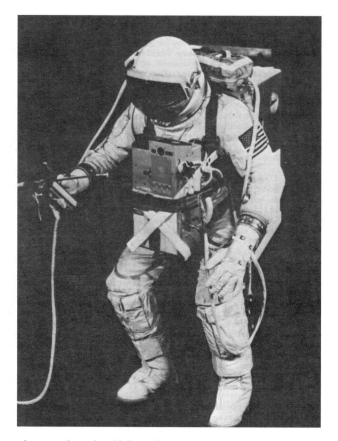

The EVA equipment that should have been worn by Scott during the planned but cancelled Gemini EVA.

forward- and downward-firing thrusters and the suit therefore incorporated several extensive modifications, mostly to the cover layer. The HT-1 high-temperature nylon was not recommended for use above 500° F, and so a stainless steel fabric was incorporated into the leg area of the suit cover layer for added protection. Other tests had indicated that internal temperatures exceeded the melting point of the aluminised Mylar, and so aluminised H-film, separated by layers of fibreglass cloth, was used to solve the excess heat problem. There were 22 layers (eleven of each) of alternating H-film and fibreglass cloth. The standard EVA cover layer was used for the upper torso assembly.

The pressure-sealing visor for Cernan's helmet was made of polycarbonate material, which had ten times more impact resistance than did Plexiglas. This eliminated the need for an impact visor in the Sun visor assembly, allowing the helmet to be redesigned to accommodate a single gold-coated Plexiglas visor to reduce both visible and infrared energy.

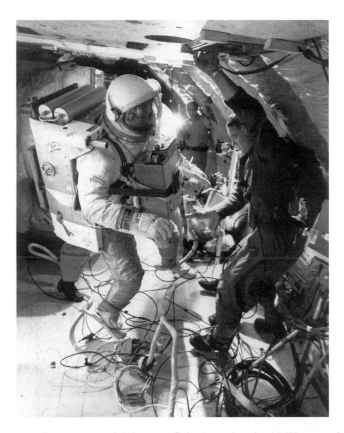

Gene Cernan practises 'zero-g weightlessness flying' wearing the AMU system he was to evaluate for the Gemini 9 mission, during one of the many parabolic flights onboard a modified aircraft during his EVA training programme. Unfortunately, he did not conduct this type of test during the mission.

The Astronaut Manoeuvring Unit

The AMU (DoD experiment D012) was a compact unit weighing 166 lbs and measuring 32 inches high, 22 inches wide, and 19 inches deep. It was located in the Adapter Module at the rear of the spacecraft. To put on the unit, Cernan had to make his way across the side of the spacecraft to the rear and remove a thermal curtain to reveal the unit, while wearing the ELSS chest-pack and connected to the spacecraft by a 25-foot umbilical tether. He then had to unfold the controller arms and nozzle extensions to protect the helmet and shoulder area from plumes of the firing jets. Finally, he had to ease into the back unit with the help of hand-rails and foot-rails, and then strap on the unit before powering it up and checking its systems, fixing a ten-foot tether, and then releasing it.

The unit consisted of independent systems for life support (by means of 7.3 lbs of oxygen stored in a supply tank at 7,500 psi and supplied to the astronaut through the chest-pack); communications (which included a telemetry and voice system), power

Ed Givens – AMU Project Officer with the USAF Space Systems Division, Detachment 2, at MSC – is assisted into a Gemini EVA suit prior to conducting vacuum chamber tests with the unit at MSC early in 1966. Givens was selection to the NASA astronaut programme as one of the nineteen Group 5 members in April 1966. Tragically, he was killed in an off-duty car accident in June 1967.

supply (supplied from two independent sets of silver zinc batteries, $+28.5$ V and ± 16.5 V, for redundancy); and an audio/visual alarm system. Propulsion consisted of hydrogen peroxide gas supplied to twelve small nozzles mounted in each of the corners of the AMU, controlled from the arm controllers for manual and automatic stabilisation. 24 lbs of hydrogen peroxide was supplied, providing 2.3 lbs thrust from each thruster. The system utilised thrusters in pairs: two each for forward, aft, up and down. The forward and aft thrusters were used in balanced pairs for movement forwards and aft and for pitch and yaw, and the up and down thrusters were used for vertical movement and roll control. The system included manual or automatic control on a three-axis stabilisation and attitude control basis, and a two-axis control in manual only.

The left-hand controller had controls for translation commands, the mode selection switch, and voice communication/VOX disable switch. The right-hand controller covered attitude control.

All fogged up

It took almost four hours to go through the eleven-page checklist to prepare for Cernan's exit into space on 5 June 1966. It was the third day of the Gemini 9 mission and the EVA had already been postponed due to the tiredness of the crew after performing three rendezvous with the ATDA during the first two days. Tom Stafford had wanted to move as far away from the target as possible before sending Cernan outside, as he did not wish to station keep with the ATDA and monitor Cernan at the same time. There was nothing that could be accomplished during an EVA that would have freed the stuck shroud, and so permission was received to delay the EVA to Day 3.

Inside the crew compartment, pre-EVA tasks included unstowing the 25-foot umbilical, which led to an obvious comment from Stafford: 'We've got the big snake out of the black box.' Working slightly ahead of schedule, the hatch was opened at 49 hours 22 minutes into the mission, and Cernan's first tasks were to stand on the seat and to retrieve the S12 micrometeorite impact package, deploy the hand-rails on the Adapter Module, and set up cameras.

Everything seemed to take longer in zero g than during training, although Cernan had no sensation of disorientation or of being lost in the black of space. Indeed, he admired the view of Earth when the umbilical stopped trying to wrap itself around him. 'Boy, it sure is beautiful out here,' he observed, looking at the view extending from San Francisco to Mexico. Stafford was holding his Pilot by the ankles, as Cernan enjoyed the view and set up a rear-view mirror so that Stafford could follow his progress to the back of the spacecraft to put on the AMU.

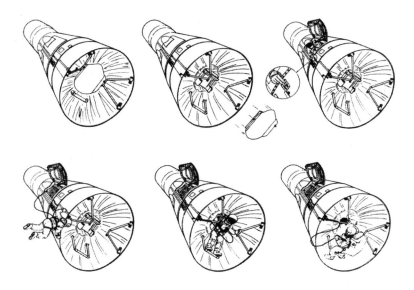

The Gemini 9 EVA profile, showing how Cernan would traverse to the rear of the spacecraft, don the AMU, and then release the unit for a period of tethered flight. The astronaut achieved all but the last part of this exercise during his mission.

Cernan's next task was to exit into space, and to begin what was called 'umbilical dynamics' to further examine what Ed White had investigated, using the tether to manoeuvre in space. Relative to Gemini, Cernan had no sensation of disorientation or of the Earth turning far below. He soon found out, however, that every motion resulted in an opposite one – as determined by Isaac Newton. The slightest movement sent him reeling towards the limit of the tether, and once there, he rebounded back again. Simply twitching his fingers would send his whole body in motion around the umbilical. With no HHMU he could not counter the movements, and began losing the battle with the umbilical 'snake'. Cernan commented that the umbilical prevented him from going where he wanted to go, and curled all over him. As he later wrote in his autobiography (published in 1999): 'I fought [the umbilical] for about thirty minutes before deciding this snake was perhaps the most malicious serpent since the one Eve met in the Garden of Eden.'

Having exceeded the record for duration outside of a spacecraft, Cernan gradually moved hand over hand to the rear of the spacraft to attach the AMU to his back. Through the small window of the spacecraft, the view of the infinity of space was limited by the 8 x 6-inch pane; but now, outside for the next couple of hours, the experience was unimaginable, as Cernan later wrote. 'Try to imagine a place with no boundaries, a room with no walls, an empty well as deep and limitless as your imagination; for that was where I was, and it was going to be my home for the next few hours ... wow!'

Moving back to the Adapter Module, he found that the installed hand-rails were inadequate and that the Velcro could not keep his body in place as he fought against the immobility of his pressure suit. The pressure had ballooned the suit, making movement difficult, and he was fighting against the pressure with every move. As all spacewalkers would discover, you do not *wear* a suit on EVA, but rather, float inside it; so it was a constant battle to set the suit in the required position. With the added thermal protection, Cernan observed that his state-of-the-art EVA suit was about as flexible as a rusty suit of armour.

At last, Cernan reached the edge of the Adapter Module, where he discovered a rough jagged edge that had been created by the separation of the Titan during launch. This had been observed before, and stand-off supports for the umbilical were provided. But it was not until he saw the actual saw-tooth jaggedness that he appreciated its potential danger, and now he had to negotiate past it without cutting the tether or slicing his suit. As Cernan progressed into the rear of the Adapter Module, Stafford lost sight of him. Cernan deployed the restraining bars to secure himself while putting on the AMU.

His heart-rate was 155 beat per minutes, and he was tired and sweating. The added exertion had begun to cloud his visor, and he needed to rest. Working to prepare the AMU took much longer than it had done during training, but Cernan slowly went through the 35-step check-list, and could not understand why he had so much difficulty in space after completing all these manoeuvres in the zero-g aircraft. Sweat floated into his eyes, and his feet kept slipping out of the foot bars, but at last he was ready to fly. He flipped the switch to receive oxygen from the unit, isolated from the spacecraft supply. Breathing from the back-pack, Cernan was the first

During his EVA, Gene Cernan took this stunning photograph of Gemini 9. The nose of the spacecraft, the umbilical, and the open hatch are clearly visible.

person to cut off connection to his spacecraft. All he had to do was release the unit, and he would be flying the AMU.

However, he had to rest to relieve the fogging of his visor; but as soon as he exerted himself, so the faceplate again fogged up. Although the faceplate was cold, the small of Cernan's back was now scalding hot. Later examination of his suit revealed that his exertions had ripped the rear seams of the suit, and that the heat of the Sun had penetrated the remaining layers and burned his back.

Communications were garbled, and Stafford had difficulty talking to his colleague at the rear of the Gemini. Stafford could not see Cernan at all, but was aware that everything was taking four or five times longer than in training, and he was about ready to terminate activities with the AMU. They agreed to review the situation after the next orbital sunrise, which provided Cernan with more time to rest.

Cernan admitted that he could not see in front of his eyes, and as he expanded energy equivalent to running up a flight of 116 steps each minute, his heart raced to three times normal. On the ground, the medics were concerned, as was Stafford inside Gemini. At the back of the spacecraft, Cernan was frustrated, and did not wish to admit that he might have to cancel the AMU flight and waste all the work he had done by being unable to complete his task. Despite his tiredness, he still wanted to release the unit. While resting, unable to wipe the inside of the fogged visor, he used his nose to clear a small gap to look through, and realised that space, as beautiful as it was, could still be very dangerous.

After reviewing the situation, Stafford ordered Cernan to reconnect to the spacecraft systems and return to the hatch. It was easier to get out of the AMU than to get into it, but it was still time-consuming. Cernan climbed back over the lip of the Adapter Module and made his way back to the hatch, leaving the $10 million AMU in the Adapter Module to burn up on re-entry. Raising the Sun visor and resting had provided about 25% vision through the faceplate – but not for long, as his exertion again fogged it. The suit's cooling system was eventually overloaded, and he began to heat up inside the suit; and his faceplate was completely fogged.

'Getting in no problem'
At the hatch, Cernan found it extremely difficult to bend the suit to get into the cramped compartment, and with the umbilical present, it was difficult for Stafford to hold on to Cernan and pull him down to shut the hatch. During the struggle to get in, Stafford radioed to the ground saying, 'Getting in no problem,' so as not to alarm the medics. Cernan had to be forced almost double to shut the hatch, and Stafford used a broomstick-like handle to reach the locking device from his position. It took 17 minutes to return Cernan to the seat and to secure the hatch.

Stafford put his visor against Cernan's, but still could not see inside. Eventually, with the cabin repressurised, Cernan removed his helmet to reveal a flushed red face, which Stafford sprayed with water from the jet gun. He had been outside for 128 minutes instead of the planned 167 minutes – almost 1½ orbits, covering 36,000 miles – and although he had not flown the AMU, he had discovered that EVA was extremely difficult.

During an interview in 1988, Cernan explained that had he proceeded to undock the unit, he thought he would have found it extremely difficult to redock, disconnect from it and return to the spacecraft systems. In one plan, he was to attempt removal of the AMU at the front of the spacecraft. Holding on to the docking bar on the nose, he would reconnect to the spacecraft umbilical and disconnect from the AMU. But he felt that trying to hold onto the spacecraft with one hand while working the connections would have been very difficult. Cernan found that maintaining position was the main problem, as he had only a bar and two stirrups for his feet, which were not held firmly. His greatest concern, however, was how to complete the job, as no-one had had sufficient experience to prepare, or act upon, what was discovered during EVA.

The Gemini 9 EVA was, however, deemed a success, and revealed the problems of maintaining position in space, even for a simple task. Gemini 9 was also a milestone in EVA development: it initiated underwater simulations that would be used for EVA training; liquid cooling for suits was investigated, and anti-fogging methods for helmets were evaluated; and the use of tethers and restraints evolved, all to the benefit of later missions and programmes.

EVA AT AGENA

The next EVA was a more straightforward affair. Gemini 10 Pilot Michael Collins would perform a stand-up EVA to conduct astronomical observations and photography. Even so, he still had a 131-step check-list to go through. The EVA suit was similar to that which Scott would have used on Gemini 8, with small amendments. The polycarbonate pressure-sealing visor of the Gemini 9 configuration was included, as was a single lens Sun visor, modified to allow attachment by Velcro instead of metal pivots. The arms and legs of the underwear were removed, the torso seams were strengthened, and fingertip lights were included on the thermal gloves, so that the hands could be seen on the dark side of the orbit. A red lens was included to avoid damage to photographic film being used in the experiments. In addition, as a result of the experience on Gemini 9, each crew-man carried an anti-fogging kit for use during EVA preparations, including wet wipes saturated with a visor anti-fog, and cleaning solution.

Irritation of the eyes

The stand-up EVA began at GET 23 hours 24 minutes, and for 30 minutes Collins followed the flight plan, taking photographs of the stellar targets. He was amazed by the sight of the stars from his vantage point of the open hatch: 'My God, the stars are everywhere, above me on all sides, even below me somewhat, down there next to that obscure horizon. The stars are bright and steady [not twinkling as a result of their light passing through the atmosphere]. This is no simulation, this is the best view of the universe,' he would write, eight years later.

Immediately after the sunrise, Collins' eyes began to water, even though he had lowered his Sun visor. He thought that the Sun had vapourised the anti-fogging solution he had applied to the inside of the helmet, causing a reaction in his eyes. When he reported this to Young, his Commander replied that his eyes were also watering. By then, both of them were experiencing difficulty in seeing the instruments and check-lists, and they therefore decided to terminate the EVA after 39 minutes, instead of the planned 115 minutes.

Back inside, Mission Control 'asked a hundred questions', but they were not sure what had caused the irritation. The crew suspected that it was caused by leaking granules of the lithium hydroxide that was used to absorb carbon dioxide. Engineers discovered that the problem was in having simultaneous use of both compressors in the oxygen supply loop to the suits. The irritation gradually subsided, and after a depressurisation check it was decided to use only one compressor during the umbilical EVA.

The second EVA of the mission began at GET 48 hours 42 minutes, immediately after Young had rendezvoused with the Agena 8 target. Upon receiving the 'go' for EVA, Young replied, 'Glad you said that, because Mike's going outside right now.' Like Cernan before him, Collins discovered that everything took much longer to perform in space than during simulations. After collecting the S12 micrometeorite collection package from the exterior of the Adapter Module, he made his way back to the nitrogen valve connection next to thruster 16, which would supply the HHMU. At the hatch area, Collins hung on while Young manoeuvred Gemini to within about six feet of the Agena.

Collins pushed off from Gemini and grabbed hold of the Agena docking collar lip – the first time anyone had manually contacted another orbiting object. Cernan had

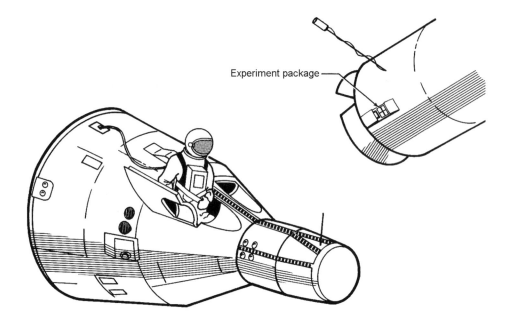

Experiment package

An artist's impression of Michael Collins' EVA at Agena 8 during the Gemini 10 mission.

told him the next stage would be difficult and, indeed it was. As he made his way around the collar, Collins lost his hold and drifted about 16 feet away from the Agena. Using the HHMU, he approached the target, and this time grabbed a handful of wires and cables on the target to secure himself as he removed the S-10 micrometeorite collection package from the Agena. The plan was to replace the package with a new one, but to do that he would have to let go of either the cables or the recently retrieved older package. Collins decided to not install the new package, but to return the older one to the Gemini. He used the umbilical to pull himself back to the hatch area, and gave the package to Young.

The umbilical had been restricted so that it would reach only 20 feet, but was now released to allow full extension to 50 feet for a full evaluation of the HHMU. Unfortunately, excessive use of Gemini's propellant early in the mission had reduced the levels to minimum for station keeping, and so the EVA had to be terminated. Collins, assisted by Young, then spent several minutes trying to get himself and the 50-foot umbilical back into the crew compartment in order to shut the hatch. With the hatch closed, the scene inside the cabin reminded Young of, 'the snake house at the zoo. It looked like a Sunday school picnic.' He added that with 50 feet of hose around Collins, 'we may have difficulty getting him out.'

Fifty minutes later, with Collins back on the spacecraft life support systems, the hatch was opened again for three minutes to jettison a duffel bag, crammed full of the ELSS chest-pack, the umbilical, empty food bags and anything else that was not needed. Collins had also reported inadvertently losing the 70-mm still camera during the EVA, and with it the photographs from the spacewalk. It had been knocked loose, and had drifted off into space.

Lessons learned
Collins had spent 1 hour 29 minutes in two periods of EVA, and had discovered, like those before him, that preparation for EVA on orbit was an important task requiring the attention of both crewmen. Trying to perform rendezvous and EVA preparations together caused them to rush, and reduced Young's ability to assist Collins. Unsecured equipment tended to drift away during EVA, and the 50-foot umbilical was a bigger inconvenience than anticipated. After stowage and handling after EVA, it was an inconvenient and undesirable item in the cramped spacecraft.

On the positive side, the tasks of crew transfer and equipment retrieval were accomplished in a deliberate way and with no increase in workload, and formation flying was a combination of thruster control by Young and EVA activities by Collins, working as a team. Despite the eye irritation and the lack of hand-holds (again), the EVA was a success, and raised confidence after the exploits of Gemini 9.

Penultimate EVA
Dick Gordon's primary task on Gemini 11 was to attach the 100-foot tether to the docked Agena, and to make more extensive use of the HHMU. A stand-up EVA was also planned, to carry out ultraviolet photography from the open hatch. The experiences of Gemini 9 had highlighted the need for adequate training for EVA. After that flight, the neutral buoyancy facility (water tank) technique was beginning to be recognised, and was used both to train the EVA pilot for Gemini 12 and to reproduce the difficulties experienced by Cernan during his spacewalk, to determine whether underwater training provided an effective simulation of real conditions in orbit. Initially, there were reservations about the use of a water tank to train for EVA, but after Cernan – as back-up Gemini 12 Pilot – evaluated the tank, and reported that the experience closely simulated his experiences in space, interest in water-tank EVA training gradually increased. However, Gordon was not scheduled to use the facility in his EVA training. Thirty five years later, the water tank – or Water Emerson Training Facility (WETF), as it is called – has become an integral tool in EVA training for American crews, and similar devices have been used by cosmonauts in their space-station EVA training.

Gemini 11 EVA equipment
The suit was basically the same as that for Gemini 10, apart from some refinements on the wrist locks, neck ring and pressure sealing zipper. A disconnect assembly was added to the suit pressure gauge to prevent fogging during EVA.

The umbilical had been shortened from 50 feet to 30 feet to reduce stowage and handling problems, and the HHMU was stowed in the Adapter Module, to be retrieved during EVA. Also in the Adapter Module was a moulded overshoe-type foot restraint that was to be evaluated during tasks performed in that area.

'Ride 'em cowboy!'

The Gemini 11 crew had allowed four hours to prepare for the first EVA, scheduled for the second day of the mission, but they had trained so much to complete this that they were ready in 50 minutes. Conrad realised that they would not need all of the allocated time, and he therefore halted preparations just prior to Gordon opening the hatch and stepping outside; and they sat there 'with all the junk on,' as Conrad later recalled. While waiting, they decided to perform oxygen flow tests, which dumped unwanted oxygen into the cabin and which then had to be vented into space. This was an ill-advised excess use of precious reserves, and so the tests were halted – which pleased Gordon. His suit was warm during the test, because the heat exchanger was designed for use in a vacuum, not under pressure.

At last, it was time to open the hatch; but, just beforehand, Gordon had difficulty attaching his visor, and Conrad could not reach the far side of the helmet to help him. Realising that they should have done this earlier, they rushed to ensure that the visor was attached, which caused the Pilot to become hot and bothered while trying to snap the visor in place. In his rush, he cracked his helmet visor at the same time.

At GET 24 hours 2 minutes the hatch was opened, with an already tired Pilot beginning his first EVA. The out-gassing of the compartment made everything float out into space, including Gordon, who had to be held down by Conrad, using a leg-strap hand-hold. A few simple activities followed; deployment of the handrail,

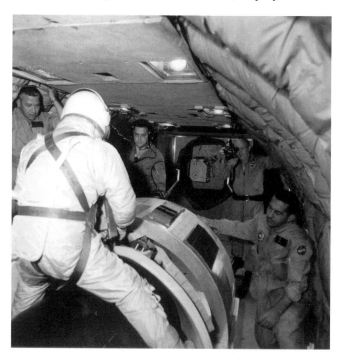

During a zero-g parabolic aircraft flight, Dick Gordon straggles the Agena by jamming his legs into the docking collar on this mock-up.

Dick Gordon's activity during the Gemini 11 mission prompted Command Pilot Pete Conrad to yell 'Ride 'em Cowboy!' as the spacecraft and astronauts flew 160 miles above the Atlantic Ocean.

retrieval of the S9 experiment; and the installation of a camera, which, with Gordon floating above it, was proving stubborn to fix in the mounting – until he hit it with his fist.

Gordon next moved to the nose of the spacecraft to install the 100-foot tether from the Agena to Gemini. On his first attempt, he missed his target and drifted of into space, in an arc above the docking adapter and towards the rear of the spacecraft. Conrad had only released a few feet of the umbilical and was able to pull on it to help Gordon to return to the hatch and try again. This time he reached the docking area and grabbed some handrails to straddle the spacecraft's rendezvous and recovery section at the nose of Gemini. Here, Gordon wedged his feet and legs between the docking adapter and the spacecraft (as he had done during simulations in the KC-135 aircraft), so that he could maintain position, allowing him to handle and attach the tether with both hands free.

'Ride 'em cowboy!' shouted Conrad, looking at the scene in front of him, as Gordon rode on the spacecraft at 17,500 mph. Once again in space, the astronaut found that the training methods used on Earth did not work so well in zero g, and for six minutes he struggled to attach the tether, and expended a great deal of energy. What had seemed a relatively straightforward task in training was quickly sapping the energy of the Pilot. His face streamed with sweat, which floated into his eyes, and he was breathing heavily, groping blindly about him while trying to secure his position on the spacecraft.

At the hatch, Gordon rested while the two astronauts evaluated the situation. The next task was the trip to the back of the spacecraft adapter for the HHMU. But Conrad knew that Gordon was still exhausted, and his right eye was still stinging from the perspiration. With an unknown expenditure of energy still required to complete the remaining tasks, Conrad elected to terminate the EVA and bring back Gordon.

When the spacecraft came back into communication range with the ground, Conrad radioed Mission Control with the news that he had, 'brought Dick back in. He's got so hot and sweaty, he couldn't see.' With the termination of the EVA, several experiments were also lost, including the D16 Power Tool evaluation, planned from the aborted Gemini 8 and brought forward from Gemini 12. The closing of the hatch went unhindered, and an hour later the hatch was again opened to jettison unwanted equipment. The umbilical EVA had lasted just 33 minutes, instead of a planned 107 minutes.

Asleep hanging out the hatch

At GET 46 hours 6 minutes, the hatch was opened again – this time for a stand-up EVA. Attached to the short tether, Gordon was able to use both hands to operate the S13 ultraviolet astronomical camera, and with Conrad maintaining the position of the docked Gemini–Agena, he was able to obtain a series of excellent photographs. When they were told they had enough oxygen, they rested until the next night-pass to take more photographs. From their vantage point they became orbital sightseers looking down at Houston and the Florida peninsula. With nothing to do they both took a short sleep, with Gordon hanging by the harness in the open hatch, and Conrad sitting inside the spacecraft. CapCom stated that they had just recorded a new space first in being the first crew to sleep in a vacuum. Inside the spacecraft, with the hatch closed after 2 hours 10 minutes, Gordon indicated that his legs were tired, due to maintaining his position for the experiment rather than from physical exertion on the first EVA.

The EVA Review Board

Prior to the final flight, the Gemini Mission Review Board became an EVA Review Board, evaluating all that had gone before and the status of the USAF AMU planned for Gemini 12. The first decision was to delete the AMU from this mission because of the problems encountered on Gemini 9 and the risks involved in its use, and on 30 September 1966 the USAF received official news of what it had feared for some time: the AMU would not fly on Gemini 12.

On Gemini 10, Collins had had relatively few problems, which provided fresh hope that the unit might after all fly. But after Gemini 11, hopes were soon dashed. Despite incorporating all the changes and recommendations from past EVA experiences, Gordon still had problems, and there was obviously more work required regarding basic EVA principles before moving on to testing advanced systems such as the AMU. George Mueller told the Air Force that despite incorporating the lessons learned from past EVAs to the subsequent, more complex activities, problems had still occurred. The last mission therefore had to be devoted to EVA fundamentals. The AMU was pulled from Gemini 12, and would not fly in space again. In 1973, a manoeuvring unit was flown inside the Skylab space station by the second and third crews, but it was not until 1984 that the much advanced Manned Manoeuvring Unit – a descendent of Gemini's AMU – finally flew in space on three separate missions. It performed flawlessly, untethered. Then in 1990, the Soviets also demonstrated their tethered YMK manoeuvring unit from Mir. Smaller rescue back-packs – SAFER units – are now used on ISS EVAs, and three decades after the AMU was assigned to Gemini, its descendants continue to be employed, proving the basic concept of the USAF device.

The finale
The EVA on Gemini 12 was planned to evaluate body restraints and workloads for a series of representative tasks, as well as stellar photography and the attachment of a tether. The Adapter Module was fitted with a new workstation, a second workstation was fitted to the Agena target docking adapter, and Aldrin was provided with several new restraints and tools.

Gemini 12 EVA equipment
Aldrin's suit was essentially the same as the suit worn by Cernan on Gemini 9, except that the stainless steel fabric on the legs was replaced by high-temperature nylon. Four layers of aluminised H-film and Fiberglas cloth super-insulation was omitted from the suit legs. The cover layer thermal layering was quilted, and was turned 90° to the first layer of the anti-micrometeorite material, so that the chance of tearing or ripping would be reduced.

During the first period of (stand-up) EVA on Gemini 12, Buzz Aldrin removes the micrometeorite package for return to the spacecraft.

Training for the EVA was highlighted by the extensive first use of the NBS water tank. Aldrin described the advantage of the water tank after Gemini 12 flew: 'A medium that has considerable advantages over the zero-g aircraft in that we can time things ... [and] look at the entire (EVA) flight plan. It has disadvantages also, in that there are buoyancy effects ... [but] these are minor in looking at the whole underwater situation. It is an excellent training device, and we should attempt to make use of it as much as we can.'

Opening the door
The first stand-up period on Gemini 12 began at 19 hours 29 minutes GET, and was similar to those conducted on Gemini 10 and Gemini 11. Aldrin conducted the S13 Ultraviolet Astronomical Camera experiment and synoptic terrain photography. He also retrieved an S12 Micrometeorite Collection package, and conducted several activities designed to allow him to become familiar with the environment and to prepare items for the umbilical EVA. Aldrin mounted the EVA sequence camera, then took it down and fixed a handrail from the hatch that would help him to reach the docking area. After 2 hours 20 minutes the EVA had been completed without any difficulty.

The final steps
The final umbilical EVA took place the following day, and began when Aldrin opened the hatch at 42 hours 48 minutes into the mission. He installed the camera as easily as on the previous day, and moved to the Agena to attach the gravity-gradient tether between the two vehicles. With hand-rails and waist tethers to secure him, he encountered none of the difficulties to which Gordon had been subjected on the previous mission. The EVA progressed to deploy the S10 Micrometeorite Collection package on the Agena, and Aldrin then moved to the rear of the spacecraft for the evaluation of tools and restraints.

Tools, tethers, hand-holds and slippers
Most of the previous EVAs had included the use of some sort of hand-rail to help the astronaut move around, but the hardware was either insufficient, inefficient, or was not fully deployed, or else the EVA astronaut encountered difficulties which hampered the correct use of the facility. Rectangular hand-rails were installed on the spacecraft Adapter Module. These lay flush at launch, and were automatically deployed 1.5 inches from the Adapter Module after spacecraft separation, although on Gemini 10, they did not fully deploy. To move into the Adapter Module, two large cylindrical hand-bars were provided in the Equipment Module – again deployed on spacecraft separation. On Gemini 9 and Gemini 11, when the astronauts had to use the nose thruster nozzles as hand-holds, it became evident that some method of traverse aid was required on the nose area of the spacecraft. Fixed hand-holds (6 inches long and an inch in diameter) were provided on the back of the Gemini 11 Agena docking cone, which helped Gordon hold on, but there were no foot restraints available. Flexible Velcro-backed portable hand-holds were evaluated on Gemini 9, when Cernan wore Velcro nylon 'eye' pads on his gloves, and attached

Buzz Aldrin's face is clearly visible with the Sun visor up during one of the Gemini 12 EVAs in November 1966. After a successful demonstration of EVA techniques, Aldrin completed the final Gemini EVAs during the last mission of the series.

Restraint devices used during EVA

Configuration	GT-9	GT-10	GT-11	GT-12
Regular handrail	x	x	x	x
Large cylindrical handbars (1.38" diameter)	x			x
Small cylindrical handrails (0.317" diameter)			x	
Telescoping cylindrical handrail				x
Fixed handhold			x	x
Flexible Velcro-backed portable handhold	x			
Ridged Velcro-backed portable handhold				x
Waist tethers				x
Pip-pin handhold/tether attachment device				x
Pip-pin anti-rotation device				x
U-bolt handhold/tether attachment device				x
Foot stirrups	x			
Foot restraints				x
Stand-up tether		x	x	x
Straps on spacesuit leg			x	x

across the spacecraft were 80 pads of 'hooks'. However, he could not exert enough force on the pad for good contact, the elastic hand attachment was insufficient and was pulled off, and launch heating damaged one unprotected patch.

It was clear, due to these experiences, that more effort was required in improving the restraint and mobility aids outside the spacecraft, and this became the focus of Gemini 12 EVA plans.

At the back of the Adapter Module, Aldrin slipped his feet into a pair of overshoe restraints – called Golden Slippers, because of their gold finish. These 'locked' the feet in a firm platform, and allowed the astronaut to sway 45° left and right. Unlike Cernan (on Gemini 9), who had to fight to keep his feet in the stirrups, Aldrin could take a break from his work and lean 90° backward to admire the view, instead of waiting for his ELSS to work overtime to cope his with physical excursions.

He next began work at the workstation (the 'busy box'), where he spent time evaluating waist tethers, tightening bolts, cutting metal and (empty) fluid lines, hooking rings and hooks, stripping patches of Velcro, and determined the efficiency of the foot restraint in maintaining his position as he worked. At one point he lost a bolt that had slipped from the workstation and had to lean back to catch it, reporting to Lovell that he 'had to do a little rendezvous there.'

He then made his way back along the spacecraft to the second workstation on the Agena. This time, Lovell could observe his Pilot as he again evaluated tethers and foot restraints and used a torque wench planned for Apollo. He also pulled electrical connections apart and then put them back together, using both waist tethers, then one on one side, then the opposite one, then none at all. He provided a running commentary and observations, comparing the work with that of the underwater simulations, and noting what worked and what did not work.

On the way back to the hatch, Aldrin wiped Lovell's window with a cloth; and as he did, Lovell asked if he would change the oil too, to which Aldrin replied that 'the air in the tyres' seemed fine. The EVA ended after 2 hours 6 minutes, and clearly demonstrated the need for efficient training devices such as the water tank. With this, and with adequate body and foot restraints, all the tasks attempted were feasible. This showed that the EVA workload could be controlled within the limits of the suit system and the astronaut, and that with proper planning, training and application, EVA could be a valuable asset to future manned spaceflight.

The final exit
The successful umbilical EVA was followed on the fourth flight day by a final hour-long stand-up EVA, during which Aldrin took UV photographs of constellations, made observations, and tossed overboard unwanted equipment in a stowage bag that contained the umbilical, empty food bags and other unwanted equipment, to burn up in the atmosphere.

Without ceremony, the hatch was closed on Gemini EVA operations. Aldrin had set a record of almost 5 hours 30 minutes in three EVAs, and had finally proven that working in a vacuum was achievable given adequate preparation and controlled execution. After the frustrations of past missions, Gemini finally pointed the way to

Extravehicular activity

Mission	Date	Astronaut	Life support system	Umbilical length (ft)	Manoeuvring device	Umbilical time (hr:min)	Stand-up time (hr:min)	Total EVA/mission time (hr:min)
GT-3	*None planned*							
GT-4	1965 Jun 3	White	VCM	25	HHMU	0:36		0:36
GT-5	*Cancelled, 1965 July*							
GT-6	*Cancelled, 1965 July*							
GT-7	*Cancelled, 1965 July*							
GT-8	1966 Mar 17	Scott	ELSS-ESP	25	HHMU	Early termination of mission prevented EVA		
GT-9	1966 Jun 5	Cernan	ELSS-AMU	25	AMU	2:07		2:0
GT-10	1966 Jul 19	Collins	Spacecraft	–	–		0:50	
	1966 Jul 20	Collins	ELSS	50	HHMU	0:39		1:29
GT-11	1966 Sep 13	Gordon	ELSS	30	HHMU	0:33		
	1966 Sep 14	Gordon	Spacecraft	–	–		2:10	2:43
GT-12	1966 Nov 12	Aldrin	Spacecraft	–	–		2:18	
	1966 Nov 13	Aldrin	ELSS	25	–	2:06		5:30
	1966 Nov 14	Aldrin	Spacecraft	–	–		1:06	
Totals: five astronauts; nine EVAs						6:01	6:24	12:35

EVA times are normally recorded as hatch opening to hatch closing, and include equipment ejection duration.

more complex operations in Earth orbit, and boosted confidence in planning for Apollo. The earlier EVAs had suffered difficulties, setbacks and disappointments, but overall the missions provided a wealth of data upon which to build for the future.

Re-entry and landing

NASA's first spacecraft featured a blunt-bodied capsule, but from Project Mercury and the designs through to Apollo they were not expected to provide long-term access to space. In the USAF, it was envisaged that routine spaceflight would be undertaken by a vehicle that could be launched either by rocket or by its own power, with the ability to use attitude control thrusters and manoeuvring engines to change orbit in the vacuum of space, and then, at the end of its mission, to control its re-entry and landing by manoeuvring to a designated landing area, to be refurbished and reused.

DEVELOPMENT OF THE PARAGLIDER

Controlled re-entry

When Mercury was authorised in 1959 it was an engineering project designed to achieve the goal of orbiting a manned spacecraft at the earliest practicable date, using the best technology available: a small ballistic re-entry capsule – cone-shaped and with a blunt body – to be launched by a missile, and after re-entry and descent by parachute, to be recovered from the ocean – a one-shot spacecraft for which there was very little capability of a controlled landing to within an accuracy of a hundred miles or so.

The development of research aircraft by the USAF had led to the X-15, which was used to study flight at the edge of the atmosphere (at an altitude of about 50 miles) up to Mach 6, and also to investigate the problems of returning such a winged vehicle to a desert landing strip at Edwards AFB, California. NASA was also studying the lifting body concept, in which the aerodynamic shape of a wingless vehicle enabled it to generate sufficient lift to allow the pilot a limited amount of control to achieve a landing. Moreover, the USAF plans for the X-20 – the Dyna Soar – envisaged a delta-winged spacecraft, launched by rocket into orbit, and then landing like a winged glider on a runway.

During the development of the ballistic Mercury capsule some thought had been given to a way to improving its lifting capability by adding an external trim flap that would allow the re-entering vehicle to 'skip' through the denser layers of the atmosphere to extend its range and lessen its re-entry loads, thereby producing a lift-over-drag capability.

Rogallo's wing
During the mid-1940s, Francis Rogallo – an engineer at the NACA (later NASA) Langley Research Center, Virginia – began to develop (at home) his own alternative controlled landing concept: a hybrid of a parachute and a wing, to be deployed above the returning vehicle. The structure would consist of an inflatable V-shaped fabric form, over which the lifting surface would be stretched. This 'paraglider' would be much lighter than some of the alternative concepts, and could be used to recover spacecraft returning from orbit under huge heavy wings.

In 1958, Rogallo's idea was brought to the attention of the Langley Committee on General Aerodynamics, which moved development from Rogallo's house to a proper laboratory; and in 1959, STG Manager Robert Gilruth suggested that further study should be directed at a proposed follow-on Mercury using manoeuvring capability for landing. The Gemini paraglider was thus born.

Land landing for Mark II
By 1960 the concept of a lifting re-entry had (in principle) won the approval of NASA Headquarters, and though funds were not available (due to the commitment to a Mercury follow-on programme), the development of lifting re-entry and land landing continued at STG, although the Mercury and newly authorised Apollo programme were of foremost importance.

Engineers at STG had expressed as interest in the paraglider, but doubted whether it could be successfully deployed in flight. They therefore decided to continue with the parachutes proven for safety and reliability for Apollo, but reserved the paraglider as of possible use. It was accepted that an ocean landing was expensive and complex, and that if routine access both too and from space was ever to become a reality, it would not be acceptable for each spacecraft to be dropped into the ocean and then taken to a museum after one flight. Continuing studies and sub-scale tests by Rogallo's team at Langley had provided a considerable amount of data, and demonstrated the capability of the lightweight and controllable paraglider to support a returning manned spacecraft. On 12 January 1961, a Langley technical liaison group on Apollo configuration and aerodynamics concluded that 'if the paraglider shows the same type of reliability in large-scale tests that had been achieved in the sub-scale tests, the potential advantages of this system outweigh other systems.' To further evaluate the system, it was decided to continue the tests with manned and unmanned drop tests.

The STG was convinced that there was still much work to be completed before the paraglider could be seriously considered for use on Apollo. How would the configuration be packaged? Would the astronauts' view from the spacecraft be good enough for them to control the landing? How would such a large structure be deployed successfully? How would the aerodynamics of the structure affect the spacecraft design? All of these questions remained largely unanswered, and STG

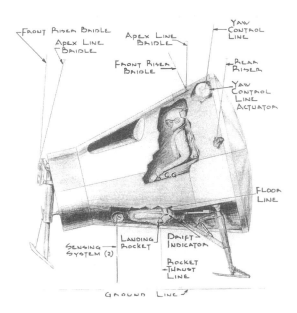

An artist's impression of the proposed Gemini parasail arrangement.

required at least six months of data-gathering to investigate the potential and the problems of such a concept before any contracts could be awarded.

The Mercury Mark II design featured the familiar parachute recovery sequence and an ocean splash-down. The capsule also featured an escape tower which, in the final design, had posed the problem of additional weight, and it was considered that the tower could be replaced with ejection seats that could be used on the proposed Titan II launch vehicle configuration; and where the advanced version of Rogallo's paraglider could be tested.

Engineers at McDonnell were already considering the incorporation of the paraglider in Mercury Mark II, and the Marshall Space Flight Center had issued two contracts for evaluating the use of the paraglider for returning spent rocket stages. With such interest, and with the prospect of a viable system, on 17 May 1961 STG awarded $100,000 study contracts to Goodyear Aircraft Corporation, Ryan Aeronautical Company, and the Space and Information Systems Division of North American Aviation. Each would evaluate the design parameters of a system for future spacecraft to provide manoeuvrability and controlled energy descent and landing by aerodynamic lift, which was soon to be designated Phase I of the paraglider development programme.

With the completion of the Phase I studies in August 1961, the paraglider proved to be a feasible system; and at the same time, work on the Mercury Mark II was sufficiently advanced that the preliminary project plan, dated 14 August, listed controlled landing as the third goal of the proposed follow-on programme. On seven manned flights of Mercury Mark II, the Pilot would be provided with a means of flying his spacecraft towards a limited landing area, by using the spacecraft offset from the centre of gravity to induce some degree of lift in the atmosphere, and by using the roll thrusters to control the amount and direction of lift through the atmosphere as well as to correct errors in the predicted landing footprint. A controlled landing required a means of softening the impact, and the paraglider was ideal for such a role. By the end of the year – as Mercury Mark II became Gemini – its use for land landing was a listed objective of the new programme.

Landing Gemini by paraglider

Early studies established that the paraglider would be stored in the recovery compartment in the nose of the re-entry module. Following re-entry, a drogue parachute would be deployed to pull away the recovery compartment from the Gemini at an altitude of 59,000 feet, and would initiate the deployment of the paraglider from its storage container at an altitude of 49,000 feet.

A 32 x 42-foot paraglider wing, attached to five gas-actuated cable reels – three in the pitch plane and two in the roll plane – would deploy and inflate at 46,000 feet, and would be controlled via differential reel-in and reel-out of the two aft pitch cables and the two roll cables. The control system would be powered by gaseous nitrogen pneumatic motors, actuated at a constant rate in response to control stick

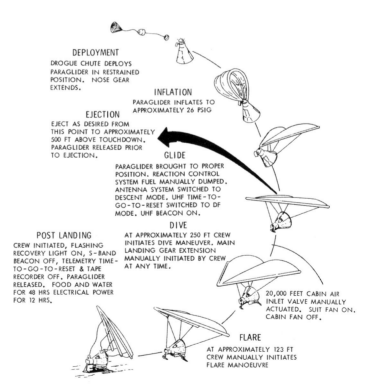

DEPLOYMENT
DROGUE CHUTE DEPLOYS PARAGLIDER IN RESTRAINED POSITION. NOSE GEAR EXTENDS.

INFLATION
PARAGLIDER INFLATES TO APPROXIMATELY 26 PSIG

EJECTION
EJECT AS DESIRED FROM THIS POINT TO APPROXIMATELY 500 FT ABOVE TOUCHDOWN. PARAGLIDER RELEASED PRIOR TO EJECTION.

GLIDE
PARAGLIDER BROUGHT TO PROPER POSITION. REACTION CONTROL SYSTEM FUEL MANUALLY DUMPED. ANTENNA SYSTEM SWITCHED TO DESCENT MODE. UHF TIME-TO-GO-TO-RESET SWITCHED TO DF MODE. UHF BEACON ON.

DIVE
AT APPROXIMATELY 250 FT CREW INITIATES DIVE MANEUVER. MAIN LANDING GEAR EXTENSION MANUALLY INITIATED BY CREW AT ANY TIME.

POST LANDING
CREW INITIATED, FLASHING RECOVERY LIGHT ON, S-BAND BEACON OFF, TELEMETRY TIME-TO-GO-TO-RESET & TAPE RECORDER OFF. PARAGLIDER RELEASED. FOOD AND WATER FOR 48 HRS ELECTRICAL POWER FOR 12 HRS.

20,000 FEET CABIN AIR INLET VALVE MANUALLY ACTUATED. SUIT FAN ON. CABIN FAN OFF.

FLARE
AT APPROXIMATELY 123 FT CREW MANUALLY INITIATES FLARE MANOEUVRE

The proposed sequence of events in the deployment of the Gemini paraglider for land landing of the spacecraft.

movements by the Pilot. This would allow the astronauts to climb, dive or bank as they descended. The Gemini would move relative to the paraglider, and the shifting centre of gravity would move forwards, aft or sidewards as the vehicle manoeuvred, thus increasing or decreasing the rate and angle of descent towards the chosen landing area. The Re-entry Module of the Gemini vehicle was designed with two forward-facing crew positions, which allowed the crew to observe in the direction in which they were heading, with the nose of the spacecraft tipped forward by the configuration of the paraglider.

At about 250 foot above the ground, a tricycle landing-gear arrangement would be manually deployed with a nose skid and two outrigger skids for stability and support. Immediately prior to landing, the nose of the spacecraft would be raised in a flare manoeuvre to increase the lift of the wing and to reduce the rate of descent, thereby allowing a touch-down cushioned by the three skids. After a sliding run-out at about 45 mph, the paraglider would be jettisoned, and the Gemini would draw to a halt.

Initial schedules proposed that the first unmanned Gemini in August 1963 would be recovered by parachute, and that the second mission – as early as September 1963 – would be the first manned mission and the first to use the paraglider. However, the schedule for developing the whole Gemini programme was very tight, and the paraglider still had to be flight-proven. Contingency plans were therefore provided for parachute recovery into the ocean should anything delay the incorporation of the paraglider.

THE PARAGLIDER DEVELOPMENT PROGRAMME

In May 1961 the development programme consisted of:

- Phase I – a series of design studies that would demonstrate the feasibility of the glider as a concept.
- Phase II-A – an eight-month research and development period during which system evaluation would identify the design with the best performance.
- Phase II-B – confirmation of the final design, fabrication of the prototype, and the beginning of unmanned and manned flight-testing.
- Phase III – the production of a flight vehicle and the building of a pilot training vehicle.

During Phase I, the first manned paraglider – called the Flex Wing – was constructed by Ryan. It was fabricated from a flexible plastic-coated wing, which was attached to a central keel and two leading-edge members that formed a V-shaped surface resembling an arrow head. This inexpensive design was mounted above a fuselage and a four-wheel landing gear system. It weighed 1,500 lbs, and was powered by an after-mounted 180-hp Lycoming engine.

Ryan's chief test pilot, Lou Everett, flew the Flex Wing in short 'hops' at the USN Brown Field, Otay Mesa, California, during Phase I tests, which were completed by August 1961. The machine was also tested in Langley's full-scale wind tunnel, which provided valuable data on stability control and the forces on the pilot stick control. On 20 November 1961, North American was awarded the Phase II-A contract after the presentation of its test programme results; but Langley would still support wind tunnel tests, and the vehicle was tested at the Flight Research Center in California.

FRC's paraglider research vehicle

Following the meeting in November 1961, engineers from the Flight Research Center in California decided that they needed flight experience with a Rogallo wing before committing it to returning a spacecraft. FRC research pilots Milton Thompson and Neil Armstrong (soon to be selected to NASA to train for Gemini missions) therefore constructed a single-seat test vehicle (initially in their spare time, but eventually with approval from FRC Director Paul Bikle) designated a Paraglider Research Vehicle (Paresev), and the completed craft – Paresev I – was rolled out in February 1962. It weighed 600 lbs and measured 15 feet in length; it had a 46-square-foot Rogallo wing attached over the top of a 10-foot vertical mast and a V-shaped metal frame; and it resembled a tricycle with a pilot seat.

The tests began by towing the vehicle behind a truck across the dry lake beds. At a speed of 40 mph the vehicle lifted by a few feet into the air, and was controlled by Thompson, who tilted it left, right, fore or aft. After a 'challenging test programme' that included several hundred ground tows, it was decided to tow the vehicle in the air to 5,000 feet, after which it would be released, to be controlled to a dry lake bed landing. In March 1962 Thompson completed the first (difficult) free flight, and succeeded in achieving a safe landing. NASA test pilot Bruce Peterson then tested the vehicle, but was injured when he lost control and it smashed into the ground. It was damaged beyond repair, but was completely rebuilt, and re-emerged as Paresev I-A, with much improved controls. The vehicle was flown by Thompson, Peterson, Armstrong, army pilot Emil Kleuver, LaRC pilot Bob Champlain, and later by astronaut Virgil Grissom.

A later version of the design included a Rogallo wing attached to the vehicle. It was then designated Paresev I-B, and went on to complete more than a hundred free flights throughout the entire programme. It produced important data on how such a vehicle handled in flight, but it differed significantly from the proposal for Gemini, in which a full-scale flexible inflated paraglider would slow a two-ton spacecraft to a controlled, safe and survivable landing.

The North American Aviation test programme

Following commencement of work on the paraglider early in 1962, North American Aviation organised a programme of tests on the proposed system in order to qualify the design for full-scale vehicle demonstrations prior to the assignment of the hardware to Gemini missions. To support the programme, the Ames Research Center began wind tunnel tests on a half-scale inflatable paraglider wing. These tests – begun on 23 May 1962 – evaluated the sequence of deployment of the paraglider, and the aerodynamic effects on the combination of wing and spacecraft during landing, and were successfully completed on 25 July.

The Paraglider Landing System Research and Development Program was carried

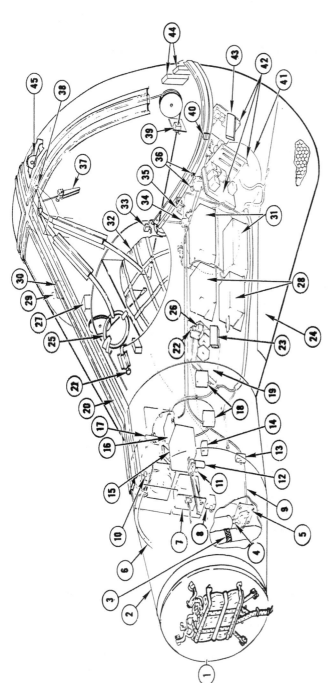

FSTV structure and components: 1) paraglider drogue parachute; 2) R&R can; 3) paraglider (packed); 4) apex fitting; 5) apex load washers; 6) MDF ring; 7) restraint load boxes; 8) camera no.1; 9) nitrogen line (high pressure); 10) forward cable cutter; 11) drogue wire cutter; 12) 'J' box antenna; 13) fuse panel; 14) T/M transmitter; 15) restraint load boxes; 16) electronic package; 17) fuse panel; 18) radio command receiver; 19) relay panels; 20) paraglider cable stowage; 21) forward main parachute attachment; 22) rate gyro; 23) radio command antennae (left- and right-hand); 24) landing pad; 25) stabilisation parachute; 26) accelerometer; 27) instrumentation battery; 28) recovery batteries; 29) camera no.3; 30) camera no.2; 31) deployment batteries; 32) main recovery parachute; 33) overboard dump; 34) inflation valve; 35) shunt; 36) redundant pressure switches; 37) pitch cable cutter; 38) stabilisation parachute attachment; 39) roll cable cutter; 40) pressure transducer; 41) pressure vessels; 42) relay panels; 43) T/M antennae (left- and right-hand); 44) fuse panels; 45) aft main parachute attachment. (Courtesy North American Aviation.)

out at North American from May 1961 through December 1964. There were two different versions of the test vehicle, designed and fabricated to physically simulate the Gemini spacecraft. The Full-Scale Test Vehicle (FSTV) provided systems and equipment that developed the deployment system. Two FSTVs were built, and were sled-launched from a C-130 aircraft, to achieve wing extraction, inflation, deployment, and glide, after which the wing was jettisoned, allowing a parachute deployment for vehicle recovery. The two Tow Test Vehicles (TTV) were used to develop the prototype control system and for unmanned and manned free-flight manoeuvres, flare manoevres and landing development. They were towed to altitude by helicopter, released, and then flown to a landing on the dry lake bed runways at Edwards AFB.

The programmes included a series of half-scale and full-scale wind tunnel tests, 25 developmental flight tests, and four free-flight manoeuvres and landing tests. Other testing included the deployment of the FSTV parachute recovery system through boilerplate drop tests, half-scale free flights, one-tenth scale drop tests, and one-twentieth scale drop tests. The development programme also included research into inflatable wing material and coatings, fabrication processes, and an extensive programme of qualification testing of the frame joints, webbing and cordage. In addition, the control actuators were designed, fabricated and qualification tested. Prior to flight tests, an extensive ground test programme centred on the packaging, stowage and extraction sequence that resulted in 100% success in this phase of paraglider development.

THE TEST VEHICLE PROGRAMME

The Full-Scale Test Vehicle (FSTV) was designed as a prototype test-bed for the development of the flight version of the paraglider wing, and to produce data on glide, manoeuvre and flared landing through remote control. These tests were initially intended to incorporate the tricycle landing gear, but MSC demanded that the expensive test hardware be provided with a system of emergency parachutes to protect the vehicles during the recovery of an unproved design, which would allow them to be used for later full-scale drop tests, and also obviate the construction of more vehicles.

The test vehicle

The boilerplate Gemini outwardly resembled a flight vehicle, with the cockpit floor, side bulkheads and substructure below the floor simulated to allow later installation of the tricycle landing gear. Frames and bulkheads were of welded tri-ten steel, with non-structural and structural doors of aluminium plate, allowing access to internal equipment. These test vehicles were constructed by North American, and not by McDonnell.

Cable and parachute riser line stowage troughs were fixed externally to the structure, while the wing pitch and roll cables were rigidly attached internally. The forward cable was designed to be fed out at a controlled rate during deployment, to reduce dynamic loads. The cables were cut prior to parachute deployment for final recovery. A representative rendezvous and recovery canister was attached on the flight vehicle where the wing stowage container was located.

Initially, a tricycle skid-landing gear was deployed to support the landing, and crushable aluminium alloy honeycomb, filled with balsa wood, was used as an energy absorber. The nose gear featured three telescoping tubes, of which the outer and middle were extended by gas pressure with spring-actuated down locks. At landing, the innermost tube telescoped into the middle tube and crushed the energy absorbing material (the balsa wood). The two primary gears were hinged arms extended by pneumatic actuators, and upon landing this arm rotated against a column of crushable material.

Four 5,000-psig vessels of 1.5 cubic feet provided a gaseous nitrogen supply for wing pressurisation, wing pressure make-up, flight control pneumatic motors, pressure for the forward cable payout control system, and deployment of the landing gear.

A change in objectives resulted in the deletion of test requirements for flight manoeuvres, flare and landing, which would be tested on the TTVs. The landing gear was excluded, and in its place an emergency recovery system was installed as the primary recovery system for the vehicles. It initially consisted of three Mercury parachutes, but was changed to only one Gemini parachute that increased touch-down sink rates from 24 fps to 32 fps. This rendered the impacting gears inadequate for the tests, and they were replaced by impact actuator pads – incorporated within the mould line of the boilerplate – that could withstand a vehicle impact weight of 4,400 lbs, a vertical velocity of 32 fps, an attitude variation at impact of 10°, a zero horizontal velocity, and a maximum deceleration to 15 g. The pads were of bonded honeycomb, mainly due to its lack of rebound characteristics. The FSTV was recovered successfully 25 times without major damage to internal systems or vehicle structure, and several pads were reworked and recycled for use on later tests.

The test profile
The sequence began with sled launch from the C-130 at an altitude of approximately 33,000 feet. The drogue parachute was deployed during the sled vehicle separation, which set the vehicle in a heat-shield-down configuration that maintained stabilisation. A timer jettisoned the rendezvous and recovery nosecone, and this also extracted the wing. The wing was attached to the vehicle at its apex and at the trailing edge by a pneumatic disconnect fitting, to form a U-shape during inflation as pressure increased, stabilising the descending Gemini. At the end of the inflation the three trailing-edge attachments were released, and after a preset delay the apex connection was released, allowing the forward cable to reel out in a controlled manned and the wing/vehicle transitions to glide. At an altitude of more than 10,000 feet the wing was jettisoned, and the 84-foot Gemini recovery parachute was deployed, allowing recovery of the test vehicle at 32-fps vertical velocity.

The drop test programme
Qualification of the emergency recovery system was completed with a half-scale and then a full-scale test vehicle over the US Navy Parachute Facility at El Centro,

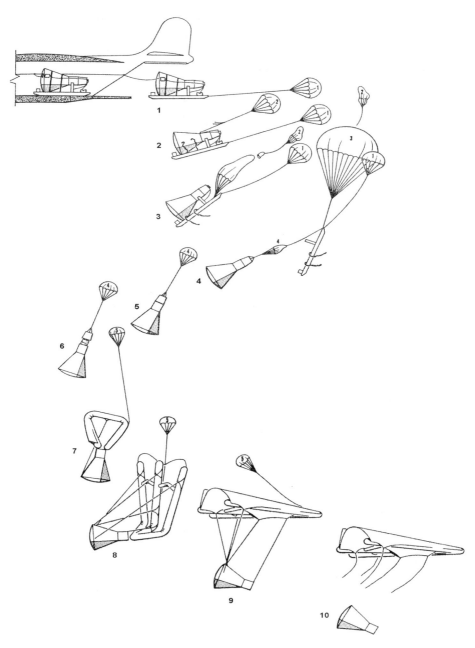

The FSTV deployment and recovery mission sequence: 1) vehicle/sled launch extraction; 2) vehicle/sled separation; 3/4/5) paraglider extraction and drogue deployment; 6) R&R can separation; 7) wing in 'U' configuration and 'Q' parachute deployment; 8) trailing edge release; 9) apex release, 'Q' parachute release, wing deployment; 10) wing jettison. (Courtesy North American Aviation.)

California. The first drop – on 24 May 1962 – used the Mercury single parachute, and was a success; and this was followed by two failures and another success. After a parachute failure on 26 June, the system was improved, but on 10 July another parachute failure entailed further repairs. The last test – on 4 September – was successful, and qualified the system for full-scale tests.

Three parachutes would be incorporated for the heavier full-scale drop tests. The series suffered numerous slips and parachute failures. One parachute was lost on 21 August, two were lost on 7 September – resulting in slight vehicle damage due to an increased landing load – and on 15 November all three failed, resulting in a crash of the test vehicle and a requirement for McDonnell to supply a replacement Gemini boilerplate. The tests were consequently suspended to allow improvements to be incorporated in the back-up system, and at the same time further tests of the paraglider concept were held on half-scale vehicles deployed beneath a helicopter.

Helicopter drop tests

These tests utilised two Half-Scale Test Vehicles (HSTV) at the FRC, where the paraglider was deployed and inflated on the ground, folded in half, tied down, and then towed aloft by an army helicopter. When the planned attitude was reached, ground command would release the tow cable and allow the unmanned vehicle to make a radio-controlled descent to evaluate control characteristics. During the first test, on 14 August 1962, the vehicle could not be released; on 17 August the wing was deployed prematurely, although the vehicle completed a stable glide; and on 23 August the vehicle descended too fast and incurred some damage during landing. The test programme was plagued by minor problems, which were delayed for several weeks; but after a halt to resolve the problems, testing was successfully completed on 23 October.

The next test series would incorporate deployment of the wing in flight, but further problems delayed the first test until 10 December. Unfortunately, the deployment system became entangled, and at 5,000 feet the paraglider was jettisoned in order to recover the test vehicle with the parachute system, thereby justifying the decision to incorporate the back-up system in the programme. Two more tests – on 8 January and 11 March 1963 – failed to demonstrate successful deployment in flight, and both HSTVs were destroyed on impact with the ground.

Delays and redirection

The failure to demonstrate full deployment in flight, using the HSTVs, led to suggestions from NASA HQ that the programme should be cancelled from Gemini. The GPO wanted to assign more time to North American, and to award the company the Phase III contract, but the Office of Manned Spaceflight objected to this proposal.

During a joint meeting on 27–28 March 1963, representatives of North American and NASA evaluated the test programme results to date. It was clear that a paraglider could not be included before Gemini 7, which was then planned for October 1965 at the very earliest. The funds for further tests on Phase II would be depleted by April, and North American awaited NASA's decision.

Suggestions for replacing the paraglider with a parasail were rejected due to its high sink rate and the necessity of firing retro-rockets to cushion the impact of landing, and several of the astronauts were not even confident that the system would even work. Walter Cunningham once observed that 'waiting until the last split second to find out if you would be around for the next second' would not promote confidence in the system. With so much money already spent on paraglider development, it was therefore decided to downgrade the programme to become a research and development project. On 5 May, North American was awarded a contract for the redesigned Paraglider Landing System Program, which would allow design, development and testing with a prototype wing to be completed before beginning the production of such a device. This programme would utilise the two Full-Scale Test Vehicles dropped from the C-135 to demonstrate deployment sequences of the wing but then recovered by Gemini parachute, and a half-scale boilerplate used in the emergency parachute test programme reconfigured as an HSTV, to be towed behind a helicopter to practise take-off techniques for the full-size trainer which was being fabricated by North American. A second trainer would also be built.

Helicopter tows and parachute drops
The HSTV was initially towed behind an automobile to gather data on wing -angle settings and attachment points. A series of 121 tows commenced on 29 July, and it was delivered to FRC in mid-August. On 20 August FRC began a series of 133 ground tows followed, in October, by a series of eleven helicopter tow tests to evaluate the consequences of deploying the wing at higher speeds and altitudes before releasing the vehicle.

Before an FSTV could be used, it was necessary to demonstrate the new emergency parachute system using the single-standard Gemini parachute from 10,000 feet, with a single second back-up system available. Between 22 May and 18 July 1963, five drop tests were successfully completed, although one of the test vehicles suffered minor damage on landing. However, on 30 July, both the prime and the back-up parachutes failed, and the test vehicle was destroyed.

North American wanted to declare the test programme completed, and had isolated the problem in moving the FSTV – but NASA disagreed, and two more HSTV drops were ordered. These were successfully completed on 12 November and 3 December, and the system was qualified for the full scale test programme/

Twenty tests were planned for the FSTV to be dropped from the C-130, and if these were successful, without problems and delays, then there was a chance that the paraglider would be installed on the final three Gemini missions The first FSTV was deployed from the C-130 on 22 January 1964, but the paraglider failed to fully deploy, and the test was considered to be only a partial success. By the time the fourth drop took place on 22 April, all had suffered problems enough to render each test a failure. NASA therefore decided that no more money should be spent on the paraglider, although the remainder of the allocated funds could be spent to complete the planned test programme.

The next two drops, on 30 April and 28 May, were successfully completed, but from the eighth to the twenty-second drop – which took place between June and October – a number of small problems were encountered. Each drop was, in the main, successful, but stable glide remained an elusive achievement. Then, on the final three tests on 23 October, 6 November and 1 December, all went well, and the principle of deployment and stable flight was fully demonstrated. But by then it was too late. On 10 August 1964, NASA had declared that the paraglider had been axed from the Gemini programme, and that each mission would be recovered by parachute and in the ocean, as were the Mercury missions.

Testing continued
Using the residual funding from NASA, together with its own money, North American supported manned demonstration flights of the paraglider after its cancellation from the Gemini programme. In November 1963, TTV No.1 had arrived at the FRC, to be followed by the second on February 1964. On 29 July 1964 a TTV was towed by helicopter to about 2,800 feet, and after a 20-minute ride the combination landed, whereupon North American pilot E. Hetzel released the cable and jettisoned the paraglider. However, on 7 August the first free flight almost ended in disaster when it spun out of control, forcing Hetzel to parachute to safety from 3,000 feet.

After a fourteen-flight radio-controlled test programme of the HSTV, two full-scale radio-controlled test flights were completed between August and December 1964. On 19 December 1964, test pilot Donald McCusker flew a successful 5-minute free flight from 9,800 feet, and despite a hard landing the paraglider design was finally vindicated.

The Full-Scale Tow Test vehicle is shown, under the control of NAA test pilot E. Hetzel, during a towed roll-out across the dry lake bed at NASA Flight Research Center in California during 1963. After the concept was dropped from the Gemini programme, tests continued to prove the concept until November 1965. (Courtesy North American Aviation.)

Meanwhile, Ryan Aeronautical Company (which had developed the Flex Wing) had proceeded with its development of the XV-8A Fleep (*Flying jeep*) as a small one-man flying vehicle powered by a small 200-hp pusher engine, and the military was beginning to show interest in developing the paraglider for cargo deployment from the back of transport aircraft in low-level air-dropped cargo delivery systems. North American detected this interest, and funded a seven-month test programme utilising redundant Gemini hardware.

By September, three manned Gemini missions had been flown, and on the 3 September 1965 McCusker was released at 8,200 feet on a 4½-minute flight – the first of a series of flights, the twelfth and last of which took place on 5 November 1965. During these tests, several of the pilots demonstrated the capability of landing the Gemini paraglider within 500 feet of a planned target point – which was the objective of the paraglider programme when it was assigned to Gemini.

Although Gemini was not a beneficiary, the paraglider proved to be a viable concept for other applications in the military – and eventually as a sport. The hang-glider derives from the Rogallo wing design.

The loss of the paraglider
The elimination of the paraglider from the Gemini programme was the result of having to construct the test hardware from scratch, of delays in the development of the hardware and the test programme, of restrictions in the budget, and of the shortage of available time left in the programme. By February 1964 it became clear that if it ever did fly on Gemini it would not be included until the later stages, and so an alternative parachute recovery system was adopted for the first missions, which would require ocean splash-down rather than land recovery. When asked by the House of Representatives if the paraglider would ever be used on Gemini, George Mueller replied that it depended on the status of the much-delayed test programme during that spring and the needs of the systems to demonstrate precise landing as part of a co-operative USAF objective, and it was unlikely before the tenth manned mission. He also added: 'We have no money included [in] 1965 or beyond for the paraglider, under the assumption we will not go into production.' The feeling at NASA was that while landing on the ground was desirable, it was far riskier than an ocean splash-down – and after three years of development and tests there were still no sustained levels of performance to warrant any further development work to rely on the system being successful before the end of Gemini. The major hurdle was not to deploy the paraglider or to recover a spacecraft on land (which the Soviets had demonstrated with Vostok and Voskhod) but to stop a two-ton spacecraft, using a paraglider, from slamming into the ground.

There was, however, some hope that the paraglider could be included in the proposed USAF MOL programme, as a military objective. On 17 March 1964, George Mueller had enquired whether the Air Force had any interest in this capability for their Gemini B spacecraft under the MOL programme, but after six weeks of evaluation the Air Force too concluded that there remained too much to be done to warrant using the system, and passed the problem back to NASA by stating: 'Should NASA qualify and demonstrate the paraglider in the NASA Gemini programme, consideration would be given to its application in Gemini B/MOL.' But by then it was too late, as NASA had already decided to eliminate the paraglider from the Gemini programme.

PARACHUTE RECOVERY

During the 1960s the paraglider was a controversial approach to landing a spacecraft. The Rogallo wing – although seemingly a good idea – was an unproved concept, and it was perhaps optimistic to expect it to be included as an operational system in an authorised programme with a tight launch schedule, specific design objectives, and a restricted budget. By May 1962 it was clear that the paraglider would not be ready until the third flight of Gemini, and that Gemini 1 and Gemini 2 would require a parachute system. By April 1963 the delays had slipped the paraglider to the seventh flight, and it appeared that the requirement for Gemini would be an ocean splash-down by parachute. (Gemini parachute deployment is discussed in the chapters on hardware (p. 27) and flight tests (p. 163).)

One of the major objectives of Gemini was to achieve accurate touch-down control with the use of a trajectory-shaping technique during re-entry, allowing for unpredictable manoeuvres during retro-fire, and compensating for atmospheric conditions and uncertainties in the spacecraft aerodynamic characteristics. Gemini 8 also demonstrated that trajectory control greatly reduced the recovery task during emergency re-entries.

The design of the Gemini spacecraft placed the centre of gravity such that the angle of attack also provided the desired lift characteristics. It was offset towards the feet of the pilots, with the spacecraft flying inverted to reduced aerodynamic heating on vulnerable areas such as hatches and windows. The spacecraft was rolled to the required bank angle in order to use the lift for both down-range or lateral range control capability. Gemini's touch-down footprint extended approximately 300 nautical miles in length and 50 nautical miles in width. Maximum range was planned as a heads-down or zero bank-angle, whereas minimum range was obtained by a 90-degree bank-angle or a rolling re-entry, which cancelled out the effects of the lifting force.

The spacecraft ability to do this required stability during re-entry. If the vehicle was not stable then the required angle could not be maintained, which would result in errors in determining the landing area. The drogue parachute was released at 50,000 feet, which avoided unstable characteristics at lower Mach numbers. During Gemini missions, two different steering techniques were employed – a rolling re-entry (used on Gemini 3, 4, 8, 9, 10, 11 and 12) or a constant bank-angle technique. (employed on Gemini 5, 6 and 7). Each used predicted range data computed from the range-to-go reference trajectory (planned) and from deviations from navigation flight conditions (actual), which were stored in the computer memory as a function of parameters that related navigated velocity and measured acceleration.

An accurate de-orbit manoeuvre had to be initiated for the guidance system to 'steer' the spacecraft using the four solid propellant retro-rockets that produced a velocity change of about 320 fps. The vehicle was held at a pre-estimated pitch attitude of 2° throughout the manoeuvre, while pitch, yaw and roll were dampened by the automatic control system. This manoeuvre produced excellent results on all missions.

SUMMARY OF RE-ENTRY AND RECOVERY

During the twelve Gemini missions, eleven re-entries were attempted, and all of them were achieved successfully. In addition, a further re-entry demonstration was achieved as part of the USAF MOL programme. (Gemini 1 was an unmanned orbital test flight, and re-entry and recovery was not planned.)

Gemini 2
The second mission was an unmanned sub-orbital test of the re-entry and guidance system. A rolling entry was programmed into the computer; but this was bypassed, and after open-loop re-entry, the spacecraft was constantly rolled from the 0.05-g point until it reached an altitude of about 80,000 feet. A pitch attitude error of 3°.2 during retro-fire caused a shift in the footprint of 14 nautical miles from the planned touch-down point.

Gemini 3
As this was the first manned flight of the series, the spacecraft propulsion system performed a pre-retro-fire manoeuvre some 12 minutes prior to retro-fire to ensure that the spacecraft would re-enter if the retro-rockets failed. The combination of pre- and retro-fire burns produced a footprint of 48 nautical miles. The plan was to fly a back-up bank-angle to cancel out the cross-range and down-range errors, and for the crew to fly by the commands generated by the onboard computer. But a decrease in lift-to-drag ratios resulted in the loss of about 160 nautical miles in down-range capability, which placed the Gemini 3 target at the edge of the manoeuvre envelope of the spacecraft, and providing very little degree of error. Gemini 3 consequently landed 60 nautical miles from the target but it was determined that if the crew had flown according to commands from the computer throughout the entire re-entry, they would have missed the target by only 3 nautical miles.

The landing took place at 14.17 on 23 March, 9½ miles from the US coastguard cutter *Diligence*, and 50 miles up-range from *Intrepid*. A helicopter was over Gemini 3 after only three minutes, but the pilots chose to await the arrival of the prime recovery vessel. The 5-foot swells did not add to the crew's comfort, and prompted John Young to remark that Molly Brown also featured the capability to 'pitch, heave and roll'. The crew arrived by helicopter on the deck of the recovery ship at 15.30, and Molly Brown was taken safely on board at 17.03.

Gemini 4
A planned pre-retro-fire manoeuvre was to be followed by a normal retro-fire about 12 minutes later. After the experience on Gemini 3, it was planned that Gemini 4

Following the cancellation of the land landing objective for Gemini, all missions were targeted to splash down in the ocean. Here, at the end of a mission, a Gemini spacecraft enters the water under a single parachute.

should use the rolling entry guidance logic in the computer, and that the crew should follow the commands manually, as a back-up, during the entire re-entry. However, McDivitt could not turn on the computer, and found that nothing would activate it. A profile that followed an open loop, by manually rolling Gemini, was therefore initiated at 400,000 feet, and was carried on throughout re-entry. As they watched the adapter burn up behind them, the crew realised that without the computer to assist them they would land short, despite some lift. It was planned to roll at about 15° per second, but because the roll-rate gyro had been turned off, the yaw thruster increased the acceleration in the roll direction – which could not be dampened – to a maximum of 60° per second. At 89,000 feet, McDivitt attempted to slow the roll rate, and at 40,000 feet he deployed the drogue parachute; but instead of stabilising the spacecraft it gyrated, as the roll was still about 50° per second. The parachutes were deployed with a jolt, resulting in cracked helmets as the astronauts placed their arms in front of their faces to protect themselves.

Splash-down was hard as the spacecraft slammed into the ocean 44 nautical miles from the intended target; but with recovery forces approaching it took only 30 minutes for the crew to be picked up by helicopter and 27 minutes later to step from the helicopter onto the recovery vessel *Wasp*. Gemini 4 was lifted onboard 1 hour 19 minutes later.

Gemini 5

Due to the large reduction in lift-to-drag ratios recorded on the first three Gemini re-entries, the sets of data stored on the computer were invalid, and therefore produced erroneous results. However, due to the short time between missions (a feature of Gemini) it was impossible to amend the computer memory in time for Gemini 5 and Gemini 6. But the crews were trained to calculate these errors, and they were able to determine the corrections required to increase accuracy in reaching the target.

The Earth rotates $360°.98$ degrees each day, but when the computer was programmed the $0°.98$ was omitted, so that when the navigation update was sent to Gemini 5 prior to retro-fire, the data were incorrect by $7°.9$ degrees ($0°.98$ x time of lift-off to time of retro-fire in days), which equated to 474 nautical miles, indicating erroneous range data displayed throughout re-entry. By the time the crew determined this there was not enough manoeuvring capability to correct it, and Gemini 5 therefore landed 91 nautical miles from the planned target. Post-flight analysis revealed that if this mistake had been corrected, the landing error would have been only $2\frac{1}{2}$ nautical miles.

During Gemini 5's eight-day flight, Hurricane Betsy threatened the landing area, and although landing conditions were more amenable for Gemini than during Mercury (wind speeds of 18–25 knots, and waves of 5–8 feet) – and overall, the

With the flotation collar attached, the Gemini 12 astronauts open the hatches and await recovery. Here, Buzz Aldrin is about to exit the spacecraft via the Command Pilot hatch, his Commander Jim Lovell having already exited the spacecraft into the lift rafts. (Courtesy British Interplanetary Society/NASA.)

weather need not be good – the spacecraft could certainly not re-enter into a hurricane. Weather bureau data were therefore heeded, and the landing area was moved so that Gemini 5 would re-enter earlier than at first planned.

As Cooper brought down the spacecraft, the re-entry gauge indicated a high rate of descent which would result in an overshoot, and he therefore turned it to the left to increase the drag and as a result reduce the landing error, increasing g-loads from 2.5 to 7.5 as he did so.

Gemini 5 landed 103 miles from the prime recovery vessel *Champlain*, and as divers were attaching flotation devices to the spacecraft, Cooper said that he wished to remain in the capsule, as it was a pleasant summer morning. With the recovery vessel so far away, however, he instead accepted the offer of a helicopter ride, and 1 hour 29 minutes after landing, both he and Conrad were on the deck of *Champlain*, to be followed by the spacecraft 2 hours 24 minutes later.

Gemini 6
On 16 December 1965, Gemini 7 was in the twelfth day of a 14-day space marathon; and onboard Gemini 6, the crew were preparing for re-entry after only one day in orbit. It was little more than a week before Christmas, and spirits were high, as the two spacecraft had achieved the world's first space rendezvous. During the penultimate orbit of the flight of Gemini 6, while flying over the United States, an unusual observation was noted by Tom Stafford: 'We have an object . . . looks like a

Tom Stafford looks up at the recovery helicopters as he and Gene Cernan relax onboard Gemini 9 and await the prime recovery vessel *Wasp*. The crew decided to remain in the spacecraft for pick-up, instead of taking the helicopter ride.

satellite going north to south, probably in polar orbit … looks like he might be going to re-enter soon … stand by one [the communications channel] … looks like he's trying to signal us … [let's] try to pick up that thing … I see a command module and eight smaller modules in front. The pilot of the command module is wearing a red suit.' Over the communication speakers on Gemini 7 and at Mission Control came the strains of 'Jingle Bells', played by Stafford ringing a bunch of small bells and with Schirra on harmonica. Mission Control broke up in laughter when, on Gemini 7, Lovell also reported seeing the object. Gemini 6 was coming home for Christmas.

Following retro-fire, Schirra placed his spacecraft in the head-down attitude to align it with the Earth's horizon, and as he neared the upper limits of the atmosphere at 328,000 feet he set the bank-angle at 55° left, and held that position until the computer took control at 279,000 feet. In order not to overshoot, Schirra turned the Gemini left and right to shorten the range. The recovery progressed normally, and with the first successfully controlled re-entry it splashed down seven nautical miles from the target, with a navigational error of 2½ nautical miles. The splash-down was the most accurate thus far, and was a significant success for Gemini. Pictures were transmitted from the prime recovery vessel *Wasp* via satellite to American TV viewers.

After splash-down, Schirra and Stafford chose to remain inside the spacecraft, although both hatches were open as the astronauts chatted to the recovery frogmen. The rendezvous and radar section that separated during descent were nearby, and were, for the first time, recovered by the frogmen. The two astronauts were onboard the carrier 1 hour 2 minutes after recovery.

Gemini 7

Two days later, Gemini 7 Commander Frank Borman almost shouted with joy when CapCom asked the crew if they were ready to come home as 'Going back to Houston' was played up to them. Waiting for retro-fire was a tense moment. If it failed, there were only 24 hours of oxygen, ten hours of electrical power, and very little water. However, retro-fire worked perfectly (as with every Gemini) and they began the long ride home. Borman likened re-entry to flying inside a neon tube, as the external temperature increased around the spacecraft.

The guidance computer had been improved, although when the bank-angle was flown, the primary crew display was still showing range error. Ground controllers had mistakenly provided the crew with a bank-angle of 35° instead of 53°. For most of the re-entry, Borman could not see outside, and he followed the descent by instruments as they banked sharply at 90°. He wondered whether the computer guidance was working correctly, as it was taking them to the left and then 180° to the right, and with his hand hovering over the manual control – in case the instruments should decide to take them to a burn-up – he remained fixed to the dials, and resisted the urge to override the machine which he still trusted.

Because they had been weightless for so long, with the increase in g-force they 'felt like a ton' as the spacecraft slowed its manoeuvres and as the needles settled down. The maximum load on the crew was 3.9 g (compared with Mercury's 7.7 from orbit)

Gemini 7 astronaut Jim Lovell is picked up by helicopter at the end of the 14-day space marathon. Borman is still in the lift raft, with the pararescue divers around him and in the water.

but it seemed to be considerably more. The descent under the parachute was normal, but when Lovell opened the snorkel, his spacesuit hood filled with smoke and an acrid smell caused his eyes to water; but even through the tears he could see the parachute deployed. Gemini 7 entered the water with a jolt. Borman and Schirra had agreed a small wager on who would land closer to the target, and when Borman could not see anything outside his window he thought that Schirra had won – when, in fact, Borman had won. Gemini 7 landed only 6.4 miles from its target, compared with the 7 miles of Gemini 6. The astronauts were onboard the recovery 30 minutes after landing, and were followed by their spacecraft 31 minutes later.

Gemini 8
Following the difficulties encountered during the early docked phase of the Gemini 8 mission, the use of the Re-entry Control System to terminate excessive roll rate in the yaw thruster of the primary system required an early termination of the three-day mission and recovery in a secondary landing area. It was a mission rule that in

The view from the lift raft, with a Gemini 4 astronaut being hoisted up.

the event of emergency activation of the RCS, the spacecraft would be recovered in the next planned landing area. Gemini 8 was therefore targeted for a Western Pacific landing after only seven orbits.

The mission flew a rolling re-entry logic that was followed by all subsequent missions that featured manual control of the bank-angle commands that were generated by the onboard computer. Due to the emergency nature of the mission, naval recovery forces in the Pacific were immobilised to head to the planned landing area 500 miles east of Okinawa and 620 miles south of Yokosuka, Japan. In the MOCR, Gene Kranz could rely on only three tracking stations (Hawaii and the tracking vessels *Coastal Sentry Quebec* and *Rose Knot Victor*) to monitor the final stages of the mission, due to the location of the spacecraft's orbital track at the time.

In the spacecraft, the crew began to wonder where they would land, and whether their wilderness survival training might after all be useful. They checked each action for re-entry, then double-checked – and then checked again. But all went well as the re-entry sequence and recovery phase followed without further incident After 10 hours 41 minutes, Gemini 8 was floating in the Pacific Ocean. Initial calls to the recovery vessels went unanswered, with caused them some concern that they were too distant for a prompt pick-up, but Mission Control in Houston informed the crew that the prime recovery vessel *Leonard Mason* would be alongside in three hours, which indicated that they had indeed achieved a landing close to the planned target.

Due to the landing area chosen for recovery, no re-entry tracking was possible, and therefore no navigational accuracy was determined for this flight, although the

The view from the helicopter. Gemini 10 astronaut John Young is about to be hoisted up, assisted by the pararesue divers. Michael Collins is already inside the helicopter.

spacecraft computer calculated a 1.4-nautical-mile planned target deviation at drogue deployment, and the spacecraft on its main parachutes was sighted by recovery aircraft. The seas were rough during the recovery operation, and it took some time for the frogmen to attach the flotation collar to the spacecraft. The swell also made the crew and the divers feel nauseous, but the floatation collar was secured 45 minutes after landing. Three hours after landing, the crew scrambled up a ladder on the side of the *Leonard Mason*, and promptly headed to the sick bay to remove their pressure garments. Soaked in sweat, the two astronauts were medically examined and found to be thirsty, but were otherwise none the worse for their ordeal.

A pinpoint landing
As a result of a 1.06% increase in retro-rocket velocity and a 2°.3 spacecraft pitch error, Gemini 9 achieved a rather larger footprint of approximately 55 nautical miles. With the crew using the rolling re-entry logic and manually controlling the bank-angle commands, the spacecraft entered the water 0.38 nautical miles from the planned landing point, with a post-flight navigational accuracy of 2.2 nautical miles. The Gemini 9 landing was the most accurate of the programme – within 0.5 nautical miles – and demonstrated the concept of controlled re-entry into the ocean. Both astronauts decided to stay in the spacecraft, and after they opened the hatch they relaxed in a gentle swell as recovery operations were completed; and as they were

close to the recovery vessel *Wasp*, they waved to the ship's crew, and indicated their success with 'thumbs up'.

The final three

The landing accuracy of Gemini 10 was 3.4 nautical miles with a 4.2-nautical-mile navigational accuracy due to yaw misalignment in the inertial platform. With the experience of Gemini 3, John Young warned Michael Collins to prepare for the shift to two-point parachute suspension. They braced their arms in front of themselves, but were disappointed when the move resulted in nothing more than a gentle forward swing. What was novel to Collins was to see clouds passing sideways by the windows as the spacecraft spun under the parachute, until the motion slowed and finally stopped, only to begin again in the opposite direction. No-one had told Collins about this, and it caused him some concern that they would be subjected to a hard landing; but they slipped gently into the Atlantic – probably on the crest of a wave.

The prime recovery vessel *Guadalcanal* was so close to the spacecraft that the sailors on deck were able to watch it descend into the water. Young and Collins chose to be hoisted up to the helicopter for a ride back to the ship, 27 minutes after landing. The spacecraft was picked up after another 25 minutes.

The penultimate Gemini was the first to employ the automatic steering mode, using the attitude-control system combined with the guidance commands. Post-flight

Gemini 11 astronauts Dick Gordon (*left*) and Pete Conrad (*right*) step onto the prime recovery vessel *Guam* after their three-day mission. Note the cameramen around the crew and inside the helicopter – the beginning of the media circus.

Recoveries

	GT-1	GT-2[1]	GT-3[2]	GT-4[3]	GT-5[4]	GT-6[5]	GT-7[6]
Date	n/a	1965 Jan 23	1965 Mar 23	1965 Jun 7	1965 Aug 29	1965 Dec 16	1965 Dec 18
Miss distance (n. miles)	n/a	14	60	44	91	7	6.4
Navigation error (n. miles)	n/a	1.2	0.8	–	474	2.5	2.3
Recovery vessel	n/a	*Lake Champlain*	*Intrepid*	*Wasp*	*Lake Champlain*	*Wasp*	*Wasp*
Recovery times (EST)							
Splashdown	n/a	09.22	14.17	12.12	07.56	10.39	09.05
Crew	n/a	n/a	15.28	13.09	09.26	11.32	09.37
Spacecraft	n/a	10.52	17.03	14.28	11.50	11.32	10.08

	GT-8[7]	GT-9	GT-10	GT-11[8]	GT-12[9]	MOL/GT-2A
Date	1966 Mar 17	1966 Jun 6	1966 Jul 21	1966 Sep 15	1966 Nov 15	1966 Oct 28
Miss distance (n. miles)	1.4	0.38	3.4	2.65	2.6	7.0
Navigation error (n. miles)		2.2	4.2	4.0	2.4	?
Recovery vessel	*Leonard Mason*	*Wasp*	*Guadalcanal*	*Guam*	*Wasp*	*LaSalla*
Recovery times (EST)						
Splashdown	10.22	09.01	16.07	08.59	14.21	n/a
Crew	01.28	09.53	16.34	09.23	14.49	n/a
Spacecraft	01.37	09.53	17.01	09.58	15.28	n/a*

Notes: 1, footprint shift; 2, lift–drag reduction; 3, footprint shift, inoperative computer; 4, invalid position update; 5, no radar below 180,000 feet; 6, lift–drag reduction; 7, emergency re-entry; 8, automatic re-entry; 9, automatic re-entry; *data not available, but the spacecraft landed seven nautical miles from the prime recovery vessel.

analysis of the profiles from Gemini 8 and Gemini 10, with that of Gemini 11, revealed only minor differences. What was found was that the automatic system responded immediately to any command to change the direction of the bank-angle, while the manual system featured a time-lapse between the command and the response, although this had no noticeable effects on the landing point of Gemini 11. At 40,000 feet, Conrad rolled Gemini to a back-up bank-angle of 44°, whereupon the computer commanded a bank-angle for full lift and a right roll to recover from Conrad's input. He was confident that the computer was working correctly, and switched to automatic for the rest of the descent – although he followed each manoeuvre with a deactivated hand-controller in case a problem should occur, in which case he would need to quickly revert to manual control.

The spacecraft splashed down – with a navigational accuracy of 4 nautical miles – 2.65 nautical miles from the planned landing point, under the eye of TV cameras on the prime recovery vessel *Guam*. Just 24 minutes after landing, the astronauts were onboard the ship, and the spacecraft arrived 35 minutes later.

Gemini 12 – the final mission of the series – was highlighted by an event during re-entry. A stowage pouch broke free of its Velcro attachment, and landed in Lovell's lap. However, he resisted the urge to grab it because he had deployed the ejection seat D-ring and was holding it between his legs. He feared that he might inadvertently grab the D-ring instead of the pouch – and he was not in a hurry to evacuate the spacecraft, especially during the height of the re-entry fireball. The spacecraft flew an uneventful re-entry – with a navigational accuracy of 2.4 nautical miles – and splashed down hard in the Atlantic Ocean, 2.6 nautical miles

The crew normally arrive ahead of their spacecraft. Here Gemini 4 is hoisted, unoccupied, onto the recovery vessel *Wasp*.

from the recovery vessel *Wasp*. It was the fifth Gemini spacecraft to splash down within sight of the recovery forces, and provided a fitting end to the recovery phase of the Gemini programme. The astronauts were on the deck of the recovery vessel 30 minutes after landing, and 37 minutes later they were followed by the final Gemini spacecraft.

The record of achievements
The re-entry phase of Gemini revealed that guidance techniques had to be insensitive to large deviations in the spacecraft lift capability; that constant bank-angle techniques depended on accurate estimates of the manoeuvring capability; and that on long-duration missions in which centre-of-gravity variations or precise aerodynamic characteristics were in question, this was not an effective method. As long as the landing target was within the footprint, the rolling re-entry technique would steer the spacecraft towards the intended target; useful and informative display for the crew allowed them to evaluate performance of guidance and navigation sub-systems and to provide back-up data for a safe in the event of a guidance system failure; and accurate navigation could be achieved during re-entry, with the spacecraft being controlled either in automatic or manual mode, or in both. It was proved that re-entry from Earth orbit required a degree of control accuracy, but did not require the astronauts to respond instantly to displayed information.

Experiments

The Gemini experiment programme featured a collection of investigations that formed a range of secondary objectives to be accomplished throughout the course of the manned missions. Fifty-four experiments were scheduled, and of these several were flown on more than one occasion, while more than forty were successfully completed. The experiments were grouped under four categories – medical, engineering, defense and scientific.

The eight medical experiments (designated with the 'M' prefix) were designed to study cardiovascular reactions, biochemical analysis of body fluids, the crews' vision and orientation, and so on. Ten engineering experiments sponsored by the Manned Spacecraft Center (MSC), and two Technological (T) experiments, would study a wide range of fields such as optical communications, the ultraviolet reflectivity of the lunar surface, the measurement of electrostatic charges, and the number of protons outside the spacecraft. A programme of fifteen Department of Defense (D) experiments concentrated on radiation, navigation, polarisation, light levels, aspects of mobility during EVA, and working environments; while the nineteen Space Science (S) experiments offered research into nuclear emulsion tests, micrometeorite collection, astronomical observations, and Earth surface and atmospheric studies. A further area that encompassed most fields, and which evolved into a separate research programme, was photography from manned spacecraft, not only of the Earth, but also of missions activities.

EXPERIMENTS ON GEMINI

Project Mercury had been focused on providing the technical effort of placing a man in Earth orbit, sustaining him there for up to a day, and then returning him safely. There had not been much allowance for anything scientific on the missions, although a few experiments – mainly of a visual or photographic nature – had been included on the later missions.

Prior to 1963 there been little thought of placing scientific experiments onboard Gemini missions, but with the prospect of a larger spacecraft flying two astronauts on longer missions, and the interest of the military, a dedicated Gemini experiment programme began to be developed.

The selection of any experiment to Gemini was mostly based on the requirement or desirability of direct astronaut participation, with most of the planning and operational management of the programme handled from MSC in Houston.

From proposal to experiment

On 21 January 1963, NASA and the DoD concluded a major agreement that defined their roles in Project Gemini, including the planning, supply and flight of several DoD experiments. By 13 September a technical development plan for proposed Gemini experiments originating from DoD PIs included thirteen USAF experiments and nine USN experiments costing an estimated $22 million (1963). The chances of their inclusion on a Gemini mission, however, still depended on a further definition of technical merits by NASA, and on the weight and volume constraints of a particular mission. However, what the USAF hoped to achieve through this co-operation was the development of its own programme of experiments with a military objective for use on its own military man-in-space programme that began with the 'Blue Gemini' concept and continued with MOL. In pursuit of this, the Air Force Systems Command established a field office at MSC to co-ordinate DoD-sponsored experiments and their integration into the NASA missions.

Initially, the Headquarters Office of Space Science and the Office of Manned Spaceflight created a Panel on In-Flight Scientific Experiments (POISE) to pursue the merits of all submitted proposals for future consideration, including those from each of the field centres. From January 1964, all experiment proposals were co-ordinated by a Manned Space Flight Experiment Board, which allocated experiment proposals to specific missions, of both Gemini as well as Apollo.

These investigations had originated from MSC, the Office of Space Science and Applications, the Space Medicine Office, the Office of Advanced Research and Technology, and the DoD, all of which had received new proposals from experimenters around the country, military installations and NASA field centres. In addition, experiments proposed but not flown on Mercury, as well as those that had flown previously, were re-evaluated for Gemini.

From the Experiment Board these proposals were forwarded to the Gemini Program Office, which then directed them to its Experiment Office for a determination of which were suitable for flight on Gemini missions. From here, the proposals were reviewed by other offices involved in the operations of Gemini – such as launch operations, flight control, astronaut training, and recovery – which forwarded their recommendations to the GPO, which at the same time had instigated a series of engineering definition studies for each experiment.

All this information was then placed before a Scientific Experiments Panel (which met for the first time on 17 December 1963) to review proposals for experiments to be flown. The initial experiments for Gemini were approved in February 1964. These were planned for the first five manned missions and were classed as Category B, or second priority items – meaning that they were not experiments that could affect design requirements, and therefore, in the event of their impeding a launch or effecting a delay on a launch, could be omitted from the mission without serious consequences for the primary mission or the spacecraft hardware. In addition to selecting experiments for Gemini, the same board forwarded any experiment that could be incorporated on future Apollo missions

Technical regulations and procedures

On 8 July 1964, Robert Gilruth issued – as part of the MSC Management Manual – a list of technical regulations and procedures for flight experiments. The purpose of the document was to establish the procedures by which the flight experiments that were proposed for manned spaceflight could be evaluated and implemented to fall within guidelines established by the Manned Space Flight Experiments Board (MSFEB). This included all engineering and operational tests of the systems, procedures and basic science studies, but did not apply to investigations carried out in ground facilities, flight qualification of standard spacecraft components, or hardware modification.

Under this plan, MSC established an Experiment Co-ordination Office (ECO) as part of the Advanced Spacecraft Technology Division within the Engineering and Development Directorate. The ECO reviewed all experiment proposals for evaluation and implementation into the manned missions. Once a proposal was received it was assigned to six groups for review and evaluation:

Flight crew operations Under the Flight Crew Operations Directorate at MSC (the Astronaut Office and training division) where the experiment's operational suitability was established. This group also ensured that mock-ups and training versions were available for establishing stowage, operating procedures and instructions the flight crew, as well as initiating check-lists and other flight plan documentation

Flight operations support Under the Director of the Flight Operations Directorate (Mission Control and Flight Planning departments), which reviewed suitability for mission compatibility with communications, suitable flight control input, and tracking and recovery support where required.

Flight preparations support Located at Cape Kennedy, and under the responsibility of Florida Operations, where flight hardware was processed, hazardous safety requirements were identified, and support equipment was developed for launch preparation and launch conditions.

Spacecraft support Associated with the Spacecraft Programme Office, to ensure that the proposed experiments, hardware and operations were compatible with spacecraft requirements, and that delivery dates for design drawings, mock-ups, training and flight hardware were met. Responsible for installation of all training, mock-up and flight hardware for crew training, simulation and the actual mission, ensured that supplied equipment was compatible with safety standards, and provided contractor documentation for implementation and installation in the spacecraft.

Engineering support The Engineering and Development Directorate ensured that specialist engineer support work was completed when required, and that the apparatus was feasible and suitable for spaceflight.

Medical support The Office of the Center Medical Programme was responsible for integration and compatibility with the health and physiological integrity of the flight crew, and also for biological specimen requirements before flight, in-flight and post-flight.

A Technical Monitor (TM), assigned for each experiment, responsible to the ECO for point of contact to and from the Principle Investigator (PI), and the PI would use the TM as his 'agent' at MSC for access to NASA policy and flight requirements, including supply of all mock-ups, training aspects, contact with the flight crew, and so on. With the PI, the Technical Monitor was also responsible for generating post-flight reports and preliminary experiment results when necessary

In support of post-flight operations a crew post-flight debriefing on the experiments would be held to benefit the PI and the Technical Monitor. This document stated that 'all equipment carried on any flight and all data resulting from an in-flight experiment will become the property of NASA. However, MSC (NASA) will protect the right of the experiment originator to the use of this data for the purpose of scientific reports.'

In June 1965 the Experiment Co-ordination Office and the Gemini Experiments Office were merged to form the Experiment Program Office (EPO), which also had responsibility for Apollo experiments and those proposed for 'future programmes'. Following the conclusion of Gemini flight operations, in December 1966 the EPO became the Science and Applications Directorate at MSC.

The selection of experiments
The experiment proposals that were submitted to the Experiment Board were passed to the ECO at the Manned Spacecraft Center for review and evaluation. From the different groups, comments were compiled and then submitted to an Experiments Review Panel consisting of the following representatives:

Chairman: Assistant Director of Engineering and Development
Secretary: Head of Experiments Co-ordination Office
Members and representatives:
  Assistant Director for Flight Operations (1)
  Assistant Director for Engineering and Development (1)
  Assistant director for Flight Crew operations (2)
  Office of the Chief of Center Medical Programmes (1)
  Gemini Program Office (1)
  Apollo Spacecraft Program Office (1)
Observers: representatives of various groups (including the DoD, Crew Systems, and others) were invited to attend as deemed necessary by the Chairman.

Following ERP review and the Center Directors approval, the experiments were resubmitted by the ECO for flight implementation. In conjunction with the appropriate departments, the ECO then devised a package of experiments for each flight, and assigned a technical monitor with all departments co-ordinating efforts to ensure smooth transition into the agreed flight plan.

All departments ensured the recovery of experiment data, and samples were acquired to provide suitable post-flight reports for analysis, review and, if required implementation for reflights and future experiments in the same or similar fields. A post-flight debriefing on experiments was held after the termination of each mission.

Operational requirements
No experiment was deleted from a Gemini mission because of flight hardware not being available at the time of launch, although three DoD experiments were removed from Gemini 4 due to the inclusion of the EVA.

Every experiment was assigned to two missions – a prime and an alternate – and as the programme developed it soon became apparent that for some experiments this was inadequate, and that a multiple assignment was required. Considerable effort was placed on ensuring that the crews were familiar with the experiments that they were to operate, although in some cases, due to the length of development time or quick reflight of experiments and constraints of crew training, this could not always be to the degree desired.

Experiment planning had to reflect not only changes in the flight plan and crew activities, but also the timing of the mission and specific days in orbit. This was revealed during the Gemini 6 mission when during the final preparations to rendezvous with Gemini 7 were underway, Comet Ikeya–Seki was discovered, and an attempt to photograph its movement through the Sun's corona was added to the mission. Although provision to do this was included in time, the delay to the launch resulted in the incorrect position of the spacecraft to accomplish the task.

During Gemini 12, ultraviolet photography of the dust entering the atmosphere was planned, along with observations of an expected meteor shower as Earth passed through the remnants of a comet's tail, and photography of the Moon's shadow moving across the Earth during a solar eclipse. The mission had been extended from three to four days, and all activities to allow the crew to achieve these objectives were completed affecting their other activities. The mission was then delayed for two days, but the objectives remained. As a result, the support personnel had to provide adjusted orbital data and rendezvous updates to allow the crew to attempt the experiments.

Public image
In the spring of 1966 – half-way through the Gemini programme – L. Day, Acting Deputy Director of the Gemini programme sent a letter to Charles Mathews, Manager of the Gemini programme, concerning the publicity and promotion of the Gemini experiments. The letter, dated 12 May 1966, indicated that the large number of experiments had generated a significant amount of data, and that in order to utilise both the experiments and the resultant data to maximum effect, it was

important that they become well known to the larger scientific and technical community. In addition, it was also important to promote the experiments as part of the public image of the programme. With the headlines featuring rendezvous, docking, EVA and long flights, it would be easy to lose sight of the experiments in the larger programme.

In his letter to Matthews, Day suggested that, whenever possible, experiments should be referred to by their titles rather than by their numbers. This would not only improve communications with other scientists, but would also enhance the public's perception of the experiments.

INTEGRATION OF EXPERIMENTS

Several of the 54 experiments were reflown, and there were 113 experiment 'missions', with an average of 11 experiments on each manned flight. Gemini 7 carried the most (twenty) on one mission, while the shorter flights of Gemini 3 and Gemini 6 carried only three each.

Summary of experiments

Sponsoring agency	Number of experiments	Total experiment mission
Scientific		
Office of Space Science and Applications	19	49
Technological		
Office of Advanced Research and Technology	2	2
Office of Manned Spaceflight;		
Manned Spacecraft Center	10	18
Department of Defense	15	26
Medical	8	18
Total	54	113

Sources for experiments flown on Gemini originated from universities, laboratories, hospitals, industry, and a variety of Government agencies. Several PIs were assignment on more than one experiment, or one experiment had more than a single affiliation. The selection of a PI and an experiment began a very close personal association with all aspects of placing experiments on-board a particular mission, from NASA management to the spacecraft contractor, mission planning team, crew training, astronauts, flight controllers, and mission evaluation teams.

Mission integration

All experiments were integrated on missions on the basis of minimum interference with crew activities, with three basic requirements:

Stowage–use–restowage Experiments (such as hand-held cameras) were stowed in the spacecraft, unstowed by the crew during the flight, operated to a preplanned schedule, and then restowed.

Hardware mounted in the crew compartment Experiments (such as Frog Egg Growth) were mounted around the cabin, and remained there for the duration of the mission, with no relocation by the crew.

Hardware integration These experiments (such as D004 or D007) had to be structurally mounted, automatically deployed, or thermally controlled, and often entailed extensive data collection.

Crew integration
Training of the crew began with the PI's instruction briefing, which included an explanation of the experiment, the method of operation, training requirements, and expected results. At this point it was determined whether the experiment could be accomplished by the crew, and how adjustments could be incorporated into the operation of the experiment. A programme of training was then established, including simulators, mock-ups, planetarium briefings where necessary, and zero-g aircraft flights. All of these were reviewed jointly by the crew and PIs in order to resolve any problems that had developed during training.

Mission planning
While the crew prepared to operate the experiments, the mission planing teams prepared to integrate the experiment operations, along with other activities and objectives, into the flight plan. Certain experiments could be conducted only at specific times during the mission, (such as photography of the zodiacal light during night passes) and any conflict with other activities (rendezvous, docking, eating, sleeping) had to be resolved The knowledge and flexibility of the experimenter was crucial in this planning. Another aspect was integration into the mission pre-launch preparations, and it was important to determine when to insert the experiment hardware into the spacecraft, as well as the timing of launch, orbital flight and re-entry.

THE EXPERIMENTS

The following is a summary of all experiments conducted during the ten manned Gemini missions between March 1965 and November 1966. Each research field is covered, with the experiment designator, name, and the missions to which it was assigned, together with details of the Principal Investigators and their affiliations. Also included is each experiment's primary objective, the equipment used, and the primary results. (For further reading, see *Gemini Summary Conference*, NASA SP-138; and *On the Shoulders of Titans*, NASA SP-4203.)

Mission time planned for experiments

Mission	Planned total mission time (hrs)*	Planned experiment activity time (hrs)**	Mission time planned for experiment (%)
3	9	0.5	5
4	140	22	16
5	288	49	17
6	66	8	12
7	392	86	22
8	90	19	21
9	90	19	21
10	90	33	37
11	90	26	29
12	122	37	30
Totals (mean)	1,377	299.5	21

* Two crew; less sleep time
** Direct crew participation time only, not including total experiment equipment operating time

Biomedical experiments
M1 Cardiovascular Conditioning (Gemini 5, 7)
PIs and affiliations: Lawrence F. Dietlein, NASA MSC; William V. Judy, NASA MSC.
Objective: To evaluate the use of pneumatic cuffs on the legs to prevent deterioration of the heart and blood distribution system during periods of prolonged microgravity.
Equipment: Each Pilot wore a pair of venous cuffs on his legs and used a pneumatic cycling system to alternatively inflate and deflate the cuffs to 80 mm of mercury.
Results: During Gemini 5 it was planned to use the cuffs for the full eight days, but they stopped working when the oxygen in the storage tank dropped below operational levels. It was determined that even with limited results, the Pilot's overall post-flight condition was better than that of the Command Pilot, with a significant reduction in the pooling in the legs indicating the effect of the use of the cuffs. The experiment was flown again on Gemini 7 for 14 days. The Pilot logged 311 hours of experiment operation, with the device being turned just three hours prior to re-entry. The significant reduction in blood pooling in the Pilot's legs (recorded post-flight) compared with that of the Command Pilot, confirmed the limited results from Gemini 5.

M3 In-flight Exerciser and Work Tolerance (Gemini 4, 5, 7)
PIs and affiliations: Lawrence F. Dietlein, NASA MSC; Rita M. Rapp, NASA MSC.
Objective: To evaluate the day-to-day physical condition of the crew.
Equipment: A pair of rubber bungee cords attached to a nylon foot strap at one end and a nylon handle at the other.
Results: This device was flown on the three longer-duration missions of four days (Gemini 4), eight days (Gemini 5) and fourteen days (Gemini 7). Data from all three

missions indicated that there was very little difference between the in-flight and pre-flight results.

M4 In-flight Phonocadiogram (Gemini 4, 5, 7)
PIs and affiliations: Lawrence F. Dietlein, NASA MSC; C. Vallbona, Texas Institute for Rehabilitation and Research.
Objective: To measure heat muscle deterioration against a simultaneous electro-cardiogram.
Equipment: Attached to each astronaut's chest was a microphone that picked up heart sounds which were then recorded on a bioelectrical recorder.
Results: No significant changes were recorded between any data obtained from the three long-duration missions and data from the pre-flight ground tests.

M5 Biochemical Analysis of Body Fluids (Gemini 7, 8, 9)
PIs and affiliations: Harry S. Lipscomb, Baylor College of Medicine; Elliott Harris, NASA MSC; Lawrence F. Deitlein, NASA MSC.
Objective: To study the astronauts' reaction to stress.
Equipment: Intake and output of body fluids were measured and analysed pre-flight, in-flight and post-flight.
Results: No gross changes were noted during Gemini 7. On Gemini 8, pre-flight and post-flight samples were obtained, but only one sample from the Command Pilot was obtained prior to the early termination of the mission. From Gemini 9 there were no gross changes recorded from the samples obtained (as with Gemini 7).

M6 Bone Demineralisation (Gemini 4, 5, 7)
PIs and affiliations: Pauline Berry Mack, Texas Women's University.
Objective: To investigate the effects of prolonged weightlessness and immobilisation associated with several days of confinement.
Equipment: X-rays were taken before and after the flight of each crew-member. Particular attention was paid to the heel bone and the end bone of the fifth finger of the right-hand.
Results: The experiment was flown on all three long-duration missions to obtain the best results. On Gemini 4 it was found that there was a distinct loss in bone mass, compared with that of the control experiment bed-rest patients for the same length of time. From the Gemini 5 results the Command Pilot showed greater changes than did the bed-rest patients, while the Pilot also revealed equivalent changes to those same bed-rest patients. During Gemini 7 the crew ate and exercised more, slept better, and rested longer than the other two crews, and as a result they were subject to a significantly smaller loss in bone mass.

M7 Calcium and Nitrogen Balance Study (Gemini 7)
PIs and affiliations: G. Donald Whedon, National Institute of Health; Leo Lutwak, Cornell University; William F. Newman, University of Rochester.
Objective: To evaluate the effects of a 14-day spaceflight on the bones and muscles of the crew.

Equipment: The intake and output of both fluid and solid matter (including perspiration) was measured and analysed pre-flight, in-flight and post-flight.

Results: This experiment was flown only on the 14-day mission, during which the crew experienced problems with leaking sample bags, one of which broke, while four others were unlabelled. From the eighth flight day, the Command Pilot showed a marked increase in calcium excretion.

M8 In-flight Sleep Analysis (Gemini 7)
PIs and affiliations: Peter Kellaway, Baylor College of Medicine; Robert L. Maulsby, Baylor College of Medicine.
Objective: To assess the crew's state of alertness, their level of consciousness, and their depth of sleep during the flight.
Equipment: The Command Pilot wore two pairs of scalp electrodes to record electroencephalographs on a biomedical recorder.
Results: The first night's record revealed poor sleep, which was expected as the Command Pilot became accustomed to the environment and to the strange sounds that disturbed his sleep. After the first night and up to GET 54 hours 20 minutes, when the sensors were dislodged, he appeared to be sleeping normally.

M9 Human Otolith Function (Gemini 5, 7)
PIs and affiliations: Ashton Graybeil, Naval Aerospace Medical Institute; Earl F. Miller II, Naval Aerospace Medical Institute.
Objectives: To evaluate the capability of the astronaut to orientate himself during the mission, and to measure changes in the functions of the otolith (the gravity gradient sensors located in the inner ear).
Equipment: Using special goggles, with one of the eyepieces containing a light course in the form of a movable white light, the astronaut used a calibrated screw to position the line to where he judged was the correct pitch axis of the spacecraft.
Results: Data from both missions indicated that given adequate cues, co-ordinated space sense existed, even in weightlessness.

Manned Spacecraft Center experiments
MSC1 Electrostatic Charge (Gemini 4, 5)
PI and affiliation: Patrick E. Lafferty, NASA MSC.
Objective: To detect and measure accumulated electrostatic charges on the surface of the spacecraft.
Equipment: An electronic field sensor (weighing 1.8 lbs) was located in the retrograde section of the spacecraft, and was controlled by a switch from within the cabin.
Results: On Gemini 4 the sensitivity of the sensor to other influences caused readings that were higher than expected. Following this experience, there was insufficient time to modify the instrument before it flew again on Gemini 5. A shield placed on the sensor had little effect, and once again, high readings were recorded. On the later rendezvous missions, this measurement became an operational procedure.

MSC2 Proton–Electron spectrometer (Gemini 4, 7)
PI and affiliation: James R. Marbach, NASA MSC.
Objectives: To measure the radiation environment immediately outside the space-craft, to correlate radiation measurements inside the spacecraft, and, from the accrued data, to predict radiation levels on future missions.
Equipment: A proton–electron measuring device (weighing 12.5 lbs) was mounted in the equipment adapter section, with its sensor facing aft, and was operated by a switch at the Pilot station in the cabin.
Results: On Gemini 4 this experiment was operated successfully, and all data was relayed to the ground. During Gemini 7, however, an erratic response indicated a failure in the proton mode, and the received data was inconclusive.

MSC3 Tri-Axis Flux-Gate Magnetometer (Gemini 4, 7, 10, 12)
PI and affiliation: William D. Womack, NASA MSC.
Objective: To monitor the direction and the maximum extent of the Earth's magnetic field with respect to the spacecraft.
Equipment: A tri-axis flux-gate magnetometer, which consisted of an electronic unit and aft-facing sensors located in the equipment adapter section. The sensors were located on a deployable boom that could be extended beyond the end of the adapter. By using two switches, the Pilot could first extend the boom, and then activate both this and experiment MSC-2.
Results: This experiment was first flown on Gemini 4, was completely successful, with all data relayed to the ground. On Gemini 7, the Z-axis detector failed before launch, but both the X-axis and Y-axis detectors performed as expected during the mission. From the Gemini 10 mission the data was not conclusive, but data collection from Gemini 12 was successfully accomplished, and was collaborated with the data from Gemini 4. During the tenth revolution of the Gemini 12 mission, the magnitudes of the geomagnetic fields were measured, and compared well with the theoretical calculated magnitudes using computer (McIlwain) codes.

MSC4 Optical Communications (Gemini 7)
PI and affiliation: Douglas S. Lilly, NASA MSC.
Objectives: To evaluate an optical communication system (laser), to check the ability of a crew to act as a pointing element, and to probe the atmosphere using an optical coherent radiator from the outside atmosphere.
Equipment: A flight transmitter and a ground-based receiver–transmitter system.
Results: This experiment was affected by unfavourable cloud-cover, as well as operating difficulties with the ground-based equipment, and yielded little useful data. However, the laser beacon was visible from orbit.

MSC5 Lunar UV Spectral Reference (Gemini 10)
PI and affiliation: Roy C. Stokes, NASA MSC.
Objective: To determine the UV spectral reference of the lunar surface at 2000–3200 Å.
Equipment: A 70-mm Maurer camera with a UV lens.

Results: When the launch date slipped, this experiment was cancelled before the flight.

MSC6 Beta Spectrometer (Gemini 10, 12)
PI and affiliation: James R. Marbach, NASA MSC.
Objective: To obtain accurate predictions of radiation doses to which the Apollo crews would be subjected, in order to assess the degree of hazard and to devise preventive measures.
Equipment: Similar to the proton–electron spectrometer used on MSC-4, but for this experiment the equipment was of a different design. It consisted of two containers, and had a total weight of 16lbs. One of the containers housed the detector, and the other housed the data-processing system. The equipment was located in the retrograde section of the spacecraft adapter, and during launch was protected by a half-hinged door that was automatically jettisoned during the separation of the spacecraft from the Titan booster.
Results: An unexpected high usage of fuel during the first two days of Gemini 10 eliminated control attitude passes, but on the third flight day – as the spacecraft passed through the South Atlantic Anomaly – one good traverse of the magnetic field was achieved. From within the Anomaly, the location of data points was good, and this provided a suitable picture of the electron distribution direction. The experiment was flown a second time on the final mission, Gemini 12, during which the equipment recorded omnidirectional flux consistent with findings recorded on Gemini 10. In addition, the equipment recorded the decay of electrons artificially injected by the July 1962 Starfish high-altitude nuclear test, which was so low that naturally trapped electrons were becoming detectable.

MSC7 Bremsstrahlung Spectrometer (Gemini 10, 12)
PI and affiliation: Reed S. Lindsay Jr, NASA MSC.
Objective: To measure breaking radiation (bremsstrahlung) flux-energy spectra inside the spacecraft while passing through the South Atlantic Anomaly.
Equipment: An X-ray detection system (weighing less than 7.5 lbs), mounted on the inner wall of the crew compartment behind the Command Pilot's seat, at shoulder height.
Results: During Gemini 10, results determined that the measurement of radiation levels is possible with this spectrometer. The experiment was flown a second time on Gemini 12, during which the crew activated the experiment four times for a total of 32 hours. From this mission, data indicated that electrons indeed penetrated the wall of the spacecraft, but that the energy distribution was within reasonable estimates.

MSC8 Colour-Patch Photography (Gemini 10)
PIs and affiliations: John R. Brinkman, NASA MSC; Robert L. Jones, NASA MSC.
Objective: To determine the possibility of obtaining true colour pictures during stand-up EVA.
Equipment: A colour patch/slate (8 × 8 × 1/16-inch) supporting four colour targets

(red, blue, yellow and grey) in a matt-finish ceramic. A three-foot extension rod held the patch 3 feet in front of a 70-mm Maurer camera.

Results: Problems with the spacecraft ECS terminated the EVA, and only four of the planned nine frames were obtained. The colour patch and rod were discarded prior to the closing of the hatch, but sufficient data were obtained from the exposed frames and back-up colour patch to determine that commercial colour film was indeed suitable for space photography.

MSC10 Two-Colour Earth's Limb Photography (Gemini 4)
PI and affiliation: Max Petersen, Massachusetts Institute of Technology.
Objective: To determine whether the Earth's limb might be used in future guidance and navigation sightings.
Equipment: A 70-mm Hasselblad camera with black-and-white film, and a special filter mosaic which allowed each picture to be taken partially through a red filter and partially through a blue filter. The experimental film magazine weighed 1 lb.
Results: Thirty good-quality pictures were taken.

MSC12 Landmark Contrast Measurement (Gemini 7, 10).
PI and affiliation: Charles E. Manry, NASA MSC.
Objectives: To measure the visual contrast of land–sea boundaries and other types of terrain for planned onboard Apollo guidance and navigation.
Equipment: A stellar-occultation photometer – a single-unit, dual-mode, hand-held, eternally powered instrument measuring $5 \times 5 \times 3$ inches, and weighing 2.5 lbs. It measured the contrast of a Sun-illuminated ground target, and to determine the extent to which the sight-line to a selected star penetrated a planetary (Earth) atmosphere.
Results: On Gemini 7 the instrument malfunctioned, and no information was obtained. The experiment was not performed on Gemini 10, due to fuel usage and limitations of time.

Technological experiments
T1 Re-entry Communications (Gemini 3)
PIs and affiliation: Lyle C. Schroeder, Theo E. Sims, William F. Cuddihy, NASA LaRC.
Objective: To inject fluid into ionised plasma during re-entry of the spacecraft to determine whether blackout would be sufficiently reduced to allow communications with the crew.
Equipment: A water-expulsion system (weighing about 85lbs) was mounted on the inside surface of the spacecraft's right-hand landing gear door. The only astronaut involvement was the use of an activation switch inside the crew compartment.
Results: Flown on Gemini 3 only. The results indicated that there was an increase in C-band and UHF telemetry signals.

T2 Manual Navigation Sighting (Gemini 12)
PIs and affiliation: Donald W. Smith, Brent Y. Creer, NASA Ames Research Center.

Objective: To evaluate an astronaut's ability to make navigational measurements with a hand-held sextant.

Equipment: A line-of-sight optical sextant, measuring $7 \times 7\frac{1}{4} \times 69$ inches, weighing 6.25 lbs.

Results: Using a combination of learning-curve data from the initial period of familiarisation and training, base-line data for comparison with flight results and data obtained during the mission determined that the standard deviation of in-flight measurements was ± 9 arcsec, which indicated that a hand-held sextant might be useful for navigational measurements during mid-course phases of lunar or even interplanetary missions. During Gemini 12, the performance of the Pilot was the same in space as it had been on the ground.

Department of Defense experiments

D1 Basic Object Photography (Gemini 5)

PI and affiliation: USAF Avionics Laboratory (H.T. Kozuma, monitor), Wright-Patterson AFB.

Objectives: To determine an astronaut's ability to acquire, track and photograph objects in space.

Equipment: A 35-mm Zeiss Contarex camera, mounted at the Pilot's window.

Results: Without problems.

D2 Nearby Object Photography (Gemini 5)

PI and affiliation: USAF Avionics Laboratory (H.T. Kozuma, monitor), Wright-Patterson AFB.

Objectives: To obtain high-resolution pictures of an orbiting object, while manoeuvring, station keeping and observing in manual control mode.

Equipment: A 35-mm Zeiss Contarex camera, mounted at the Pilot's window.

Results: The experiment could not be performed, due to the abandonment of the Radiation Evaluation Pod (REP).

D3 Mass Determination (Gemini 8, 11)

PI and affiliation: AFSC Field Office (Rudolph J. Hamborsky, monitor), NASA MSC, DoD.

Objective: To assess the techniques and the accuracy of a direct contact method of measuring the mass of an orbiting object.

Equipment: There was no special equipment required for this experiment. Following the docking with the Agena, the Gemini would push the docked combination with a known thrust. From the change in velocity from the orbiting object, its mass could be computed.

Results: The early termination of Gemini 8 prevented docked manoeuvres. During Gemini 11 the experiment was successfully completed, and proved the method feasible. However, it was determined that additional statistical data should be obtained from other 'future missions' before such a system could be adopted as an operational mode to determine the mass of an orbiting object.

D4 Celestial Radiometry (Gemini 5, 7)
PI and affiliation: AF Cambridge Laboratory (Burden Brentnall, monitor), USAF Hanscom Field.
Objective: To provide information on the spectral analysis of star fields, the principle planets, the Earth, the Moon, and other objects such as satellites and the REP.
Equipment: Common mirror optics radiometric measuring devices measured radiant intensity from the ultraviolet through the infrared as a function of wavelength; radiometer, interferometer, and cryogenic interferometers. Minor variations were incorporated for Gemini 7.
Results: On Gemini 5 the crew gathered 3 hours 10 minutes of data from twenty-one measurements of thirty objects. This demonstrated the advantages of manned missions to obtain basic data in the identification and selection of a target, the choice of equipment mode, the ability to track effectively, and the augmentation, validation and co-ordination of data through real-time voice comments. During the second flight, Gemini 7, the astronauts took 37 separate measurements obtained in 3 hours 16 minutes 19 seconds of data-gathering. (See also *D7 Space Object Radiometry*.)

D5 Star Occultation Navigation (Gemini 7, 10)
PIs and affiliation: Robert M. Silva, Terry R. Jorris, AF Avionics Laboratory.
Objective: To investigate feasibility and operational value of stellar occultation measurements in the development of a simple, accurate and self-contained navigational capability.
Equipment: A stellar-occultation photometer (see *MSC12*).
Results: On Gemini 7 the instrument malfunctioned, and no useful data was obtained. On Gemini 10, due to a difficulty with attitude control while docked to Agena, only five stars could be tracked to total occultation (six were required). In the undocked configuration, while seven stars were tracked this time, problems were encountered in entering visual occultation data into the computer. It was determined, however, that the technique was accurate and flexible, and could be useful for automatic, semi-automatic and aided manual navigation applications.

D6 Space Photography (Gemini 5)
PI and affiliation: AF Avionics Laboratory (H.T. Kozuma, monitor), Wright-Patterson AFB.
Objective: To study the problems associated with acquiring, tracking and photographing terrestrial objects.
Equipment: A 35-mm Zeiss Contarex camera, muonted at the Pilot's window.
Results: Weather hampered much of this experiment, with several of the designated areas covered in cloud, although parts of the experiment were performed successfully.

D7 Space Object Radiometry (Gemini 5, 7)
PI and affiliation: AF Cambridge Laboratory, USAF Hanscom Field.
(See *D4 Celestial Radiometry*.)

D8 Radiation in Spacecraft (Gemini 4, 6)
PIs and affiliation: M.F. Schneider, J.F. Janni, G.E. Radke, AF Weapons Laboratory, Kirkland AFB.
Objective: To Measure radiation levels and distribution inside the spacecraft.
Equipment: Seven sensors inside the spacecraft – five on the wall of the pressure vessel, and two inside the cockpit, one of which was shielded to simulate the amount of radiation which the crew would receive beneath the skin. The shield was removed during passes through the South Atlantic Anomaly. On Gemini 6, an additional removable brass shield was placed on the tissue – the equivalent of the ionisation chamber of the Command Pilot's hatch.
Results: On Gemini 4, recorded radiation doses were within acceptable levels. On Gemini 6, during the first run the survey was performed by the Pilot, but the Command Pilot was performing station keeping activities and inadvertently failed to remove the shield from the sensor. During the second run, both crew-members were busy station keeping with Gemini 7. Additional data were obtained, but the primary objectives of this experiment were not achieved.

D9 Simple Navigation (Gemini 4, 7)
PIs and affiliation: Robert M. Silva, Terry R. Jorris, AF Avionics Laboratory.
Objective: To gather information on phenomena that could be used for autonomous space navigation.
Equipment: A hand-held (8-lb) sextant, containing filters of natural density, blue haze, and green emission.
Results: During Gemini 4 the results were good, but a lack of statistical data made evaluation difficult. During Gemini 7, crew performance and equipment operation were classed as excellent. The results included 37 star-to-horizon, five planet-to-Moon (or star-to-Moon) limb, six star-to-star, and eight zero measurements.

D10 Ion-Sensing Attitude Control (Gemini 10, 12)
PI and affiliation: Rita C. Sagalyn, AF Cambridge Laboratory.
Objective: To Investigate the feasibility of an attitude control system using environmental positive ions and an electrostatic detection system to measure the spacecraft pitch and yaw.
Equipment: Two sensors, each $11 \times 6.5 \times 6$ inches, were mounted on 3-ft-long booms, and weighed 7 lbs. There were seven computed data-points, and it operated at an angle of $\pm 15°$.
Results: From the results obtained during Gemini 10, a comparison of the system with the inertial guidance system showed agreement in the measurement of both pitch and yaw angles, and the response of the system to variations of position was very rapid – in the order of milliseconds. The Gemini 12 data proved that it was possible to measure both pitch and yaw to within a fraction of degree, which could significantly reduce the crew-time required for such manoeuvres as docking, photography and re-entry. The Gemini 12 crew was able to reduce the time to align the inertial platform from forty minutes to five minutes by using the pitch and yaw sensors as a reference. It was also determined that the addition of an horizon sensor

would provide complete data on spacecraft position and attitude, and that the addition of a servosystem would allow a complete automatic attitude control system from the lowest orbital altitudes to at least ten Earth-radii.

D12 Astronaut Manoeuvring Unit (Gemini 9, Gemini 12)
PI and affiliation: AFSC Field Officer (Edward G. Givens, project officer), NASA MSC, DoD.
Objectives: To provide EVA mobility and control in attitude and translation, and to provide oxygen supply and communications.
Equipment: A rectangular (32 × 22 × 19-inch) aluminium back-pack (weighing 166 lbs fully loaded) with a form-fitting cradle in which the EVA pilot was positioned during operation. There were four forward-firingand four aft-firing thrusters, as well as two up-firing and two down-firing thrusters. A total of 24 lbs of hydrogen peroxide was provided for thruster supply, controlled via two side arm supports. The left-hand assembly was used for translation in four directions, a switch control for the selection of manual or automatic stabilisation, and a volume control for voice communications. The right-hand assembly contained the controls for positioning the unit in pitch, roll and yaw, and contained 7.5 lbs of oxygen and a battery-powered UHF transceiver to provide communications with the Command Pilot inside the spacecraft.
Results: The AMU exercise was terminated early during Gemini 9, when the Pilot's suit became overheated and his face-plate repeatedly fogged. He had turned on internal power, but had not released from the stowage location. The experiment was cancelled from Gemini 12 prior to the flight, to allow more evaluation of basic EVA techniques before again attempting such an exercise (inside Skylab in 1973).

D13 Astronaut Visibility (Gemini 5, 7)
PI and affiliation: Seibert Q. Duntley, University of California.
Objective: To test the crew's visual performance during the flight, and their ability to detect and recognise objects on the Earth's surface.
Equipment: An In-flight Vision Tester – a small, self-contained, binocular optical device with a transilluminated array of 36 high-contrast and low-contrast rectangles, half of which were orientated vertically and the other half horizontally. The size of the rectangles, contrast and orientation were random, with presentation sequential, and with non-repetitive sequences. The visual acuity equipment consisted of an in-flight photometer to monitor the spacecraft window, test patterns at two ground observation sites, and instrumentation for measuring the atmosphere, lighting and patterns.
Results: This experiment was first flown on Gemini 5, during which the crew showed no degradation of visibility during the eight-day flight. Land observations were partially obscured by weather conditions, and the opportunity to perform observations was hindered by fuel cell troubles. When the weather was good the crew was unable to properly orientate the spacecraft due to the thruster problems. Ground smoke markers were observed on each pass, on the 92nd pass the Texas site was glimpse and photographed, and it was reported that on the 107th pass the crew

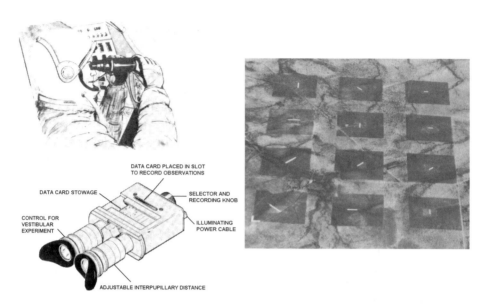

The visual acuity experiment carried out on Gemini 5. At right is the ground-marker test near Laredo, Texas.

again saw the site. The second opportunity to fly this experiment was on the 14-day Gemini 7 mission. The crew reported no apparent changes in visual performance and they saw patterns during the 17th pass and the 31st pass. (See also *S8 Visual Acuity*.)

D14 UHF–VHF Polarisation (Gemini 8, 9)
PI and affiliation: Robert E. Ellis, US Naval Research Laboratory.
Objective: To obtain information on communications systems through the inosphere.
Equipment: A UHF/VHF transmitter with an 8-foot extendible antenna, was mounted on top of the centre line of the retrograde adapter section.
Results: This experiment was not performed during Gemini 8, due to the early termination of the mission, but during Gemini 9 the experiment was performed six times. A further three planned measurements were not possible, as the antenna was inadvertently broken off by the Pilot during EVA. Only a limited number of measurements were obtained, and the experiment was only partially successful.

D15 Night Image Intensification (Gemini 8, 9)
PIs and affiliation: Thomas J. Shopple, George F. Eck, Albert R. Prince, Naval Air Development Center.
Objective: To develop a system for night surveillance of the Earth's surface.
Equipment: An image-orthicon camera, a portable viewing mirror, a recording monitor, a 16-mm camera, a TV camera control unit and an equipment control unit. One of the astronauts observed the scene directly, while the other observed through the viewing monitor. The observations were afterwards recorded, to be compared with the scenes recorded on film.

Results: This experiment could not be performed during Gemini 8, due to the early termination of the mission. During Gemini 9, 42 sequences were recorded – thirteen of medium to heavy cloud formations, and fourteen over open ocean areas. The conclusion was that lights easily identified cities, and that cloud formations were prominent even at night, as were lightning flashes, stars, and the airglow. Coastlines contrasted poorly compared with peninsulas, which were the most significant geological features observed. The Pilot also stated that the scenes viewed on the monitor were superior to the recorded film sequences of the same features.

D16 Power Tool Evaluation (Gemini 8, 11)
PI and affiliation: AF Avionics Laboratory (Victor L. Ettredge, monitor), Wright-Patterson AFB.
Objective: To determine an astronaut's ability to perform specific work tasks under zero-gravity conditions and in a pressurised suit.
Equipment: A minimum-reaction battery-powered tool, measuring 10.7 inches long and weighing 7.6 lbs; a hand wrench; a workplace with seven non-detachable bolts (four on the face and three on the reverse side); and a knee tether. The equipment was mounted in the retro-adapter, to be operated by the Pilot during EVA.
Results: Due to the early termination of Gemini 8, no EVA was performed during this mission. During Gemini 11, early termination of the umbilical EVA prevented the operation of this experiment.

Space Science experiments
S1 Zodiacal Light Photography (Gemini 5, 8, 9, 10)
PI and affiliation: Edward P. Ney, University of Minnesota.
Objectives: To photograph the Zodiacal Light (in the west after twilight and in the east before sunset), in order to determine its origin, and to determine the minimum angle from the Sun at which it could be studied without the interference of twilight; also to determine whether the Gegenschien could be detected and measured above the airglow layer.
Equipment: A 35-mm Widelux camera with high-speed colour film. On Gemini 9, the experiment was planned for operation during EVA.
Results: Gemini 5 obtained fourteen usable frames. There were no results from Gemini 8, due to the early termination of the mission. The Gemini 9 EVA operation was terminated early due to face plate fogging, although seventeen pictures were taken from inside the spacecraft. During Gemini 10, twenty pictures were obtained, but were difficult to use quantitatively, as the film was only half as sensitive as that used on Gemini 9. The crew found that observations of the same star-field in various exposures revealed that dirty spacecraft windows caused variations in light transmission by a factor of six. The horizon of the Earth could not be seen on any of the exposed frames.

S2 Sea Urchin Egg Growth (Gemini 3)
PI and affiliation: Richard S. Young, NASA Ames Research Center.

Objective: To evaluate the effects of gravitational fields on cells exposed to low-gravity conditions.

Equipment: A metal cylinder (3.25 × 6.75 inches, and weighing 25.4 oz) containing eight separate samples of sea urchin eggs, sperm, and a fixative solution. The crew-member operated a handle at one end, and this activated either fertilisation or the fixative.

Results: This experiment was flown only on the first manned mission, but could not be completed, as the handle broke near the end of the flight.

S3 Frog Egg Growth (Gemini 8, 12)

PI and affiliation: Richard S. Young, NASA Ames Research Center.

Objective: The study of the effects of sub-gravity on the development of a gravity orientated biological system.

Equipment: Two units, each mounted on a hatch sill structure, and each having four two-celled chambers – one for frog eggs, and the other for fixative. Each weighed 4 lbs. At GET 40 minutes, the Pilot turned a handle that allowed fixative (formalin) into the right-hand Chambers 1 and 2, killing the eggs and preserving them for post-flight microscopic study. Fixative was released into the right-hand Chambers 3 and 4 at GET 2 hours 10 minutes. The Command Pilot released fixative into two of the left-hand chambers immediately prior to re-entry, with the contents of final two chambers left alive for comparison. During Gemini 12, only one unit was flown, and was attached to the Pilot's hatch.

Results: During Gemini 8 the first and second chambers were correctly activated, and the second chamber was activated fifteen minutes late. The termination of the mission prevented the activation of the rest of the chambers. During Gemini 12, all phases of this experiment were completed, with good results. It was determined that gravity was not necessary for eggs to divide normally or for the latter states of their development.

S4 Radiation and Zero- g on Blood (Gemini 3, 11)

PI and affiliation: Michael A. Bender, Atomic Energy Commission.

Objective: To measure changes in human blood samples in order to examine the biological effects of known quantities and qualities of radiation.

Equipment: A phosphorous-32 radiation source, housed in an hermetically sealed aluminium box (3.7 × 1.3 × 3.8 inches, and weighing 1 lb), and located inside the cabin at the right-hand (Pilot) hatch. An identical control package was kept in a laboratory at the Cape during the flight.

Results: During Gemini 3 there were no apparent effects. When this experiment was reflown on Gemini 11, the bread mould *neurospora* and a thermoelectric cooler were added. Neither orbital spaceflight nor stress associated with launch and re-entry produced any significant, unpredicted genetic abnormalities on the samples flown. The ground control package also agreed with mission results, and it was concluded that there was nothing which linked radiation with the effects of spaceflight.

S5 Synoptic Terrain Photography (Gemini 4, 5, 6, 7, 10, 11, 12)
PI and affiliation: Paul D. Lowman Jr, NASA Goddard.
Objectives: To obtain high-quality pictures of large areas of land that have been previously well-mapped by aerial photography – for comparison, and to serve as a standard for the interpretation of pictures of unknown areas of the Earth, the Moon, and other planets; to obtain high-quality photographs of relatively poorly mapped areas of the Earth, to investigate continental drift, the structure of the Earth's mantle, and the overall structure of the continents.
Equipment: A modified 70-mm Hasselblad 500C camera, with 55 frames per roll of film. Gemini 11 and 12 also carried the 70-mm Maurer general-purpose camera.
Results: This (and *S6*) was one of the most frequently flown experiments on the Gemini series, and produced 895 photographs of usable quality, but varying from poor to fair to excellent. One hundred usable terrain study photographs were obtained on Gemini 4; Gemini 5 added 170 usable pictures, with a large portion being of excellent quality; Gemini 6 added a further 25 fair-to-excellent pictures; and Gemini 7 obtained 250 useful pictures, with cloud cover over many areas, altough dirty spacecraft windows accounted for some of the poor-quality frames. Gemini 10 obtained 75 pictures – again mostly of good quality, although dirty spacecraft windows and cloud cover affected some of the frames. During Gemini 11, all designated areas and some additional photographs were obtained, resulting in 145 photographs of excellent quality. Finally, 130 usable photographs were taken – mostly with the Hasselblad camera on Gemini 12.

S6 Synoptic Weather Photography (Gemini 4, 5, 6, 7, 10, 11, 12)
PIs and affiliations: Kenneth M. Nagler, US Weather Bureau; Stanley D. Soules, Environmental Science Service Administration (ESSA).
Objective: To augment information from meteorological satellite photographs from normally 400 nautical miles (or more) altitude. Photographs could be taken from Gemini from 100 nautical miles altitude.
Equipment: A 70-mm Hasselblad camera, with colour film.
Results: A wealth of usable photographs were obtained from the seven missions, totalling approximately 1,370 frames. Gemini 4 provided 200, of which about half were useful for weather studies; Gemini 5 added a further 250 excellent frames; and Gemini 6 obtained 100 high-quality pictures; Gemini 7 obtained 240 pictures, although some of them were unusable due to the coating on the windows;. Gemini 10 obtained 200 high-quality pictures; Gemini 11 obtained 180 good-quality pictures; and Gemini 12 obtained 200 excellent pictures of cloud patterns.

S7 Cloud-Top Spectrometer (Gemini 5, 7)
PI and affiliation: Fuad Saiedy, University of Maryland.
Objective: To determine the altitudes of clouds.
Equipment: 35-mm cameras with diffraction gratings and infrared film.
Results: The results obtained from Gemini 5 were good enough to warrant the design of a second-generation series of weather satellites. Unfortunately, the early termination of Gemini 8 prevented further investigations.

S8 Visual Acuity (Gemini 5, 7)
PIs and affiliations: Seibert Q. Duntley, University of California.
(See *D13 Astronaut Visibility*.)

S9 Nuclear Emission (Gemini 8, 11)
PIs and affiliations: Maurice M. Shapiro, Naval Research Laboratory; Carl E. Fichtel, NASA Goddard.
Objective: To study cosmic radiation at orbital altitudes.
Equipment:A nuclear emission package, 8.5 × 6 × 3 inches and weighing 13 lbs, stowed in the spacecraft retrograde adapter. A spring-loaded fairing jettisoned at orbital insertion, exposing the package. It was to have been retrieved by the Pilot during EVA.
Results: Early termination of the Gemini 8 mission resulted in non-retrieval of the experiment pack, but, during Gemini 11 the Pilot successfully retrieved the package. The results obtained agreed with data from high-altitude balloon flights into the upper reaches of the atmosphere.

S10 Agena Micrometeorite Collection (Gemini 8, 9, 10, 12)
PI and affiliation: Curtis L. Hemenway, Dudley Observatory.
Objective: To retrieve collector plates exposed to micrometeorite impacts, and to return them to Earth for study.
Equipment: The package measured 5.5 × 6.25 × 1 inch and weighed 4 lbs, and was mounted on the Target Docking Adapter (TDA) of the Agena. It was hinged to fold open, to expose eight plates of highly polished surfaces – plastic and glass – to be opened by the Pilot on EVA, and left for retrieval on a later mission.
Results: Gemini 8 was terminated before any EVA could be attempted. Gemini 9 EVA was postponed until the third flight day; and no EVA was attempted in the vicinity of the ATDA due to the attached launch shroud, and so the equipment was not retrieved. The Gemini 10 Pilot was able to retrieve the package from the Agena 8 target during rendezvous with the vehicle. Only four outer panels were exposed, as the package had remained in a closed position. The micrometeorite flux values generally agreed with known values from other experiments. Micro-organisms on exposed areas were dead, but those inside had a good survival rate. The plan was to have the Gemini 10 Pilot leave a new package on Agena 8 for a future retrieval, but this was not achieved, as there was a risk of his umbilical becoming entangled on the target vehicle. During Gemini 12 the EVA Pilot removed the protective fairing and exposed both the interior and exterior collection surfaces. As this was the final flight of the series, the package was lifted on Agena 12, for possible retrieval during a later (Apollo ?) orbital flight, allowing prolonged exposure to the environment. However, in December the unit re-entered after approximately forty days in orbit.

S11 Airglow Horizon Photography (Gemini 9, 11, 12)
PI and affiliation: Martin J. Kooman, Naval Research Laboratory.
Objective: To Photograph in the atomic oxygen and sodium spectra of the Earth's airglow, in order to study the characteristics and dynamics of the upper atmosphere.

Equipment: A 70-mm Maurer camera, with an extended exposure timer, an illuminated camera sight, and two-point variable pitch brackets for mounting the camera at the Pilot's window.

Results: Gemini 9 obtained 44 pictures including three pictures of dayglow; Gemini 11 obtained 23 useful pictures showing variations in altitude and intensity of the airglow; and Gemini 12 obtained 23 good pictures of sunlight and night airglow.

S12 Micrometeorite Collection (Gemini 9, 10, 12)

PI and affiliation: Curtis L. Hemenway, Dudley Observatory.

Objectives: To determine micrometeoroid activity in the near-Earth environment; to expose microbiological specimens to determine their survivability in a vacuum, at extreme temperatures, and to radiation; to search for any organism capable of living on mircometeoroids in space.

Equipment: An aluminium box measuring 11 × 5.5 × 1.25 inches and weighing 6 lbs 8 oz. It contained two collection experiments, an internal electric motor, and thermally insulated batteries. One compartment was sterilised to determine whether any non-terrestrial organisms were present, and the other contained the bacteria, moulds, and spores for an assessment of survival.

Results: During Gemini 9 the collection box was successfully recovered after 16 hours' exposure. Penetration holes were recorded, and some fractions of biological specimens survived, but there was no evidence of non-terrestrial organisms. The experiment was flown a second time of Gemini 10, and was retrieved during EVA. However, it apparently floated out of the cabin later in the EVA, and was lost. The Gemini 12 package was recovered after 6 hours 20 minutes exposure, and recorded fewer penetration holes than did Gemini 9. Here there was no living organism from space on the sterile collection surfaces. This confirmed sounding-rocket findings that solar ultraviolet radiation and soft X-rays are responsible for the death of micro-organisms exposed to space conditions.

S13 UV Astronomical Camera (Gemini 10, 11, 12)

PI and affiliations: Dr Karl G. Henize, Dearborn University and North Western University.

Objectives: To obtain data on UV radiation for hot stars, and to develop and evaluate basic techniques for the photography of celestial objects from manned spacecraft.

Equipment: A 70-mm Maurer camera with a UV lens. The spacecraft did not have UV windows, and the photographs had to be taken through an open hatch during stand up EVA by the Pilot. The Gemini 10 photographs had included spurious streaks and on Gemini 11 a carbon dioxide canister was included to obviate this effect.

Results: During Gemini 10, 22 photographs of the southern Milky Way were obtained. There were four problems: twelve frames were marred by a vertical streak probably caused by static electricity from the camera operation in vacuum; the image quality was poor at the centre of the field and was good away from the centre – possibly as a result of the film bowing towards the end in vacuum, and being too close to the lens when exposed; a cable release was broken during assembly of the camera and bracket; and the crew backed out, preventing proper insertion into the mount. The experiment was considered successful, however, as it provided useful scientific data and demonstrated the need for better equipment on future flights. During Gemini 11, 39 frames were exposed: five were excellent, six were good, eight were fair, thirteen were poor, two were bad and five were useless. On Gemini 12, 30 frames were exposed: three excellent, seven good, nine fair, eight poor, one bad and two light-struck. The problems continued with focus, static marks and light streaks, but the centre of the image improved, indicating that increased tension of the film-retaining spring eliminated warping (bowing) of the film.

S26 Ion Wake Measurement (Gemini 10, 11)
PI and affiliation: David B. Medved, Electro-Optical Systems Inc.
Objective: To investigate the structure of ion and electron wake.
Equipment: Inboard and outboard ion detectors, an electron detector and a data programmer on the Agena adapter. The inboard detectors collected data when the Agena was parallel to the flight path, while the outboard detector collected data when the vehicle was yawed at right angles to the flight path. The programmer sent real-time telemetry to the ground during the undocking period. This was critical, as the GATV time-delay tape recorder inadvertently cut off during undocking.
Results: During Gemini 10, limited results were obtained, due to fuel usage. Electron and ion temperatures were higher than expected, and shock effects were registered during docking and undocking. During Gemini 11 the radar, onboard voice tapes recorder (for recording start and stop times) and auxiliary receptacle (to provide time markers) were not operating. The thruster firing in adapter-south configuration decreased the ion flux to the outboard sensor, increased it to the inboard ion sensor, and enhanced the electron concentration on the outboard electron sensor. The strip-chart data revealed that definitive wake-cone angles could be determined. In many cases, electron distribution followed ion depletion effects, which indicated that the wake is plasma rather than ions.

S29 Libration Regions Photography (Gemini 11, 12)
PI and affiliation: Elliott C. Morris, US Geological Center.
Objectives: To investigate the L4 and L5 libration points of the Earth–Moon system in order to investigate the possible existence of clouds of particulate matter orbiting the Earth in these regions.
Equipment: A 70-mm Maurer camera.
Results: Because of Gemini 11's three-day mission delay, the experiment could not be carried out as planned, and the crew instead took pictures of the Gegenschein and two comets. During Gemini 12, of the eleven pictures of L4, only three were properly exposed. The mechanical failure of the shuttle mechanism in the red-lens assembly caused overexposure. The unknown amount of double exposure was caused by the failure of the film-advance at the end of the first roll. No conclusive results were obtained.

S30 Dim-Light Photography/Orthicon (Gemini 11)
PI and affiliation: Curtis L. Hemenway, Dudley University.
Objectives: To obtain pictures of faint and diffused astronomical phenomena, such as the airglow layer in profile, the brightest part of the Milky Way, the Zodiacal Light at 60 elongation, the Gegenschein, and the libration points of the Earth–Moon system.
Equipment: A low-light TV system (see *D15*) plus a spacecraft optical sight.
Results: Flown only on Gemini 11, during which 400 frames were obtained. About 30% of film for *D15* and *S30* was exposed. The camera recording cathode rays shorted out and failed during the final sequence.

S51 Sodium Vapour Cloud (Gemini 12)
PI and affiliation: Blamont Centre National de la Research Scientifique, France (the only non-US experiment flown on Gemini).
Objective: To measure the daytime wind velocity of Earth's high atmosphere as a function of altitude between 30 and 80 nautical miles, with the use of rocket-produced vertical sodium clouds.
Equipment: A 70-mm Maurer camera.
Results: Flown only on Gemini 12. The crew did not see the firings, but took 26 pictures of the area during the firings. The camera shutter locked in the open position, and all of the pictures were overexposed.

S64 Sunrise Dust UV Photography (Gemini 12)
PI and affiliation: Curtis L. Hemenway, Dudley Observatory.
Objective: To provide UV photographs of dust in the Earth's atmosphere.
Equipment: Used black-and-white film in a 70-mm camera with a UV lens.
Results: Carried on Gemini 12 as an operational experiment only. A series of sunrise photographs was obtained in the UV region, although electrostatic marks in the camera fogged nearly all of the exposures, and very little useful information was obtained.

Eclipse photography (Gemini 12)
Objective: The crew was to have initiated a phasing manoeuvre to place the spacecraft in a position to photograph the solar eclipse. With the original launch date of 9 November, eclipse photography would have been undertaken during high-apogee orbit on November 12, and magnitude and direction would be evaluated in real time. The experiment was instead revised, and was performed at 16 hrs GET, 12 November.
Equipment: A 70-mm general-purpose camera, a 70-mm superwide camera, and a 16-mm-sequence camera.
Results: Originally delayed because the Moon was out of phase, the experiment was rescheduled to take photographs of the eclipse and the resulting shadow on Earth. During totality, four exposures were made with the general-purpose camera, two with the superwide camera, and two series with the 16-mm camera. The 70-mm exposures were to overexpose to provide any useful data, and the sequence camera secured useful frames during totality.

The Red Sea area, as seen from Gemini 11.

Space photography

The ten manned Gemini mission provided more than 2,400 photographs in the fields of geology, oceanography, agriculture, hydrology, urban planning, atmospheric structure, environmental pollution control, meteorology, land management and aerospace engineering, and contributed to the creation of the awareness of the environment that led to a global environmental movement in the late 1960s. In addition, the photographs had applications in military objectives and in gathering data from celestial objects, and offered the public images of space exploration. Some of the images of astronauts and spacecraft in orbit became some of the most widely used images of the space programme in books and magazines, and in portraying the image of man in space. Several images from Gemini have entered classic space history, and are still regularly used almost four decades after they were obtained. Arrays of lenses and films were used to achieve the results, and many cameras were used in a variety of roles. (Typical camera equipment was mostly adopted from standard photographic equipment, and representative images from the missions feature throughout this book.)

Gemini photographs are of great scientific, academic and engineering value, and many are visually stunning. The successful use of camera; on the Gemini missions ensured their inclusion in the Apollo and follow-on programme, which provided even more spectacular images. This has continued to the current Space Shuttle and ISS programmes, and not only offers the scientist and engineer photographic documentation of the Earth and its environment, but also provides the general public with a tantalising glimpse of life in space.

Experiment performance status

Mission	Number of experiments	Experiments accomplished	Problem source
3	3	2	Experiment
4	11	11	
5	17	16	Mission
6	3	3	
7	20	17	Experiment
8	10	1	Mission
9	7	6	Mission
10	15	12	Mission
11	11	10	Mission
12	16	12	Experiment and mission (1)
Totals	113	90	

More than 79.6% were accomplished overall. More than 13.2% were not accomplished due to mission problems, while more than 6.1% were not accomplished due to experiment equipment problems

Experiments – flight assignments

No.	Experiment	Flight 3	4	5	6	7	8	9	10	11	12
		No./Flight (3)	(11)	(17)	(3)	(19)	(12)	(7)	(15)	(11)	(16)
Biomedical experiments (8)											
M1	Cardiovascular conditioning			x		x					
M3	In-flight exerciser		x	x		x					
M4	In-flight phonocardiogram		x	x		x					
M5	Bioassays; body fluids		x				x	x			
M6	Bone demineralisation			x		x					
M7	Calcium balance study					x					
M8	In-flight sleep analysis					x					
M9	Human otolith function			x		x					
Manned Spacecraft Center experiments (10)											
MSC1	Electrostatic charge		x	x							
MSC2	Protonelectron spectrometer		x			x					
MSC3	Tri-axis magnetometer		x			x					
MSC4	Optical communication					x			x		x
MSC5	Lunar UV spectral reflectance								x		
MSC6	Beta spectrometer								x		
MSC7	Bremsstrahlung spectrometer								x		x
MSC8	Colour patch photography								x		x
MSC10	Two-colour Earth-limb photography		x								
MSC12	Landmark contrast measurement					x			x		
Technology experiments (2)											
T-1	Re-entry communications	x									
T-2	Manual navigation sightings										x
Department of Defense experiments (15)											
D-1	Basic object photography			x							
D2	Nearby object photography			x							
D3	Mass determination						x			x	
D4	Celestial radiometry			x		x					

Experiment		1	2	3	4	5	6	7	8	9	10
D5	Stellar occultation navigation					x	x				
D6	Surface photography			x				x			
D7	Space object radiometry			x	x			x			
D8	Radiation in spacecraft	x				x					
D9	Simple navigation	x				x					
D10	Ion-sensing attitude control				x				x		
D12	Astronaut Manoeuvring Unit		x			x					
D13	Astronaut visibility		x	x		x					
D14	UHF/VHF polarisation			x		x					
D15	Night image intensification		x			x	x				
D16	Power-tool evaluation					x	x				
Space science experiments (19)											
S1	Zodiacal Light photography		x		x	x	x				
S2	Sea urchin egg growth	x									
S3	Frog egg growth	x				x	x				
S4	Radiation and zero g on blood	x									
S5	Synoptic terrain photography	x	x	x	x		x	x			
S6	Synoptic weather photography	x	x	x	x		x	x			
S7	Cloud-top spectrometer	x		x							
S8	Visual acuity	x	x								
S9	Nuclear emission	x				x	x				
S10	Agena micrometeorite collection	x	x		x	x	x				
S11	Airglow horizon photography	x	x		x	x	x				
S12	Micrometeorite collection	x	x		x	x	x				
S13	UV astronomical camera	x	x		x	x	x				
S26	Ion wake measurement		x			x					
S29	Libration region photography	x				x					
S30	Dim-light photography/orthicon		x	x		x					
S51	Sodium vapour cloud	x	x			x					
S64	Sunrise UV photography	x	x			x					
	Eclipse photography	x	x			x					

Military Gemini

During the mid-1950s the USAF began the evaluation of a manned spaceflight programme, including space stations and manned lunar landings. With the creation of NASA in 1958, most of these plans were transferred to the civilian agency or were cancelled, allowing the USAF to concentrate on the development and deployment of ICBMs and responses to the Soviet missile threat – the Cold War.

THE X-20 DYNA SOAR

Despite the reduction of manned space activity by the Air Force, there remained one project that required funding and support: the X-20 one-man winged space-aeroplane known as Dyna Soar. Dyna Soar was to have been launched on top of a military (Titan) rocket, have the capacity for manoeuvring in space to perform satellite inspection, interception and reconnaissance missions, to manoeuvre during re-entry, and to land on skids on a runway, to be refurbished and used again.

The disadvantage of the system was the price tag of $1,000 million (1963). This constantly affected the mission profile, objectives and capabilities, which changed the systems from short sub-orbital missions to much higher altitudes to increase re-entry speeds and heating, to full orbital missions. Every change induced a further delay that pushed the cost ever higher. But the hardware and programme difficulties represented only half of the problem. The then Secretary of Defence, Robert S. McNamara, considered that it was much too soon to develop an operational space-plane concept, and it was decided, after a cost/performance analysis, that as an experiment, X-20 was just simply too expensive.

When NASA announced the development of the Mercury Mark II in 1961, the final fate of X-20 was sealed. As plans for the new NASA programme evolved it took some of the objectives of Dyna Soar, including rendezvous and docking, and with little modifications to the basic design, but required major system adjustment to the X-20 in the USAF programme. For most of 1963 the requirement for the X-20 programme gradually decreased, and its benefit to NASA was questioned.

There were questions into whether the funds could be better spent other than in supporting the further development of the X-20. Studies into the possible military use of Gemini were underway, and plans for a military space station were evolving. Therefore, on 10 December 1963 McNamara announced the termination of Dyna Soar without any flight hardware built or missions flown. Dyna Soar was ahead of its time, and would have included everything in one programme, including precise re-entry and land landing, vehicle turnaround and reuse, and the demonstration of the capability of man in space over automated spacecraft. The vehicle was to be launched on a Titan I, but soon outgrew this to be launched instead on a Titan II, and finally, as its launch weight increased, on the larger Titan III.

On the very same date, 10 December, came the announcement of plans for the USAF Manned Orbiting Laboratory, which featured the orbiting of a two-man Gemini, crewed by USAF personnel who would occupy a pressurised laboratory attached to the rear of the Gemini for missions up to 30 days. This programme was expanded by a co-operative agreement with NASA on 25 January 1965. The concept was founded on proven technology. The USAF could use a reconfigured NASA Gemini on a Titan III launcher, while NASA, under the agreement, could use the Titan III instead of the Saturn 1B to launch Apollo Command and Service Modules into Earth orbit. A subsequent study determined that the Apollo/Saturn combination was far more versatile than the Gemini/Titan III combination, and the Apollo/Titan III never progressed further than paper proposals.

USAF CO-OPERATION IN GEMINI

The idea of the Air Force utilising Gemini for its own needs had evolved over the previous two years. On 7 December 1961 NASA officially announced plans to develop the two-man Mercury capsule, and also announced that the launch vehicle would be the USAF Titan II. On the same day, Robert Seamans, NASA Associate Administrator, and John Rubel, DoD Deputy Director for Defense Research and Engineering, offered recommendations to McNamara on the division of effort and responsibility for the new programme between NASA and the DoD. It was stressed that NASA would remain in control of the management and direction of the programme, and that the DoD would provide contractor resources (mainly USAF) in attaining programme goals. Primary NASA contractor for DoD procurement of the Titan and Atlas was the Air Force Space Systems Command, which would also handle the preparation and launch of the vehicles for the civilian agency, providing the Air Force and DoD personnel with experience in the design development and operation of a manned spaceflight programme. DoD support would continue in areas of launch operations, astronaut training, recovery techniques, and the tracking and communications network.

By 21 January 1963, Jim Webb, NASA Administrator, and Robert McNamara completed a major policy agreement which defined the roles of NASA and the DoD in Project Gemini The result was a joint NASA–DoD Gemini Program Planning Board which would evaluate potential experiments, conduct flight tests, and interpret results. NASA would continue to manage and direct the programme, but the DoD would participate in the development of the spacecraft, crew training, pre-flight checkout, launch, and flight operations, be mainly responsible for the Titan II and Atlas–Agena vehicles, and also allocate funds toward completion of Gemini programme objectives.

This co-operative agreement reached a point of conflict in the spring of 1963 when McNamara testified in the DoD 1964 fiscal budget statement, and in an article *Missiles and Rockets*, published in April 1963. He expected that the USAF would take over the role of management and direction of Gemini towards the end of the flight programme, to meet USAF objectives, while NASA would move to operational aspects of Apollo. In response, Seamen stressed NASA's primary management responsibility in Gemini, and stated that although the role of the DoD was vital in programme management, development and operations, NASA would retain the final responsibility for programme success to the last flight.

USAF Gemini plans, 1963
These comments had evolved from USAF studies into the use of Gemini for DoD objectives apart from the NASA stated role. On 8 February 1963 Colonel Kenneth Shultz, of Headquarters, AF Office of Development Planning, had outlined the DoD objectives for Gemini at the first meeting of the Joint Planning Board.

Three general objectives were established as a range of orbital experiments:

1. Relating to such possible future missions as the inspection and interception of both co-operative and passive or non-co-operative objectives in space under a variety of conditions, the logistics support of a manned orbiting laboratory, and photoreconnaissance from orbit.
2. Gaining military experience and training in all aspects of manned spaceflight.
3. Assessing the relationship between man and machine from the aspect of potential military missions.

Some of these were adapted from the cancelled X-20 Dyna Soar, but also aimed at a future longer-term presence of military man in space, and stated the desire to establish a military space platform for photoreconnaissance from orbit. Between 1956 and 1960 the USAF conducted U-2 high-altitude spy-plane missions over foreign territory – notably the Soviet Union. The purpose of the flights was to access the missile capability of the Soviet Union, and to update intelligence of what became known as 'the missile gap' between the USA and the USSR. Estimates of Soviet missile strength varied considerably as the data was collected and interpreted by six bodies: the USAF, the USN, the US Army, the CIA, the Joint Chiefs of Staff, and the State Department. The information was often confusing, misinterpreted, or simply incorrect.

However, the over-flights continued until Gary Powers was shot down by a SAM ground-to-air missile on 1 May 1960. The U-2 flights were cancelled (at least over the Soviet Union) and although the data failed to reveal the number of Soviet missile; they offered valuable information on the infrastructure and facilities of missile and submarine bases, airfields, mineral mining and railways. Each flight was a risk–as demonstrated by the Gary Powers incident – and to prevent a repetition, the USAF was developing an unmanned military satellite programme – Discovery.

Discovery and Zenit

Discovery was a programme of unmanned spy-satellites, each carrying a recoverable capsule into orbit, and which, after the photographs and data had been recorded, would separate, re-enter and be air-recovered over US territory. Originally called Corona, this series of satellites became publicly known as Discovery. The first launch occurred in early 1959 and, with mixed results, continued through December 1962, encompassing the Cuban missile crisis of October that year.

The year 1962 also saw the first flights of the Soviet unmanned spy satellites called Zenit, which had evolved from the early design studies of the Vostok manned spacecraft. The derivatives of Zenit continued flying for more than forty years, while various generations of American spy satellites were launched by unmanned launch vehicle and Shuttle vehicles during the same period.

In order to evaluate the capabilities of unmanned operations, as opposed to the stated potential of also using USAF personnel in space, in 1963 the USAF proposed flying military-orientated Gemini missions in addition to the NASA series.

The Martin Marietta Report, April 1963

On 5 April 1963 the primary contractor for the Titan II – Martin Marietta – reported on the results of an investigation into improvements on the Gemini launch vehicle. The restricted study covered both NASA and USAF mission requirements, the performance of the Titan launch vehicle, improvements to the propulsion systems, and the structure of the Titan. The document also provided an early indication of the types of mission that the USAF hoped to perform with Gemini spacecraft.

The study assumed that launches would occur between 1964 and 1967, and centred on improvements to the Titan II to allow the use of new materials for construction, improved engine uprating, and advanced guidance's systems (and also removing redundant research and development equipment). A high-energy third stage could be incorporated on to the Titan vehicle, allowing a payload of 7,300–10,500 lbs to be placed in an orbit of 90–400 nautical miles. For the purpose of the study, four two-stage and five three-stage Titan vehicles were defined. The two-stage vehicle encompassed a Stage 2 design increase, ullage volume reduction, engine uprating, and second-stage restart and guidance modifications. The three-stage vehicle used an Agena D, Centaur or a 'liquid injection stage' (LIS) for the third stage. The LIS was the optimum choice, using GLV propellants and an ablatively cooled engine. These feasibility studies depended heavily on uprating the YLR87-AJ-5 engine.

The USAF mission profile envisaged for this programme included six specific missions, grouped in three categories, and defined as 'unclassified orbital mechanics considerations'. Initially, the missions would have a principle objective of training personnel and developing orbital equipment and techniques Secondly, operational reconnaissance and surveillance missions would proceed with interim satellite interception and rendezvous missions (SAINT). Launches would occur from the Atlantic Missile Range (Cape Canaveral) in Florida, although the Pacific Missile Range (Vandenberg) in California was also considered. It was conceded that this was by no means the full extent of the missions that could be achieved, and other possibilities could include 'shuttle (ferry) vehicle tasks, multiple launches with subsequent rendezvous, and joining of vehicles in orbit'.

With little change to the objectives of the NASA Gemini missions, additional experience could be gained by USAF personnel in preparing, launching, controlling, flying and recovering Gemini DoD missions with little or no adjustments, thus providing experience for more advanced missions.

Possible DoD Gemini missions
The report indicated that development of orbital techniques would be conducted with maximum payload capability. If reconnaissance and surveillance sensors were to be carried on the Gemini, a minimum payload of 400 lbs would be necessary – but up to 1,500 lbs desirable – in an orbit of 90 × 150 nautical miles.

For a reconnaissance mission, the definition was the mapping and gathering of intelligence data over large areas on a relatively routine basis, while surveillance was determined as gathering high-resolution or detailed data from a few small areas. For the purpose of the study, the geographical locations were confined to 40° and 80° north latitudes. Taking into account safe consideration from the AMR, a mission profile of between 43° and 61° was optimum, and similar targets could be achieved from PMR, with primary interest on due south maximum coverage per pass it was suggested an orbit of 125 to 150 nautical mile was optimum but could be as low as 90 nautical miles and circular for sensor considerations. Thus the report predicted an AMR launch between 43° and 61° azimuth yielding a 53° and 40° inclination while from PMR a 180° azimuth related in a 90° orbital inclination.

The AMR orbit was particularly useful for photoreconnaisance missions during winter months, when northern regions were insufficiently illumined, and for low-latitude targets. Lighting conditions over target areas would see launch windows of only one hour in the winter, to as much as four hours in the summer, and required either day or night launches, depending on the target and the time of year. The first orbit was within tracking range of the 'territory of interest', while the next five orbits would not pass over the target territory, and would allow tracking, orbit adjustments and communications with US territory tracking stations. Orbit seven would take the Gemini over its target area in a northerly trajectory, while southern passes, beginning on the ninth pass, would pass over 'recon territory'.

The PMR allowed almost every launch to pass over the recon territory, since the orbital ground-track covered 170° of longitude. If daytime coverage was desired, then only half of the orbits could be considered operationally useful, while tracking could be performed from Alaska, Hawaii and Western Canada to support communications in these orbits. The first southern operational passes occured on the third and fourth revolutions while the first northern operational passes occurred on the tenth or eleventh orbit.

Landing at the end of the mission would be in what was termed a 'fenced-in area on the continental United States' – in other words, a military base, enabling land-based recovery of Gemini within controlled conditions. Landing could take place from a north-westerly or south-westerly direction during the day or the night, and included the capability to recall the vehicle frequently at least once a day as the

CONFIDENTIAL

Mission	Mission Objectives	Launch Site	Launch Azimuths (deg)	Launch Window (hrs)	Orbit Altitude Band (n.m.)	LV or SC Changes Required Spacecraft Propulsion	LV or SC Changes Required Guidance	LV or SC Changes Required Additional Payload* (lb)
Personnel Training	Training in conditions of orbital flight, rendezvous & docking.	AMR	81 to 98 for rendezvous	2 to 3 for rendezvous	90 perigee	None	None	None
Equipment Development	Develop & evaluate orbital equip. & techniques				90 to 150 circular for sensor testing	None	None	300 up to maximum vehicle capability
Recon & Surveillance	Systematic coverage of large areas & high resolution data on specific targets	AMR	43 to 61	For photo: summer (4 to 5) winter (approx 1)	90 to 150	None	None	1800 (R or S) 600 (R only)
		PMR	180				RGS marginal	1500 (R or S) 600 (R only)
Minimum Interception	Pass near the target satellite, moving in the same general direction	AMR	50		100 to 200 for recon satellites & up to 400 for comm & other high altitude satellites	None	None	Min of 400
		PMR	180	Several minutes			RGS marginal	Min of 100
Interception with period synchronization	Like minimum but with orbital periods matched for longer & recurring contacts	AMR	50			Additional 300 to 400 ft/sec	None	Approx 850
		PMR	180				RGS marginal	Approx 550

*AMR weights have been converted to equivalent weights on a 105° Azimuth plot.

A summary of Blue Gemini objectives. USAF document, 5 April 1963, declassified 1975. (Courtesy USAF.)

mission and real-time situation dictated. The proposed landing areas for the military Gemini mission were Edwards AFB (California), Ewendover AFB, White Sands Missile Range (New Mexico), and the Las Vegas Bombing Range (Nevada).

The NASA/DoD Joint Ad Hoc Study Group report, May 1963
This evolved from the 21 January 1963 agreement between NASA and the DoD on joint participation on Gemini. It convened at MSC, Houston, on 25 March 1963, and completed two weeks of study into the feasibility of a joint programme, identified the experiments that could be performed under DoD investigators within the NASA programme, and indicated the maximum benefit if USAF participation was attached to a defined goal.

On 9 April an interim report was presented to the Gemini Programme Planning Board, upon which the Study Group was instructed to address further attention to defining DoD experiments to be carried on the NASA programme, and to evaluate an additional add-on programme of between two and six Gemini flights sponsored and funded by the DoD, which would further expand experiment development for future DoD spaceflight equipment for DoD mission objectives and training for DoD astronauts.

In order to achieve this within a short time (the final report of just 100 copies was published on 6 May 1963) a policy board and five working committees were established within the Study Group, and were comprised of both USAF and NASA personnel, in areas of schedules, manpower, costs, facilities, hardware, trade-offs and benefits.

Objectives of the Air Force programme
The Study Group recognised the important of NASA effort in using Gemini objectives as a stepping stone to the Moon, and the need for NASA to retain programme direction, management and control until those objectives were realised.

In addition, the USAF believed at that time that a contribution to national security could be obtained by the service participating in Gemini after the lunar landing development objectives had been secured and when NASA moved on to Apollo Applications. These stated objectives were:

1. Inspection, identification and neutralisation of potentially hostile spacecraft.
2. Observation of surface activities to provide intelligence, warning and targeting data.
3. Retaliation from a space platform by bombardment or by other forms of attack of hostile activities.
4. Development of protective measures to safeguard satellite activities from hostile action, and other developments of reliable space hardware through in-space experiments conducted under active, manned control.

The general USAF Gemini objective was 'to increase the fund of experience and information necessary to permit adequate decisions related to possible manned military space systems, and to develop, test and evaluate space equipment for the accomplishment of military mission'.

The plan was also to expand Air Force organisational experience with knowledge in Gemini systems and experience in manned spaceflight, training Air Force individuals to gain working experience, allowing the Air Force 'to further exploit Gemini experience, to conduct other manned spaceflight activities as required.' The plan also suggested that a component flow of data from NASA to the Air Force would allow utilisation of experience gained during the NASA programme, to prevent duplication in the AGF programme and to allow time-critical adjustment from civilian to military aspects of the programme.

It was also stated that limiting Air Force participation in Gemini to just a programme of experiments would not solve defence objectives, and in order to fully develop the capabilities, authorisation of two to six additional military Gemini flights was proposed. This additional programme became known as 'Blue Gemini'.

BLUE GEMINI

The study group also evaluated what the NASA programme could contribute to the DoD objectives in addition to possible future dedicated DoD flights. In each of the major objectives of the NASA programme, the application to DoD objectives was summarised:

Long duration Recognised as a joint objective with NASA to better understand the limits of man's capabilities in space for up to 14 days. A baseline of data would be provided, which could be useful in assigning military objectives in recognising capabilities, restrictions and requirements.

Rendezvous and docking Although the NASA experiments were geared towards Apollo's lunar requirement, the Group indicated that the use of an Agena target vehicle could contribute to the evolution of techniques. Other rendezvous experiments could develop techniques for the orbital inspection and interception programmes.

Post-docking manoeuvres The development of this capability was useful for the military flights, in its ability to changing orbit during surveillance missions or inspection missions. In addition, significant manoeuvring capability for a quick recall might be desirable for an inspection mission or for 'orbital adjustment of observation or bombardment craft'.

EVA USAF interest in this area was limited to manned inspection and repair. Expansion of this programme would be limited by the then current design of EVA suit, and anything more might require removal of the ejection seats to provide more room, which was recognised as highly undesirable for Gemini.

Point landing The lifting and manoeuvring of Gemini was recognised as a potential benefit to the military programme, and the land landing capability using the paraglider was of more benefit than in the NASA programme. It was suggested that after the initial NASA trials, the land landing capability be applied to all future DoD Gemini missions, to allow landing in a secured military air base or facility.

In the review there was no suggestion of major modifications to the NASA programme; rather, the adaptation of a range of additional and alternative experiments to meet DoD objectives. The areas to address included hardware, facilities, scheduling and manpower, but the overriding obstacle (apart from the ever-present budget limitations) was launch weight. In recognising this, the Study Group suggested a programme that ranged from a series of small experiments of only a few pounds, to fly on NASA missions, to adapting the Titan IIIC vehicle to launch a 14,000-lb Gemini payload.

The proposed extensions of NASA Gemini experiments and objectives were:

Piggy-back experiments Every NASA Gemini mission revealed potential for carrying a military payload that had no impact on the NASA mission to which it was assigned (which became the series of DoD experiments actually flown on Gemini). To support these experiments – which were of lower priority than the NASA experiments assigned to support the lunar programme – the Group suggested the creation of a DoD experiment support office at MSC, to be managed by ten to fifteen USAF personnel, with members of the USAF involved in the In-flight Experiments Panel at MSC.

Additional flights of the NASA Gemini The study evaluated a minimum conversion of the NASA Gemini to support a basic military mission. A 1,000-lb payload could be assigned if a Gemini was launched on a low-altitude (for surveillance objectives) *one-man*, two-day mission with no rendezvous, no docking, and parachute recovery – presumably in the ocean, although land landing could be achieved using the ejection seat system for the lone astronaut, and the DoD experiment package from the second crew position.

This type of mission could, according to the 1963 report, involve an experiment and hardware lead-time that resulted in a launch in January 1966. One mission could be inserted into the NASA schedule, but this would be detrimental to Gemini operations for the lunar programme. A second flight could be added and launched by NASA at the end of the civilian programme, but this would restrict the release of the NASA Gemini team for up to six months after the end of the civilian programme, which again would affect the implementation of Apollo full-scale operations. Without a dedicated USAF programme, only two flights could be supported in the NASA Gemini schedule, but with various implications for Apollo. The report concluded that during 1967 and 1968 a USAF team could launch and operate Gemini military missions from LC 19, but this was also when the development of more advanced satellite inspection and reconnaissance missions could be expected, and flying marginal Gemini missions at this time was not attractive.

The report suggested the conversion of LC 16 at the ETR for the launch of military Gemini spacecraft concurrently with NASA Gemini operations from LC 19. This provided the advantage of a back-up launch facility, and also offered the possibility of a NASA Gemini and USAF Gemini rendezvous in space (as was achieved with Gemini 6 and Gemini 7). The report conceded, however, that without modification to Gemini and its launcher, there was very little to be gained that would not be achieved by piggyback experiment or from the NASA programme. From two basic USAF Gemini missions there would be mission planning, training, launch operations, flight control and recovery for later application, but, as the report stated, 'the benefits appear [to be] not worth the cost'. The report concluded that apart from training or as preparation for a dedicated later programme, this type of USAF Gemini programme should not be considered as a sole effort.

Add-on flights of a modified Gemini and launch vehicle A more flexible programme would be possible if one Pilot position were to be removed from the spacecraft, and this was studied, together with crew-activated booster stages to place heavier payloads into orbit. The cost and technical hurdles to achieve this pointed to the possibility that this type of configuration could be achieved, and that certain objectives not possible on the NASA Gemini orientated programme could also be achieved. But the modifications seemed to be too severe. Because of the development of the Titan III, further studies in upgrading the NASA Gemini as a one-man vehicle, and improvement in the performance of Titan with upper stages, were not considered in this report. The add-on NASA-type Gemini missions were small in value but high in cost, and the major modification of upgrading the NASA Gemini was highly expensive for very little return. The study therefore opted for development of Gemini on the Titan IIIC, and the development of an additional add-on module.

Add-on flights of the basic Gemini on a Titan IIIC Recognising the physical aspect of mating the Gemini to the top of a Titan IIIC, and man-rating the booster, it was found that a little-modified Gemini with a two-man crew could be easily launched, with a payload capability of 22,000 lbs, into a low easterly orbit, which afforded a 14,000-lb payload potential.

On these missions DoD experiments could be carried in the spacecraft's Equipment Module or a special 'transition type' attachment also carried into orbit by the Titan III, with access either by control links from the crew compartment or by EVA. A significant amount of payload could be used in additional resources to adjust trajectory (fuel reserves), offering multiple rendezvous for surveillance and inspection missions, and high-inclination orbits for photoreconnaissance missions.

The report offered the suggestion that an unmanned launch would be possible on the fourth Titan IIIC development shot, and a second could be used for a manned test flight within the development programme. Costing for such flights would be offset as part of the Titan development programme already funded, and would gain immediate experience and application to later Titan IIIC Gemini missions. In addition, the study also suggested that it was 'certainly possible that a *small station* may be possible using Titan IIIC, in which other developments relating to bombardment and command and control could occur. The Gemini experience is again directly applicable ... The concept seems so clearly attractive that further study is warranted.'

Air Force astronaut crew participation

At the time of the study report, NASA was planning a dual-launched Saturn 1B Apollo rendezvous mission in 1966, at the same time as the final Gemini missions were flown. It was suggested in the report that if the Gemini rendezvous and docking experiments were achieved early, and if there was success with this proposed dual Apollo mission, one or two later Gemini missions could be released from the NASA programme to allow early flight-training for USAF astronauts.

At the time, NASA policy was that any astronaut with a potential for crew assignment on a lunar landing mission must have had previous experience on rendezvous and docking in Earth orbit. This would be a priority for crew training until sufficient numbers had been trained in the technique (a minimum of two men on each lunar mission). Therefore, *if* everything went according to plan, USAF astronauts could be inserted into later stages of NASA Gemini training after the eighth manned mission, during which at least ten NASA astronauts (prime and back-up) would have received various levels of rendezvous training and flight experience, or were in line to fly forthcoming missions. The report concluded that even at the most optimistic schedule this would not be before early 1966.

It was shown that six months could be sufficient for mission training, and prior to this there was required a period of 'astronaut' training in aspects of technical and space sciences, and Gemini systems. In addition, such 'military astronauts should participate, as had NASA astronauts, in the input into design development and testing of flight hardware for the DoD programme.' At least six months' general spaceflight training, followed by a further six months' support and technical assignments, would require the first DoD astronauts being selected in the summer of 1964, under the supervision of an Operations and Training Officer from USAF staff.

An initial group of at least two or three USAF astronaut designees would be assigned to the USAF Aerospace Research Pilot School at Edwards AFB, with temporary duty at NASA MSC, Houston. This allowed for early full-time assignment at MSC if required, to take advantage of the earlier flight opportunity in real-time conditions. The selection of an initial group was limited by the availability of NASA training facilities, and the announcement of additional USAF crew-members would be made as requirements dictated.

The Group reviewed several options for assigning USAF personnel as flight crew:

Air Force crew participation in the civilian NASA programme If no dedicated programme was forthcoming, the Air Force crew-members should be available to support the Gemini 8 mission as part of support of the civilian Gemini programme, in anticipation of crew assignment on later missions. At least four Air Force officers would need to be selected by September 1964, to support the anticipated 18-month training programme.

Air Force add-on two-flight programme This would require the assignment of three crew-members by July 1964, and the assignment of an 18-month training programme for two one-man flights using existing NASA simulators and training facilities.

Air force adds-on four-flight programme This required the selection of eight Air Force astronauts as crew-members by July 1964. In addition to the use of the Titan II or Titan III launch vehicles, this programme required the purchase and fabrication of an additional Gemini mission simulator (in addition to the two planned by NASA) for crew training.

Air Force add-on six-flight programme This required an Air Force cadre of eight astronauts to support a one-man mission based on the use of the Titan II, and fourteen if the Titan III was the launch vehicle. Candidates would need to be selected by July 1964, to complete 18 months training, and two additional simulators would be required.

Whichever programme was adopted, the Group indicated that the first USAF Gemini crew-members should be selected as soon as possible, be assigned to the USAF ARPS school for Gemini preparation, and use existing and operating Gemini training facilities and procedures at NASA. Additional crew-members would be selected as and when required. If there was no requirement for assignment to NASA for flight-crew participation, or if non-participation was not established, then there would be no need to assign astronauts for training.

Study group summary, collusion and recommendations
In review, the study stated that DoD experiments included in the twelve mission NASA Gemini programme would offer early experience, and would aid in the progress towards participation in a larger manned military mission capability. The estimated cost of experiment programme was $17 million dollars (1963), and it would be managed from Houston.

The inclusion of two additional launches in the Gemini programme for DoD objectives would generate a minimum additional expense over the basic experiment programme, but was the least desirable of all options due to its stretching the NASA Gemini manifest to the point that it would affect the early implementation of the Apollo lunar effort.

If more than two DoD missions were to be authorised, then it would be an advantage to activate LS 16 as an alternative and additional launch facility, to support the DoD programme and to not impact on NASA Gemini launch operations. However, other limitations would enforce continued use of one-man spacecraft: duplicate check-out facilities and operating teams already prepared for Titan III launches, and a limitated use of the facility beyond this programme.

The value of DoD participation in NASA Gemini appeared to be in the sharing of facilities, to ensure the success of the lunar programme and to establish a springboard capability to conduct independent manned space operations using the proven Gemini systems, hardware and infrastructure adapted to military objectives and operations, for proposed experiments, the development of rendezvous and docking techniques, duration flight, and EVA. Gemini could also be used for inspection and reconnaissance missions with a two-man crew, and effect a land recovery using paraglider technology.

Further study (beyond the account of this report) suggested additional consideration of follow-on mission and non-reoccurrent costs, the most effective being a six-shot programme using the Titan IIIC, and LC 41 as the complex. There was time to amend facilities and insert programmes into the schedule, and to also allow selection and training of adequate crew-members to man these flights.

A minimum of seventeen non-flight-crew military personnel was required to complete the basic experiment piggyback programme; while a two-flight add-on programme, managed by NASA from LC 19, required 47 support personnel, and a four- to six-flight programme from LC 16 (Gemini Titan II) or a four- to six-flight Titan III programme from LC 40 or LC 41 each required 127 military support personnel.

The management command and control facilities from both the NASA and the USAF programmes were studied, and it was concluded that they should be kept separate. USAF operations facilities at AMR were capable of all aspects of launch preparation and launch orbital insertion, and those at PMR were also expected to be adequate. In addition, only one tracking station and one tracking ship would be sufficient for the planned military programme, to support the existing USAF tracking structure and network.

Of most concern were the security and public information aspects of such a programme – particularly the proposed two-shot flight under the open NASA programme. Control of public information could be contained, and for dedicated DoD flights of four or six missions, normal Air Force security, classification and public information would be enforced.

Recommendations by the Group included:

- Implementation of a joint NASA/DoD experiment programme presented to the Gemini Program Board, and funded from the Fiscal Year 1964 budget.
- AFSC should establish an office at MSC under the jurisdiction of the Space Systems Division.
- The Air Force should be asked to consider a six-shot experiment test programme for the Titan IIIC, for inclusion in Fiscal Year 1964, and to prepare long-lead plans for procurement of equipment and facilities to support an expanded military manned space programme.
- When an in-flight experiment is selected, or when specific experiments and add-on programme alternatives are finally selected, then specific security and public information plans should be implemented.

The demise of Blue Gemini and the emergence of the MOL

Blue Gemini was never a defined or official programme. Several USAF officials requested more participation in NASA's Gemini programme, while others wanted nothing to do with the civilian agency, as this would threaten pure DoD programmes. At the Pentagon there was still not a clear objective for a US military

A model of the USAF MOL configuration, built in 1967 after the merger of McDonnell and Douglas aircraft companies. Gemini B is at left, and behind it are the forward compartment containing the attitude control units and the internal transfer tunnel. The laboratory is the black and white section in the centre, and the Titan III aft section is in grey and black at right. (Courtesy Curtis Peebles via the British Interplanetary Society.)

man in space programme. With the December 1961 agreement between NASA, the DoD, launch vehicle contractors and ocean recovery forces, participation in Gemini was expanded to included small experiments that had DoD applications.

By November 1962, MSC staff presented details of Gemini to SSD representatives in order to lay groundwork for Air Force planning, and to open future channels of communications between the two bodies, as the idea of co-operation seemed a good idea. But Secretary of Defense Robert McNamara wanted more USAF control than at first proposed. He suggested the merger of the NASA programme with the Air Force project, and the placing of both under the direction of the DoD! This was not unusual for the military, as there was a belief, in the USAF, that anything in Earth orbit should be under the auspices of the military, and that NASA should only manage programmes that operated beyond Earth orbit, such as Apollo.

NASA, however, did not approve. Gemini was a step to the Moon, and the transfer of the programme to the USAF was a threat. In addition, Earth orbit was not the sole territory of the USAF. At this time NASA was also beginning to review Apollo hardware to develop a space station, but planners were told that any new project could not, in fact, be called a 'new' programme, but rather, an extension of Apollo, in order to receive funding and not jeopardise the Apollo lunar programme. To NASA, the threat of the USAF take-over of Gemini had the same effect of risking the lunar effort as had the extended Apollo missions.

The objections from NASA staff offices were clear. In addition to delaying the Apollo programme, there were other problems: the appearance of even more military influence in a civilian agency, at a time when the American public questioned the value of a civilian space programme, foreign nations questioning America's peaceful exploration of space; and the fear of being barred from operating tracking stations abroad.

There were even some factions of the USAF that were opposed to the take-over of

Gemini, as they feared that it would threaten the X-20, and the cost of an expanded Blue Gemini would far outweigh anything that could be gained. By 9 January 1963, NASA's arguments seemed to have influenced top Pentagon officials, who were not wholly convinced that the USAF should take over Gemini from NASA. McNamara reluctantly agreed, but thought that the Air Force was wrong in not securing a larger slice of Gemini. In return for NASA's endorsement of the X-20, the USAF gave up its demands on Gemini. A series of joint management meetings were suggested to discuss the DoD role in Gemini, including the participation in a series of experiments.

The joint NASA/DoD agreement of 21 January 1963 restricted the USAF to 'pre-flight checkout, launch operations and flight operations to assist NASA and meet DoD objectives', and obliged it to 'utilise that national interest [in Gemini] and to avoid unnecessary duplication of efforts.' In effect, the DoD role remained unchanged, NASA still administered Gemini, and foreign tracking stations would be civilian in manning, with the only military personnel being from the medical services.

By then, the X-20 was on the decline, and the new Manned Orbiting Laboratory was gaining support. After the 1963 cancellation of the Dyna Soar, the MOL became the primary USAF programme for its military man-in-space effort, with full Presidential approval on 25 August 1965. The plan included two unmanned test flights and a series of five two-man Gemini spacecraft (adapted from the NASA design and termed Gemini B), attached to a pressurised laboratory on a Titan IIIC launched from Vandenberg AFB, California, into high-inclination orbits. On missions of up to thirty days in duration, the crew would live and work in the laboratory, conducting classified military experiments, observations and research. The MOL would use the NASA Gemini tracking network and Gemini capsules from McDonnell, as well as certain phases of NASA astronaut training – but nothing else. It was a DoD project.

Justification for the MOL was seen to be questioned. With the achievements of the NASA programme, the escalation of the war in south-east Asia, the development of unmanned reconnaissance satellites, and a delayed and reduced budget, the MOL faced severe diffiluties. Congressmen in Washington asked whether the USAF MOL and NASA's space station plans could be merged into a NASA-managed programme – reversing the idea of a NASA and a USAF Blue Gemini – but it was soon realised that this could not be easily (or economically) accomplished.

As NASA's space station programme grew, so the MOL programme came under even more scrutiny as the cost of supporting the Vietnam War escalated and the use of unmanned reconnaissance satellites continued to evolve. After six years of planning and development, the MOL was cancelled in June 1969, and some of the hardware developed for the programme was incorporated in the development of what became NASA's Skylab space station.

Although none of the five manned flights were launched, one of the two planned unmanned test flights left the pad, and gave a former NASA Gemini capsule a second flight into space to enter the history books as the first 'reused' ballistic spacecraft, fifteen years before the Space Shuttle *Columbia* claimed the same record for a manned orbital spacecraft.

The only MOL-related launch took place on 3 November 1966, a few days before the launch of the last NASA Gemini mission, Gemini 12, on 11 November. Gemini 2 was refurbished to ride on this launch vehicle (Titan IIIC) as a development test flight of the heat shield and mock-up MOL laboratory. The five operational missions were to use the more powerful Titan IIIM version. (Courtesy USAF.)

Gemini 2 flies again

The single MOL test flight was flown on 3 November 1966 (1966-99A) with a mock-up laboratory on a Titan IIIC launch vehicle. The launch was also scheduled for an MOL heat-shield test using the refurbished NASA Gemini 2 re-entry module, which had flown the unmanned ballistic test in January1965. Its heat-shield had been fitted with an access hatch, which in orbit would allow the astronauts to gain access to the MOL (just one of several alternative designs). It was the sixth research and development Titan IIIC to be launched from the Cape. When the vehicle reached 125 miles, the Gemini was separated to fly a maximum lift/drag re-entry profile for maximum heat load. At 17,500 mph it travelled 5,500 miles and splashed down only seven miles from the prime recovery ship *La Salla*. The MOL canister entered orbit after capsule separation, to continue its orbital test programme under the MOL programme. It re-entered the atmosphere on 5 January 1967.

Gemini 2 went on display at the Kennedy Space Center as the first American spacecraft to make two spaceflights – both of them unmanned. In November 1965, NASA and the USAF agreed the transfer of a number of excess items of equipment to the MOL. Shortly thereafter, several Gemini boilerplate test vehicles and support and handling equipment were transferred to the MOL programme, and MOL astronauts were briefed on the Gemini programme.

In March 1967 the Gemini 6 re-entry module was transferred to the MOL for astronaut training. The Static Test Article that was modified for altitude chamber tests and designated Gemini 3A was brought to manned flight status and assigned as MOL 1 Gemini capsule. This capsule was later placed on display at the USAF Museum in Dayton, Ohio. In order to fly MOL missions, the Gemini B spacecraft required a number of modifications to the basic NASA spacecraft

Gemini B

In June 1964 a declassified document reviewed the status of the programme and described the Gemini B spacecraft. The design of the Gemini B spacecraft as a recovery vehicle and for MOL crew ascent was somewhat different from that planned for NASA Gemini missions. The re-entry capsule was close to the NASA version, with modifications introduced as required to change it to the military

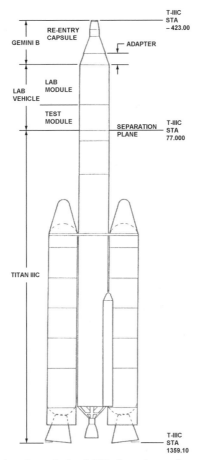

A declassified USAF drawing of the MOL launch configuration, *c*.1964. (Courtesy USAF.)

mission, including the insertion of a land landing gear, and provision, if required, for an escape tower instead of the ejector seats.

The adapter section (shown in the illustration below) was a new feature, and contained the retrograde propulsion system for de-orbit burns, power and ECS for use during ascent and initial orbital operations, and initially an inflatable crew tunnel for crew access to the MOL from the crew compartment. This was later changed to an internal transfer tunnel via a hatch that passed through the re-entry heat shield in the base of the recovery spacecraft module.

It was proposed that Gemini B remain an integral part of the MOL complex, having no independent manoeuvring capability, and self-sufficient only during ascent and re-entry. Sub-systems not required for MOL were to be deleted, and additional modifications would allow man-rating on the Titan IIIC launch vehicle, interfacing with the laboratory module, and capacity for thirty-day orbital storage while the mission was completed – double the orbital lifetime of the longest Gemini mission flown by NASA.

It was expected that an escape tower would be mounted on the vehicle, provisions for crew transfer, and the monitoring of Gemini systems from inside the orbital laboratory. Replacement of cryogenic storage bottles were expected to be with high-pressure bottles, and the replacement of fuel cells would be with chemical batteries. It was also suggested that the Gemini inertial guidance system might have to be replaced by a much simpler device as there was no rendezvous and docking planned, with a system capable of supplying attitude reference only for retro-fire and re-entry. The NASA G&N system would therefore be used as a primary or back-up system only during the launch phase.

By attaching an escape tower or the tunnel exterior to the vehicle, the structure of Gemini would have to have been strengthened In addition, a 20-lb increase in the

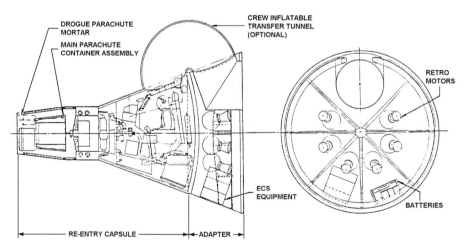

An artist's impression of the Gemini B vehicle, here with the optional inflatable transfer tunnel and off-set entry hatch in the shortened adapter section. Later versions cut a hatch-way through the heat shield of the re-entry spacecraft. (Courtesy USAF.)

thickness of the re-entry heat shield allowed for a wider range of re-entry conditions. NASA's water parachute system was likely to be used initially, with a new adapter section incorporated as a fairing for the laboratory module.

Additional provision had to be made to activate and monitor the status of the laboratory from the spacecraft, and to monitor and check the Gemini from the laboratory. During the time that the spacecraft was dormant in orbit, the laboratory was to have provided a power supply and environmental control to the spacecraft. There was apparently no provision for Gemini to interface with MOL experiments as a carry vehicle only to provide a return of small data packages at the end of the mission although this could have changed if the missions were flown.

Design considerations
Launch abort Changing the Titan II to Titan IIIC greatly increased the propellant load to lift the facility, but also increased the risk of launch vehicle explosion. In addition, altered flight characteristics affected acceleration levels and dynamic pressures. These breached the operational use of ejector seats, and escape towers were instead considered for Gemini B.

Crew transfer A variety of transfers into the MOL were considered. Only EVA had no impact on Gemini B or MOL design, and followed the NASA Gemini profile of EVA by one crew-member from the crew hatch (not blocked by the inflatable tunnel!). Considerable equipment relocation in the Re-entry Module, the addition of smaller adapters and access tunnels, and the breaking of the thermal seal to insert a hatch in the re-entry heat shield, were all evaluated.

Reactivation After thirty days of 'semi-dormant' storage (similar to those which the Russian Soyuz achieved in Salyut/Mir, or those of Apollo at Skylab), the spacecraft systems had to be reactivated. A review of procedures and systems was required to lead to redundancy or replacement by simpler and more reliable systems.

Gemini B preliminary weight breakdown

	lbs	lbs
Re-entry module		4,372
Structure	1,407	
Heat protection system	318	
Attitude, stabilisation and control	359	
Retrograde, landing and recovery	217	
Navigation equipment	138	
Tele-instrumentation	113	
Electrical power system	277	
Communication system	58	
Crew and survival equipment	940	
Miscellaneous and contingency	220	

	lbs	lbs
Adapter module		1,628
Structure	290	
Crew transfer system	180	
Retrograde system	509	
Electrical power system	200	
Environmental control system	257	
Tele-instrumentation	88	
Miscellaneous and contingency	104	
Escape tower (effective on-orbit mass of 100 lbs)		2,500
Total weight of Gemini B at launch		8,500
Total mass of Gemini B on orbit (equivalent)		6,100

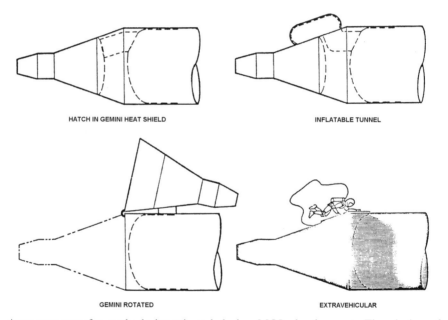

HATCH IN GEMINI HEAT SHIELD

INFLATABLE TUNNEL

GEMINI ROTATED

EXTRAVEHICULAR

Astronaut transfer methods investigated during MOL development. The placing of Gemini at the nose of the laboratory required an ingenious method of moving the crew from Gemini B to the laboratory. (Courtesy USAF.)

MOL astronauts

Seventeen men were chosen to train for the USAF MOL missions, in three selections. To be selected, each candidate had to be a qualified military pilot, a graduate of the ARPS course, a serving military officer, and recommended by a commanding officer and a US citizen by birth.

The USAF originally announced that up to twenty 'aerospace research pilots' would be selected to MOL as potential crew candidates. There were to be at least five

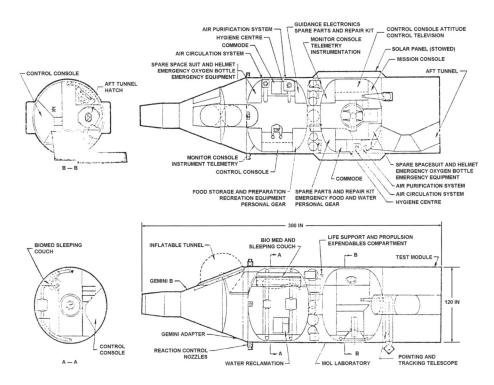

The dual pressure compartment of the MOL is shown in this declassified USAF document. A single compartment configuration was also considered. In one of the final configurations the dual compartment was believed to be the one closest to that finally chosen, alhough the heat shield hatch was positioned over the inflatable tunnel, and the aft tunnel was removed from the design. (Courtesy USAF.)

two-man one-month missions, and the selection therefore took into account attrition and possible non-flying support roles.

The first selection was named on 12 November 1965:

Major Michael J. Adams, 35, USAF
Major Albert H. Crews Jr, 36 USAF
Lieutenant John L. Finley, 29, USN
Captain Richard E. Lawyer, 33, USAF
Captain Lachlan Macleay, 34, USAF
Captain Francis Gregory Neubeck, 33, USAF
Major James M. Taylor, 34, USAF
Lieutenant Richard H. Truly, 28, USN

The second selection was announced on 30 June 1966:

Captain Karol J. Bobko, 28, USAF

Lieutenant Robert L. Crippen, 28, USN
Captain Charles Gordon Fullerton, 29, USAF
Captain Henry W. Hartsfield Jr, 32, USAF
Captain Robert F. Overmyer, 29, USMC

The third and final selection was announced on 30 June 1967:

Major James A. Abrahamson, 34, USAF
Lieutenant Colonel Robert T. Herres, 33, USAF
Major Robert H. Lawrence Jr, 31, USAF
Major Donald H. Peterson, 33, USAF

A fourth selection was in progress when the programme was cancelled in June 1969.

Of the above, Adams left the MOL on 20 July 1966, and transferred to the X-15 programme. He died on 15 November 1967, while flying his seventh X-15 mission, during which he attained the USAF Astronaut rating awarded to pilots who flew to more than 50 statute miles altitude. Finley was reassigned to the Navy at his own request in April 1968. Lawrence was the first African American selected for astronaut training, and was killed on 8 December 1967 in the crash of an F-104 at Edwards AFB.

When the programme was cancelled there were fourteen MOL astronauts in training, and all but Herres expressed interest in transferring to the NASA astronaut programme. Deke Slayton said he would only accept those under 36 years of age, as he had more astronauts than flight seats. Therefore, on 14 August 1969, Bobko, Crippen, Fullerton, Hartsfield, Overmyer, Peterson and Truly became the seventh NASA astronaut class, and all went on to fly on the Shuttle. Albert Crews, who had been an X-20 candidate, joined NASA at MSC as a research pilot and CB ground support duties on Shuttle missions for the next thirty years. The rest returned to their military careers (see Appendix).

The USAF MOL astronaut group in 1968: (*left–right*) Herres, Hartsfield, Overmyer, Fullerton, Crippen, Peterson, Bobko and Abrahamson; (*front row*) Macleay, Lawyer, Taylor, Crews, Neubeck and Truly. Not present is Adams (deceased), Finley (returned to the USN), and Lawrence (deceased). (Courtesy USAF.)

Robert Lawrence (second from left) was one of the four USAF pilots selected for the MOL in June 1967, but he was killed in an air crash in December 1967. He would have been eligible to transfer to NASA in 1969, and could have flown on an early Shuttle mission. In the centre is a model of the Titan IIIM laboratory and Gemini spacecraft. (Courtesy USAF.)

Crew training

The training for MOL astronauts in relation to Gemini spacecraft operations included:

- Academic training, both additional and refresher (based on the NASA academic training syllabus).
- Environmental: high and low g, vibration, noise, low pressure and pressure suits.
- Contingency: water, tropical, desert areas; Gemini ejection seat, egress, parachute and rescue.
- Program Office participation: design and development systems and experiment check-out; integrated testing and acceptance activities.
- Simulations: launch, re-entry and recovery; on-orbit supervision of Gemini while inside MOL; EVA through Gemini hatch; underwater EVA training.
- Proficiency: maintenance of flying proficiency in high-performance aircraft.
- NASA Gemini 'over the shoulder' participation 'to the maximum extent possible'.
- Final participation in pre-launch check-out of spacecraft at launch site.
- Medical support: medical evaluation and monitoring teams; recovery and rescue team training co-operation and integration.

Exact details of MOL crew training are still difficult to obtain, but it is known that since MOL Gemini missions were launched on one booster, the Gemini was already attached to the laboratory, and remained there until separation at the end of the mission for re-entry and landing. No rendezvous and docking training simulations were performed as part of the MOL astronaut training programme. Specific crew assignments for the planned five manned missions are also unclear, but according to MOL programme managers, James Taylor was the leading candidate for the command seat on the first MOL manned mission.

A place in history

The flight operations of Gemini were completed with the recovery of the Gemini 12 crew in November 1966. Preparations to fly the USAF MOL missions continued until June 1969, when the programme was cancelled without completing a single manned spaceflight. The following month, Apollo 11 achieved the goal of landing a man on the Moon using rendezvous and docking techniques evaluated during the Gemini programme. All three Apollo 11 astronauts were Gemini veterans, as were eight of the twelve astronauts who flew Apollo 7–10, and two of the Apollo 12 crew had also previously flown on Gemini. Of the eighteen astronauts selected to crew the first six Apollo missions, thirteen had had previous experience on at least one Gemini flight, while the other five had performed Gemini support and technical assignments prior to moving to Apollo.

From Apollo 13 onwards, several Gemini veterans worked on each of the remaining Apollo missions, with four of them going on to make a second Apollo flight. In addition, Pete Conrad moved over to Skylab and Tom Stafford moved to Apollo–Soyuz, while John Young remained in the Astronaut Office to take command of two early Shuttle flights, including the coveted first flight. (See Appendix)

Although no Gemini hardware flew after 1966 there was a range of proposals and studies that existed on paper that could have resulted in other missions and could be termed the lost legacy of Gemini.

LOST MISSIONS OF GEMINI

Some of the earliest proposals for adapting Gemini hardware to other missions (in effect, a Gemini Applications Programme) – similar to those suggested for Apollo missions (Apollo X, Extended Apollo, Apollo Applications) – were those to send Gemini spacecraft to the Moon.

Lunar Gemini
As part of the August 1961 proposal for Mercury Mark II, the use of a Centaur

upper stage as a docking target was considered to allow a high-apogee orbital mission as well as circumlunar flyby missions (similar to that performed by Soviet Zond unmanned lunar probes between 1968 and 1970), but without entering lunar orbit. Members of the Space Task Group proposed that the Centaur would be launched on a Titan II, followed by the Gemini carrying about 600 lbs of additional equipment mainly taken up with a back-up inertial navigation system and additional heat-shield protection from the higher temperatures encountered during re-entry.

Emphasising the reduced cost of this programme over Apollo, suggestions were made that Americans could fly around the far side of the Moon by May 1964. Although officially dropped from the plans of Mercury Mark II within weeks of the report, the suggestion of sending Gemini to the Moon in some form would not go away. (Proposed Gemini missions to the Moon have been analysed by space author Mark Wade (http://www.astronautix.com/), and include the development of a one-man open cockpit lunar landing module at 1/20 the overall cost of that proposed by the Apollo landing programme.)

A 1960s version of the 1990s NASA philosophy of 'cheaper, faster, better', it would certainly cost less, and by changing the launch schedule it could in theory place Americans on the Moon by early 1966; but whether it was better is open to question. The plan would require every previous Gemini mission (fourteen of them) to be launched at 45-day intervals, and without encountering any of the problems that the flown Gemini programme had to overcome in respect of hardware, testing, flight operations, launch failures or flight emergencies.

The plan also involved the use of the Titan II-launched 9,600-lb lander for three Earth orbital-docking tests, followed by two Centaur docking missions that would allow a lunar fly-by. The programme would be completed by two Saturn C-3s (essentially a Saturn V but with only three F-1s in the first stage, and different propellants loading in the upper stages) carrying the Gemini on a lunar orbital mission, and finally a Gemini and lander for a landing attempt.

When approached with these plans, NASA Headquarters management was cautious, recognising that if the Mercury follow-on programme was promoted to include lunar missions then this could seriously threaten the plans for Apollo, which were far beyond the initial feat of landing on the Moon, and that Gemini could only theoretically ever achieve. Although the lunar plans were dropped, the ideas for the spacecraft, boosters, rendezvous and docking and lunar duration missions were approved, and in December 1961 Mercury Mark II became the Earth-orbital stepping stone to the Moon – and in January 1962 was named Gemini.

Rescue and logistics

Although officially Gemini would not fly to the Moon, NASA considered utilising the design of Gemini in the role of a crew rescue vehicle or logistic resupply craft. With Apollo planned to go far beyond the initial lunar landing and possibly become crucial to creating a lunar base as well as Earth-orbital space stations, a reliable system of crew resupply and rescue would have to be incorporated into the infrastructure. It was felt that Gemini might be the spacecraft to fill that role.

An eight-week study awarded by NASA HQ to the Space Technology Laboratory

An artist's impression of a Gemini-type spacecraft approaching a space station on a resupply mission. This was one of a range of proposed missions that might utilise the resources of the Gemini spacecraft, including lunar distance flight and astronaut rescue, as well as USAF objectives and space station transfer vehicles. (Courtesy British Interplanetary Society/Douglas Aircraft.)

and McDonnell in the summer of 1962, examined the use of the Gemini spacecraft as a 'lunar logistics and rescue vehicle' and considered what changes might be required to support this new role. The costing for such an amendment was far higher than any return that such a system would offer, and with Gemini itself under financial scrutiny and restraint, anything that placed more strain on the completion of just the basic Gemini programme was soon rejected.

Power to the cause
Another short-lived idea, in the spring of 1964, was to use Gemini boosted by a Saturn 1B to achieve circumlunar flights after the end of the core Gemini missions and before the mainstream Apollo flights commenced, or as a contingency plan to beat the Russians to the Moon. At Marshall, Wernher von Braun and others were opposed to anything that threatened the full Congressional support of Apollo, and on 8 June 1964 a NASA HQ memo indicated that anything relating to Gemini lunar

proposals would be limited to in-house studies, thus preventing the issuing of contracts to pursue the matter further.

One of these 'in-house studies' in 1965 proposed using a Titan IIIC transtage to boost the spacecraft into a higher orbit, under the mission scenario Gemini Large Earth Orbit. In this proposal, the solid-fuel retrograde rockets used during re-entry would be delivered, and the liquid OMS upgraded, to increase reliability and to save 50% of the 1,200 lbs weight reduction proposed. A modified trans-stage launched by a Titan III-C would be docked with the Gemini to boost the spacecraft to a high Earth orbit, or by the addition of a second trans-stage in a circumlunar trajectory. Following the end of the scheduled Gemini missions, a December 1966 test of the system would be flown. In this mission, a 5,400-lb unmanned Gemini circumlunar spacecraft would be flown to test the 36-fps re-entry velocity in a verification of the heat-shield. A manned qualification in Earth orbit would follow in February 1967, during which a two-man Gemini would rendezvous and dock with a double trans-stage, to be boosted to a high-apogee orbit and re-entry speed. The lunar circumlunar flight was proposed by April 1967.

However, even the author of this report suggested time and effort would be better spent on developing techniques learned on Gemini–Agena, to demonstrate space station concepts using Gemini spacecraft in Earth orbit instead of sending them to the Moon. NASA management agreed, and once again, for fear of losing funds from the Apollo programme, any future studies of this concept were quietly discouraged. Despite this, Pete Conrad successfully lobbied for flying a high-apogee orbit on his Gemini 11 mission in September 1966, and set an altitude record of 975 statute miles that was not surpassed until Apollo 8 left for the Moon in December 1968. This was the closest approach of a Gemini spacecraft to the Moon.

Space rescue
Studies in using Gemini as a potential rescue craft first appeared as early as 1962, but it was not until after the Apollo fire of 27 January 1967 that NASA re-evaluated the safety aspects of Apollo, including the potential for the rescue of stranded crew-men. Several concepts were evaluated using Gemini as the core rescue craft, including sending a stretched re-entry module, unmanned, to the Moon by Saturn V, with the capability of returning three Apollo crew-men stranded in lunar orbit. Transfer would be via EVA – if, of course, the crew was still capable of accomplishing this feat. This design was rejected in favour of the Lunar Surface Rescue Vehicle.

This scenario featured two Gemini spacecraft that included the Lunar Surface Survival Shelter and the Rescue Spacecraft. Initially prior to the Apollo landing attempt, the Gemini on top of an LM descent stage would be landed unmanned near the intended Apollo landing site, and in the event of a failed LM ascent stage lift-off, two LM astronauts, with supplies for up to 28 days, could transfer to the Shelter and await the rescue craft. The CM Pilot, in lunar orbit, would make the trip home alone – a contingency for which each CMP trained on the Apollo landing missions. The unmanned Gemini rescue craft would land near the stranded LM crew, using an extended two-man re-entry module for the crew return, and three LM descent stages for lunar orbit insertion, lunar landing and lunar ascent. In an alternative design, the first two stages were replaced by modified service modules to bring the Gemini to the Moon to rescue the stranded astronauts. Each rescue return profile featured direct ascent from the lunar surface towards Earth.

In reflection, it was concluded that it would be more advantageous to develop a larger Universal Lunar Rescue Vehicle that could not only recover all three astronauts from lunar orbit but, if required, could also be used to rescue a stranded crew from the Moon. But with cut-backs in Saturn V production for mainline Apollo, there was neither the funds nor hardware for anything resembling a rescue capability with Gemini or even Apollo.

Logistics spacecraft
The Gemini design was also studied for use as a space station logistics vehicle in addition to primary transport craft Gemini B for the USAF MOL programme. Stretching the Re-entry Module or the Adapter Module could facilitate the use of additional storage volume. Studies into using Gemini as a payload on the 1960s studies of Single Stages to Orbit designs were also popular at the time, and featured in the book *Frontiers of Space*, by Philip Bono and Ken Gatland (Blandford Press 1969, illustrations on pp. 66 and 68). In addition, in the Manned Orbital Research Laboratory feasibility study the use of Gemini as well as Apollo logistics vehicles implied a capability similar to that developed in the Soviet space station programme by adapting the manned Soyuz as a space station ferry vehicle (Soyuz T and TM) and an unmanned resupply craft (Progress).

McDonnell's Big G study
In August 1969 – three years after the end of Gemini flight operations, and weeks after the cancellation of MOL – McDonnell Douglas, under an MSC contract, issued an eight-volume final report on Big G, in which an extended Gemini re-entry module provided a preliminary definition study of a logistics spacecraft that would be used to resupply an orbiting space station. Design requirements featured the capability of land landing and spacecraft reuse, and generated two concepts using three launch vehicles – the Saturn 1B, the Titan IIIM, and the Saturn V 1C/S-IVB. The Saturn 1B was discarded in the final report.

Mini-Mod Big G was a nine-man minimum modification of the USAF Gemini B spacecraft proposed for MOL, while the Advanced Big G was a twelve-man concept with the same exterior configuration but with a much improved state-of-the-art subsystems incorporated.

The design featured an extended crew module to incorporate a 156.6-inch (13.05-foot) diameter heat-shield and a cargo propulsion module. Big G would be recovered by the parawing and skid landing gear, in an advanced version of that proposed on the original Gemini spacecraft seven years earlier. Spacecraft propulsion was via a single liquid-fuel supply system that provided for orbital transfer, rendezvous, attitude control, and retrograde manoeuvres. Crew launch escape would be via an Apollo-type escape tower.

Behind the crew module would be the propulsion module, with the cargo module at the rear featuring an internal access tunnel from the crew module through the

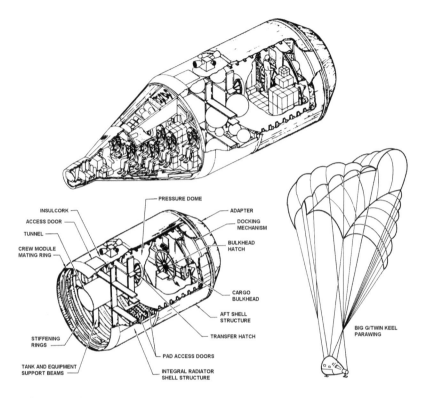

INSULCORK
ACCESS DOOR
TUNNEL
CREW MODULE
MATING RING

PRESSURE DOME
ADAPTER
DOCKING
MECHANISM
BULKHEAD
HATCH

CARGO
BULKHEAD
AFT SHELL
STRUCTURE
TRANSFER HATCH

STIFFENING
RINGS

TANK AND EQUIPMENT
SUPPORT BEAMS

PAD ACCESS DOORS
INTEGRAL RADIATOR
SHELL STRUCTURE

BIG G/TWIN KEEL
PARAWING

Three features of the McDonnell Big G study of 1969, showing the internal arrangement of the crew, propulsion and cargo modules, and the twin-keel parawing recovery for land landing.

heat-shield into the cargo module. The cargo module also housed a rear docking mechanism and controls for attaching to a space station. Once again this was only a paper study, with no firm plans to produce the hardware, but to define the design aspects of any future space station logistics vehicle based on reusable capsule technology.

In the event, of course, the only US space station that NASA orbited was Skylab in 1973, and a proposed Skylab B that could have featured Apollo CSM resupply capabilities or the hoped for twelve-man space station were not pursued.

A Gemini observatory

The Big G concept seems to have evolved from a 1967 concept of Advanced Gemini designs by McDonnell that featured the capability to orbit astronomical equipment. Launched by a Saturn 1B, an enlarged re-entry module resembled that of the later Big G design, and housed solar or astronomical telescopes for observations from low Earth orbit. The overall length of this Gemini was more than 28 feet from nose to rear of the extended adapter, which had a maximum diameter of 21.6 feet and a mass of 41,000 lbs.

A satellite retrieval mission

This author also discovered a further analysis of a possible mission for Gemini while researching for this book in 2000. During a search of Gemini archived documents at NARA Fort Worth, the report, dated 28 July 1965, was entitled 'Preliminary Analysis of a Satellite Retrieval Mission for Gemini.' It was categorised as NASA working paper No.5033, and was compiled by the Mission Feasibility Branch at MSC Houston.

The study evaluated the use of a Gemini–Agena combination to rendezvous with a Pegasus C to retrieve a micrometeroid detection panel from the satellite. The study was initiated *before* the third and final Pegasus was launched later that month. Pegasus was a series of three satellites derived from the upper stage of the Saturn 1 and Apollo Service Modules, fitted with two 49.2 × 15.1-foot extendible wings consisting of 208 panels of extremely thin sheets of aluminium, polyurethane, mylar and copper, that when punctured by a micrometeorite created an electrical short circuit that was detected by analysing the strength of the impact, its direction and its frequency in relation to the risk of Apollo lunar spacecraft. Pegasus 1 (A) was launched on 16 February 1965, into a 308 × 455-mile 31.73 orbit; Pegasus 2 (B) on 25 May 1965, into a 312 × 460-mile orbit with the same inclination; and Pegasus 3 (C) on 30 July 1965, into a 332 × 352-mile orbit at 28°.8 inclination.

Pegasus 3 was designed to be compatible with a Gemini–Agena rendezvous, and included a panel that could be retrieved during an EVA attempt. Once Pegasus had been orbited, then mission planning would be completed to place the Agena in the same orbital plane as the satellite. No sooner than 23.5 hours later, Gemini would be launched to perform a Hohmann transfer to Pegasus after docking to Agena. The spacecraft/Agena combination would approach Pegasus in a way similar to that which Gemini 10 approached the Agena 8 target without docking to it. The Pilot would perform an EVA to retrieve the sample panel.

Mission planning called for a 2–5-day duration, with the first day taken up with rendezvous and docking with the Agena as a trade-off between the late Agena launch window to match the orbital plane of the satellite and the earliest Gemini launch window. A total of 28 hours was reserved for phasing with the Pegasus and for non-rendezvous procedures such as retro-fire and recovery. The rendezvous with Pegasus was planned to occur five hours after Gemini–Agena docking.

The report did not include an intended launch date, but it would probably take place in late 1966 (Gemini 10–12? – possibly as an alternative for Gemini 10 instead of Agena 8?). The report did, however, highlight principal problem areas that required further study.

- Although the propulsion systems could theoretically handle such a mission, docking with an Agena had yet to be achieved. Therefore, with no data on orbital restart of an Agena with a Gemini docked, no accurate prediction of the capability of the system to support this type of mission was possible. This scenario was to include no less than *seven* restarts, and though fifteen were theoretically possible, the Agena could guarantee only five.
- At that time, Gemini 6 planned to test the compatibility of the Gemini–Agena

docked configuration. In reality, with the loss of the target vehicle these tests were slipped to Gemini 8. In the event, Gemini 8's early return and Gemini 9's inability to dock with ATDA indicated that such compatibly tests would not be available until Gemini 10 at the earliest, and this would seriously delay any satellite rendezvous attempt.

• The accurate tracking of pegasus was another problem area. Preliminary indications revealed that ground stations could provide an accuracy of only 200 yards at a slant range of 1,500 nautical miles.

• Forward vision would be limited by the docked configuration of the Agena, and it was suggested that the addition of a transponder on Pegasus would be a navigational advantage

• Finally, if Pegasus was tumbling it would present a serious hazard for a closest approach and EVA, and further study into a tumbling satellite was suggested.

In the event, delays in docking with the Agena and performing higher-priority mission objectives pushed this possible mission to the end of 1966 and finally into 1967, to be addressed in Apollo missions, which in turn were cancelled due to the pad fire.

Despite a range of studies for utilising Gemini hardware, none were utilised for flight operations beyond MOL, and even under this programme only one unmanned sub-orbital flight was achieved. In a prediction of Apollo, what was planned for Gemini was not to materialise. Gemini had achieved more than was originally intended, but as with the lunar programme, its full potential was not to be realised.

A PLACE IN HISTORY

Gemini was conceived as a bridge to the Moon – a platform to develop techniques, to test equipment and procedures, and to gain experience that helped NASA move from the pioneering one-man short-duration and essentially experimental Mercury series to the much more demanding Apollo lunar programme. It was successful in all of its stated goals except land landing – and even the ocean splash-downs proved that pinpoint accuracy was possible. The record was impressive, with ten manned missions in 1 year 8 months, and the whole programme completed in less than 1,900 days from official approval on 7 December 1961 to the recovery of the Gemini 12 astronauts, 5 years 8 days later, on 15 December 1966.

But it was not all plain sailing, as the programme struggled with budget and management issues in its early years, with a fear that Gemini was draining valuable resources from Apollo. Disputes with the USAF, difficulties with the Titan, the Agena, land landing, fuel cells and EVA – all added stress to the programme, but also provided hidden learning curves that helped to create the NASA 'machine' that took men to the Moon in the often stated 'can do' attitude which the world came to recognise as NASA's own.

The primary objectives and accomplishments were as follows:

Long duration The four-day flight of Gemini 4, the eight-day mission of Gemini 5, and the space marathon of Gemini 7, provided not only biomedical data proving that the human body could withstand spaceflights of up to two weeks in duration, but also helped to finally place the US manned space programme on an equal and eventually in a leading role over the Soviet Union, which had dominated the headlines with its achievements in space since 1957.

The flights provided lessons in crew compatibility, mission planning, extended mission control operations, spacecraft stowage and post-flight recovery from longer missions that would be essential not only to Apollo but also to follow-on programmes. The record of Gemini 7 would last until the Soyuz 9 cosmonauts landed after 18 days (and not in particularly good condition) in June 1970, and until Skylab in 1973. Indeed, Gemini 7 remained the fourth longest US manned spaceflight until the early 1990s.

Medical data collected during the Gemini missions included:

Peak heart rates during launch and re-entry (beats/minute)

Mission	Astronaut	Launch	Re-entry
3	Grissom	152	165
	Young	120	130
4	McDivitt	148	140
	White	128	125
5	Cooper	148	170
	Conrad	155	178
6	Schirra	125	125
	Stafford	150	140
7	Borman	152	180
	Lovell	125	134
8	Armstrong	138	130
	Scott	120	90
9	Stafford	142	160
	Cernan	120	126
10	Young	120	110
	Collins	125	90
11	Conrad	166	120
	Gordon	154	117
12	Lovell	136	142
	Aldrin	110	137

In his autobiography, *Carrying the Fire* (published in 1974), Michael Collins suggested that the higher beat-rates of the Command Pilots were a reflection of other NASA aviation medical tests that suggested higher rates of the heart (anxiety) in the crewman who had the responsibility of command rather than the task of 'doing the flying'.

Radiation doses

Mission	Mission Duration d:h:m	Mean cumulative dose in millirads Command Pilot	Pilot
3	00:04:52	20	42
4	04:01:52	42	50
5	07:22:55	182	170
6	01:01:51	25	23
7	13:18:35	155	170
8	00:10:41	10	10
9	03:00:20	17	22
10	02:22:46	670	765
11	02:23:17	29	26
12	03:22:34	20	20

Gemini 10 passed near the inner Van Allen radiation belt in the South Atlantic Anomaly, where the phenomenon dips to lower altitudes, but even these doses were well within the safety limits. The higher-altitude Gemini 11 and Gemini 12 missions did not pass through this region to record higher-radiation levels.

Weight loss (pounds)

Mission	Command Pilot	Pilot
3	3	3.5
4	4.5	8.5
5	7.5	8.5
6	2.5	8
7	10	6
8	No data (mission terminated early)	
9	5.5	13.5
10	3.0	3.0
11	2.5	0
12	6.5	7

Generally, the longer the flight the greater the weight loss. In addition, Cernan's exertions during his EVA contributed to his recorded weight loss.

EVA heart rates

Mission	Average	Peak
4 (White)	Not available	
8 (Scott)	No EVA attempted	
9 (Cernan)	150	180
10 (Collins)	118	165
11 (Gordon)	140	170
12 (Aldrin)	105	155

Again, excursions by Cernan and Gordon during their EVAs increased their heart rates. In comparison, by following a controlled and paced EVA flight plan, Aldrin managed to complete his task with a lower recorded rate. This was evidence of the value not only of EVA planning, suitable tethers and restraints, but also of the use of the water tank in EVA training.

Biomedical Up to Gemini 7, biomedical attention focused on attaining the 14-day duration objective, while the missions flow in 1966 concentrated on performance and life support relating to EVA activities. Standard clinical methods were adopted for obtaining data (pre- and post-flight clinical evaluations and in-flight monitoring and sample gathering). Major emphasis was placed on the cardiovascular system, and the data revealed that the human body exhibited constant and predicable changes during and after flights of up to 14 days, but also that such changes were reversible. Most importantly for Apollo, none of these changes demonstrated degradation of human performance or safety during the missions, and with the lunar flight lasting no more than 10–12 days there was nothing to suggest that Apollo astronauts would not be capable of medically achieving their goal.

Despite scientific predictions of adverse reactions to long spaceflight, no such effects were recorded during the Gemini missions. The fears included reduction in cardiovascular response to exercise, nausea, renal stones, urine retention, muscular non-cordination and atrophy, reduced exercise capacity, euphoria, impaired physiomotor performance, hallucinations, and the need for sedatives.

The Gemini flights demonstrated adequate performance of the vision on long flights, minor variations in sleep, and the adequacy of the astronauts to perform in-flight manoeuvres and to respond to in-flight emergencies.

The noted losses in bone density skeletal calcium and muscle nitrogen were recorded on the longer flights, but all returned to pre-flight levels within 50 hours of landing. The peak levels observed during Gemini 5 were significantly lower on the Gemini 7 mission, which indicated to scientists that adaptation was occurring.

It was recognised that since each flight was recording different data, this could only be a possibility, and that exercise, fluid intake, diet, and so on, were contributary factors. It was decided to evaluate these findings before extended-duration missions of several months were attempted (as approached during the Skylab missions) but without affecting Apollo on missions within fourteen days.

What was of concern was that the longer the flight the higher the weight loss, despite an adequate diet. Studies into recording the complete intake of food and liquid, and the collection of body wastes, in conjunction with measuring the mass of the human body during long flights to determine weight loss, were beginning to be identified as a way to measure this trend. One man with a special interest in these studies was Dr William E. Thornton, who in the mid-1960s was on active duty with the USAF at the Aerospace Medical Division at Brook AFB, San Antonio, Texas. Thornton was working on human adaptation to spaceflight, and was involved with the development of experiment and exercise equipment for the USAF MOL programme. The Gemini results were of direct application to Thornton's studies, and after 1967, when he was selected for astronaut training by NASA, he continued his work on space adaptation research with related assignments in the Skylab and Shuttle/Spacelab programmes.

Rendezvous and docking Perhaps the most significant achievement of Gemini was in the area of rendezvous and docking. During the missions, several techniques were applied on almost all of the missions flown, and many of the astronauts who participated in Gemini performed similar rendezvous and docking operations in Apollo spacecraft from 1968 to 1975.

Ironically, those that pioneered American docking activities in the astronaut office from 1965 to 1975 came from the astronauts selected in the first three groups. However, for twenty years after ASTP, no American spacecraft *docked* to another vehicle until Space Shuttle *Atlantis* linked with Mir in 1995. Since then, docking has become a regular feature of Shuttle missions and it is worth remembering that these techniques were first evaluated during Gemini, and have continued well beyond the Apollo era.

EVA Although plagued with difficulties, the Gemini EVA programme was an outstanding success in that it proved that man could work outside his spacecraft, and despite serious restrictions in physical effort and lack of restraint could overcome the difficulties and return safely to the spacecraft. In addition, adequate training coupled with suitable body restraints and a paced work plan revealed that EVA operations were not only possible, but also pointed to a high degree of success, which was clearly demonstrated on Skylab and on the Shuttle, as well as by cosmonauts outside Salyut and Mir.

Experiments An opportunity to fly small experiments on each Gemini mission as secondary objectives was an added bonus that was gladly accepted. A combination of scientific, engineering, technical, medical and DoD investigations were carried out, and pointed the way to other experiments flown on Apollo, the student experiments flown on Skylab, and the smaller experiments flown on the Shuttle under the Getaway Specials, Hitchhiker, and Middeck programmes. For the first time, small, compact and useful scientific investigations could be performed alongside the more spectacular objectives, and could significantly contribute to the gathering of scientific data from manned spaceflight.

Launch operations Gemini provided experience in almost routine and regular launch operations that were not witnessed by Americans again until the advent of the Shuttle two decades later. Meanwhile, the Soviets initiated their routine space station operation programme in 1971, and continued it for thirty years until the demise of Mir and the introduction of ISS rotating crews early in 2001. The demonstration of a reliable and routine resupply mission – often hinted at during Gemini flight operations and paper studies – was realised with the Progress resupply craft that began flying in 1978, and has continued for almost 25 years to support space station operations – an overlooked legacy of Gemini. The experiences of Gemini also provided launch operations with experience in handling payloads and launch vehicles to a tight schedule. This has resurfaced during Shuttle and ISS operations, and once again highlights the legacy of the pioneering programme.

Flight control Gemini was controlled from Mission Control Center, Houston, for the first time, and used an expanded worldwide tracking and data communications network that became essential for Apollo and the Shuttle. The flight control teams that pioneered the Mercury missions were to provide the core of experience during Gemini that helped create the 'stuff of legends' during Apollo. The recognition of 'Houston' to the world during Apollo 11, Apollo 13, and so many other historic missions, the challenge of adversity, and to mission success, began during the summer of 1965 and continued through 1966, underlining the flight controllers' lore that second only to crew safety, 'the mission comes first' – a code that continues to the present day.

Landing and recovery Although land landing was not demonstrated during the Gemini programme, significant success was achieved in aiming the spacecraft to a pinpoint ocean landing. Gemini 8 demonstrated the flexibility and capability of real-time mission planning and contingency operations in the safe return of the crew just ten hours after launch. After the paraglider was deleted from Gemini, it was decided that tests should continue to develop the system that achieved a part demonstration of the original goal, and so essentially, even in land landing, Gemini was successful.

Public awareness Gemini was a media event in the mid-1960s: launches were covered live, regular reports from orbit were headline news, and landings were transmitted on national TV. The public was still wild about space, and this was the programme that would help take Americans to the Moon before the Soviets. Indeed, there were no cosmonauts placed in orbit from March 1965 until April 1967. Throughout this period of Gemini, astronauts caught up with and surpassed Soviet achievements in many fields. However, serious questions were being asked about whether there was still a race to run. The escalating Vietnam War, the continuing domestic problems of hunger, housing, health, education, old age, and the new identification of global pollution – all were beginning to attract as much attention as the manned space missions. It was a spectre that would continue to haunt the argument for manned spaceflight from that time.

The end of the programme
On 1–2 February 1967 a two-day end-of-programme conference was held in the auditorium of the Manned Spacecraft Centre at Houston. Twenty-one papers reviewed the achievements in rendezvous and docking, EVA and in-flight experiments, although the media paid little attention to the event as it was not headline news. The conference took place less than a week after the loss of the three Apollo 1 astronauts in the pad fire, and while a nation mourned and once again questioned the value of manned spaceflight, the Gemini conference helped play a part in refocusing attention on the success of the programme and to the future.

The full Moon – the next target – seen from Gemini 12. Lovell later flew on the first manner lunar mission (Apollo 8), and Aldrin flew on the first manned landing mission (Apollo 11).

Current locations of Gemini spacecraft (2001)

Gemini	Location
GT-1	Destroyed during re-entry April 12
GT-2	USAF Museum, Cape Canaveral Air Force Station Florida (reused for MOL test)
GT-3	Grissom Memorial, Spring Mill State Park, Mitchell Indiana
GT-4	National Air and Space Museum, Washington DC
GT-5	Space Center, Houston adjacent to Johnson Space Center
GT-6	St Louis Science Center, St. Louis, Missouri
GT-7	National Air and Space Museum Washington DC
GT-8	Neil Armstrong Air and Space Museum, Wapanoneta, Ohio
GT-9	NASA KSC Visitors Centre, KSC, Florida
GT-10	Norsk Teknisk Museum, Oslo, Norway
GT-11	California Museum of Science and Industry, Los Angeles, California
GT-12	NASA Goddard Space Flight Center, Greenbelt, Maryland

Steps to the Moon

Twenty-three months after Gemini 12 splashed down, Apollo 7 took the next Americans into orbit in October 1968 – and just nine months later, Neil Armstrong stepped onto the Moon. But could Apollo have reached the Moon without Gemini? It is a difficult question, because the programme was already in development when the lunar landing goal was set in 1961. Certainly the techniques undertaken by Gemini – rendezvous and docking, EVA and long duration – could have been performed during Apollo Block 1 missions, but not before 1967. Hardware delays prevented anything from flying prior to that year, the difficulties in launching and qualifying the Saturn V continued into 1968, and, most importantly, the development of the LM was to constantly push the first manned flight to the spring of 1969. Technically, therefore, Apollo could have proceeded without Gemini, but what it provided was much-needed operational experience and the unmeasured feeling of success and achievement that continued with Apollo in overcoming the tragic fire. It has often been recorded that the fire did not delay the Apollo landing on the Moon, and neither did Gemini. In fact, Gemini created the confidence to advance the Apollo programme at a rapid pace and in respect to fallen comrades.

The overall achievement of Gemini did not pass unnoticed by President Lyndon B. Johnson, who said, after Gemini 12 landed: 'Ten times in this programme of the last twenty months we have placed two men in orbit ... ten times we have brought

Jim Lovell and Buzz Aldrin carry the notices 'THE' and 'END' to signify the completion of Gemini flight operations, as they head out to the launch pad on 11 November 1966. As they walk to the spacecraft, they begin to take the last of Gemini's steps to the Moon.

them home. Gemini 12 was the culmination of a great team effort, stretching back to 1961, and directly involving more than 25,000 people in NASA, the Department of Defense, and other government agencies, in the universities and other research centres, and the American industry. Early in 1962 John Glenn made his historic orbital flight, and America was in [orbit] ... Now, nearly five years later, we have completed Gemini, and we know America is in space to stay.'

Gemini provided the necessary experience to boast confidence in Apollo's ability to reach the Moon, and although it was still recognised in 1966 that the road would not be an easy one, NASA, the astronauts, and America, were ready to proceed. When, almost three years later, Neil Armstrong performed his 'giant leap for mankind', many of the essential small steps for Apollo to reach the Moon five months within President Kennedy's commitment date were achieved with a programme called Gemini.

Appendix

THE ASTRONAUTS

The thirty NASA astronauts selected between April 1959 and October 1963 were all eligible for assignment to the Gemini missions. Of the seven members of Group 1, Cooper, Grissom and Schirra flew on Gemini, Shepard and Slayton were grounded for medical reasons, Carpenter took an assignment in the US Navy, and Glenn retired in January 1964. All nine members of Group 2 were assigned to Gemini missions, but See was killed before making a flight. All fourteen members of Group 3 received technical assignments in support of Gemini, but Freeman was killed prior to a flight assignment, and Bassett was killed shortly before his scheduled flight. Anders, Bean and Williams served on back-up crews, while Chaffee, Cunningham, Eisele and Schweickart worked mainly on Apollo assignments.

The following are brief biographical sketches of the Gemini astronauts, with rank at the time of selection to the astronaut traing programme, and information current in April 2001. Biographical sketches of the seventeen Manned Orbiting Laboratory astronauts are included separately.

Aldrin, Edwin E. ('Buzz') (Major USAF), *back-up Pilot Gemini 10, reassigned back-up Pilot Gemini 9, Pilot Gemini 12*, was born on 20 January 1930 in Montclair, New Jersey. He graduated from the US Military Academy at West Point in 1951, gained his pilot wings in 1952, and received a PhD in astronautics in 1963. He flew 66 combat missions with the 51st Fighter Interceptor Wing during the Korean conflict, and is credited with two MIGs destroyed and one damaged. He served as an aerial gunnery instructor at Nellis AFB, and then as Dean of Faculty USAF Academy before assignment as an F-100 Flight Commander with the 36th Tactical Fighter Wing, in Bitburg, Germany. He was assigned to the Gemini Programme Target Office, AF Systems Division, Los Angeles, California, as a member of a USAF group of specialists making recommendations for participation in the NASA Gemini programme. He was then transferred the USAF Field Office at MSC, Houston, which had responsibility for integration of DoD experiments on Gemini flights, and became instrumental in developing orbital rendezvous techniques. He was selected to

NASA Group 3 in October 1963. Following Gemini assignments he worked on the AAP (Skylab) and was then assigned to Armstrong's Apollo crew for what eventually became the back-up crew for Apollo 8. Aldrin was Lunar Module Pilot on Apollo 11, and was the second man to walk on the Moon. He logged 21 hours 36 minutes on the lunar surface, including 2 hours 31 minutes EVA. He left NASA in 1971 to return to the USAF as Commander Edwards AFB. He retired from the USAF in 1972, entered private business, and became a successful author. He logged more than 289 hours in space on two spaceflights, including eight hours on EVA.

Anders, William A. (Captain USAF), *back-up Pilot Gemini 11*, was born on 17 October 1933 in Hong Kong. He was a graduate of the US Naval Academy, gained an MSc degree in nuclear engineering, and served as a nuclear engineer and instructor pilot at the Air Weapons Laboratory, Kirkland AFB, New Mexico, holding technical management responsibility for the radiation shielding and radiation effects programmes. He was selected with the Group 3 astronauts in October 1963, and worked on AAP (Skylab) assignments before assignment as LMP on Borman's Apollo crew. In December 1968 he flew as LMP (without an LM) on Apollo 8 (and was one of the first three men to orbit the Moon) and as back-up CMP for Apollo 8, before resigning from NASA in 1969 to become executive secretary of the National Aeronautics and Space Council. In August 1973 he became Atomic Energy Commission ambassador to Norway (1976–1977). He afterwards joined General Electric Company, and from 1989 he was with General Dynamics until his retirement in 1994. Anders logged 147 hours on one spaceflight.

Armstrong, Neil A. (civilian), *back-up Command Pilot Gemini 5, Command Pilot Gemini 8, back-up Command Pilot Gemini 11*, was born on 5 August 1930 in Wapakoneta, Ohio. He eaned a BSc degree in aeronautical engineering in 1955. He flew 78 combat missions as a naval aviator during the Korean conflict,and after graduation from Purdue he joined the NACA Lewis Flight Propulsion Laboratory (later the NASA Lewis Research Center) subsequently transferring to the High Speed Flight Station at Edwards AFB, California, for flight-test work, flying the F100, F104, B47, F102, X5 and X-15 rocket planes – the last of them, six times. He was a candidate for the USAF X-20 Dyna Soar programme, but joined the NASA astronaut programme with the second selection in September 1962. After Gemini assignments he was assigned to command an Apollo crew, and became back-up Commander for what eventually became Apollo 8. In training for Apollo 11 he survived a crash of the LLRV, ejecting safely. Little more than a year later he became the first man to walk on the Moon, spending 21 hours 36 minutes on the surface, including 2 hours 31 minutes EVA. After a public world tour and administrative roles, he resigned from NASA in 1971 to become Professor of Engineering at Purdue, and later entered private business. He logged more than 206 hours in two spaceflights, including 2 hours 31 minutes on EVA.

Bassett, Charles A. (1931–1966) (Captain USAF), *original Pilot Gemini 9*, was born on 20 December 1931 in Dayton, Ohio. He gained a BSc degree in electrical engineering from Texas Technology College, and carried out graduate work at the University of Southern California. He graduated from the USAF Test Pilot School and ARPS, and served as an experimental test pilot and engineering test pilot in the Fighter Projects Office at Edwards before joining NASA with Group 3 in October 1963. He was killed in the crash of a T-38 in February 1966.

Bean, Alan L. (Lieutenant USN), *back-up Command Pilot Gemini 10*, was born on 15 March 1932 in Wheeler, Texas. Following a year as an electronics technician with the Naval Reserve in Dallas, he gained a BS degree in aeronautical engineering in 1955, and received his naval commission. After completing flight he completed a four-year assignment with Attack Squadron 44 at the Naval Air Station, Jacksonville, Florida. He was a 960 graduate of the US Navy Test Pilot School, Patuxent River, Maryland, and remained there as a test pilot until 1962. He was with Attack Squadron 172 at Cecil Field, Florida, when selected by NASA as one of fourteen astronauts of Group 3 (14 October 1963). He received technical assignments in spacecraft recovery systems, and during the Gemini programme he served as a CapCom for Gemini 7/6 and Gemini 11. He became Chief of the AAP Branch of the Astronaut Office (CB) in September 1966, but when astronaut C.C. Williams was killed in a flying accident in October 1967, Bean was reassigned to Pete Conrad's Apollo crew to replace him as Lunar Module Pilot. Bean served as back-up LMP on Apollo 9 before flying to the Moon as LMP on Apollo 12, and on 19 November 1969 he became the fourth man to walk on the Moon. On the lunar surface he completed two EVAs totalling 6 hours 24 minutes, during a surface stay of 31 hours 31 minutes. Following Skylab 3, he served as the back-up American Apollo Commander for the 1975 Apollo–Soyuz Test Project with the Soviets, and in August 1975 was assigned to the Space Shuttle programme, before retiring from the US Navy in October 1975 with the rank of Captain. Bean subsequently served as Acting Chief of the CB, supervising the training of the 1978 and 1980 Classes of astronaut candidates. Although widely expected to take the Commander's seat on STS-9, carrying the first Spacelab mission, he resigned from NASA on 26 June 1981 to pursue, full-time, his hobby of painting. Since leaving NASA he has developed a highly successful career as a space artist, focusing on the Apollo lunar missions, with his work being exhibited around the world and featured in several books. He logged 1,671 hours on two spaceflights, including 10 hours 30 minutes during three EVAs.

Borman, Frank (Major USAF), *back-up Command Pilot Gemini 4, Command Pilot Gemini 7*, was born on 14 March 1928 in Gary, Indiana. He graduated from the US Military Academy in 1950, and earned an MSc degree in aeronautical engineering in 1957. Following graduation from West Point, Borman completed pilot training in 1951, and during the next five years completed a variety of assignments with various fighter squadrons in the US and the Philippines before returning to West Point as an instructor of thermodynamics and fluid mechanics. A graduate of the 1960 class of ARPS, he remained an instructor there until selected for the astronaut programme in September 1962, as a member of Group 2. Following completion of his Gemini assignments, he was assigned as Commander of the third manned Apollo mission, and in the autumn of 1968 was reassigned to the second manned flight, Apollo 8. He was one of the first three astronauts (with Lovell and Anders) to ride on a Saturn V and fly to the Moon, and during Christmas 1968 they completed ten orbits of the Moon. Borman resigned from NASA and from the US Air Force in 1969, to enter private business, and eventually became Chairman of the Board for Eastern Airlines. He left Eastern Airlines in 1986, but continued to hold executive positions in a number of companies. He logged more than 477 hours in two spaceflights.

Carpenter, M. Scott (Lieutenant USN), was born on 1 May 1925 in Boulder, Colorado. In 1943 he joined the US Navy, and after the end of World War II he attended the University of Colorado, majoring in aerospace engineering, although he did not receive his BS until 1962. On re-entering the US Navy he received flight training, and was assigned to Patrol Squadron 6, flying anti-submarine, ship surveillance and aerial mining missions in the Yellow Sea, the South China Sea and the Formosa Straits during the Korean conflict. In 1954 he entered the US Navy Test Pilot School at Patuxent River, and after graduation he remained there until 1957. He attended Navy General Line School in 1957 and Navy Air Intelligence School in 1958, and was an intelligence officer onboard the *Hornet* when selected as one of America's first astronauts in April 1959. He served as back-up to John Glenn on Mercury 6, replaced Deke Slayton as Pilot on MA7 when he was grounded, and flew as pilot on MA7 (Aurora) in May 1962. After the flight he worked on Apollo LM development, and served as executive assistant to the director of MSC. In the spring of 1965 he took leave of absence to participate in the US Navy's Man in the Sea programme, and spent thirty days in Sealab II on the ocean floor – the first man to live in inner and outer space. He returned to NASA MSC to liaise with the US Navy for underwater EVA training, but later suffered a motorcycle injury. He left the astronaut programme on 10 August 1967, and returned to the US Navy to serve as Assistant Operations Sealab III, until he retired with the rank of Commander on 1 July 1969. Carpenter has since been an engineering, aeronautical and environmental consultant, has been involved in breeding wasps, and is a successful author. He logged more than four hours in space.

Cernan, Eugene A. (Lieutenant USN), *back-up Pilot Gemini 9, reassigned Pilot Gemini 9, back-up Pilot Gemini 12*, was born on 14 March 1934 in Chicago, Illinois. He gained a BS degree in electrical engineering, and a MS degree in aeronautical engineering. He entered flight training after graduation from university, and was assigned to Attack Squadron 126 and the 133 at Miramar NAS, California, before selection as a member of NASA's Group 3 astronauts in October 1963. After Gemini he worked briefly on AAP (Skylab) technical assignments, and was assigned to Stafford's Apollo crew as LMP, serving as back-up Apollo 7 LMP and prime LMP for Apollo 10, flying the LM to within nine miles of the Moon. He was back-up Commander for Apollo 14, and as Commander of Apollo 17 he was the eleventh man on the Moon and the last man to leave the lunar surface. He logged 74 hours 59 minutes, and 22 hours 3 minutes on three EVAs. He also worked on technical assignments for the joint US/USSR Apollo–Soyuz Test Project. He left NASA in

1976 to enter private business, and set up his own company, The Cernan Corporation – an aerospace and energy management consultancy. He logged more than 566 hours in space.

Chaffee, Roger B. (1935–1967) (Lieutenant USN), was born on 15 February 1935 in Grand Rapids, Michigan. He graduated from Purdue University in 1957. He obtained a BSc in aeronautical engineering, and carried out graduate work at the Air Force Institute of Technology. A Navy ROTC student, he entered active duty upon graduation from Purdue, and completed pilot training. In March 1960 he was assigned to Heavy Photographic Squadron 62, Jacksonville, Florida, and flew 62 classified photographic missions through October 1962, including several over Cuba during the missile crisis. He was at AFIT at Wright-Patterson, working on his MSc degree, when selected by NASA as one of the Group 3 astronauts in October 1963. After general training he was assigned to Apollo development issues and technical assignments involving deep-space communications. He was assigned to the Apollo 1 crew in March 1966, and was killed in the pad fire on 27 January 1967.

Collins, Michael (Lieutenant Colonel USAF), *back-up Pilot Gemini 7, Pilot Gemini 10*, was born on 31 October 1930 in Rome, Italy. He graduated from the US Military Academy, West Point, and earned a BS degree in 1952. He completed a four–year assignment with the 21st Fighter–Bomber Wing at bases in the US and Europe. After a period as a maintenance and training officer, he graduated USAF Test Pilot School and remained there as a test pilot and student at ARPS until selection by NASA. He was a member of the Group 3 selection of October 1963. Following his Gemini assignments Collins was assigned to Borman's Apollo crew as CMP, and was eventually assigned to Apollo 8, but was removed a few weeks before the flight due to surgery for a bone spur on his neck. Lovell replaced him on the historic first flight to the Moon. Collins was reassigned to Apollo 11 as the CMP on the historic first landing mission in July 1969, and became the first man to fly solo behind the Moon – out of contact with the two astronauts on the other side of the Moon, and also out of contact with all humanity on Earth. He was offered the back-up Commander seat on Apollo 14 that would have placed him as Commander of Apollo 17, with a chance to walk on the Moon; but he declined, and in 1969 he left NASA to enter private business and to become Director of the National Air and Space Museum. He became a successful writer of books documenting the American space programme. Collins logged more than 266 hours on two spaceflights, including 2 hours 46 minutes on EVA.

Conrad, Charles ('Pete'), Jr (1930–1999) (Lieutenant USN), *Pilot Gemini 5, back-up Command Pilot Gemini 8, Command Pilot Gemini 11*, was born on 2 June 1930 in Philadelphia, Pennsylvania. In 1953 he gained a BS degree in aeronautical engineering. Following pilot training he attended the US Navy Test Pilot School, from where he graduated in 1957, and remained at Patuxent as a test pilot, flight instructor, and performance engineer. Prior to selection to NASA, he was assigned to Fighter Squadron 96 at Miramar, California. Conrad was one of nine Group 2 astronauts selected by NASA on 17 September 1962. He had been short-listed for the

Mercury astronaut selection in 1959, but was not selected. One of his early assignments involved the development of the Apollo Lunar Module, but it was during the Gemini programme that he made his first flights into space. He also served as back-up Commander for the second and then the third Apollo (which became Apollo 9) before flying as Commander of Apollo 12 in 1969. On 19 November 1969 he became the third man to walk on the Moon, completing two EVAs (with Al Bean), logging 6 hours 24 minutes during a stay of 31 hours 31 minutes. Conrad replaced Cunningham as Chief of the Skylab Branch of the Astronaut Office in the summer of 1970, and after Skylab 2 he retired from NASA and from the US Navy, with the rank of Captain, in December 1973. He became Vice President (Operations) and Chief Operating Officer for American Television Communications (ATC) – a Denver-based cable TV company – until March 1976. He then joined McDonnell Douglas Corporation as Vice President, working in various marketing and business development roles until 1993, when he joined the company's Delta Clipper launch vehicle technology development programme. He had also continued his association with the space programme, as a consultant and test subject for underwater EVA simulations for the ISS, during his years as an executive at McDonnell. From 1993 he worked on McDonnell's revolutionary space launch system for three years, including serving as a remote pilot and CapCom for a series of flight tests at the White Sands Missile Test Range in New Mexico. On 31 March 1996 Conrad retired from McDonnell to head a private space launch firm called Universal Space Lines, based in Irvine, California. He died from injuries sustained in a motorcycle accident in California on 8 July 1999. Conrad logged more than 1,179 hours in four spaceflights, including 11 hours 33 minutes on four EVAs.

Cooper, L. Gordon ('Gordo') (Captain USAF), *Command Pilot Gemini 5, back-up Command Pilot Gemini 12*, was born on 6 March 1927 in Shawnee, Oklahoma. After three years of study at the University of Hawaii, he was commissioned in the US Army. He transferred this commission to the USAF, and was placed on extended active duty in 1949. He received flight training, and was assigned to the 86th Fighter–Bomber Group, in Munich, Germany, flying F-84s and F-86s for four years. He received a BS degree in aeronautical engineering in 1956. Cooper is a 1957 graduate of the USAF test pilot school at Edwards AFB, where he was assigned to the Performance Engineering Branch of the flight test division, performing flight tests on experimental fighter aircraft. He was one of America's first seven astronaut, selected on 9 April 1959, and was back-up to Walter Schirra on Mercury Atlas 8 before flying the 22-orbit one-day Mercury 9 mission – the longest and final flight of the programme – in May 1963. He was assigned to the Apollo Branch Office of the CB, and following his Gemini assignments was assigned to the Apollo Applications programme, to become the Branch Chief for a short while, replacing Al Bean. He was reassigned as back-up Commander Apollo 10, and was expecting to be Commander Apollo 13 and to set foot on the Moon. When he was not assigned nor offered another landing flight, and with no prospect of achieving his goal of being the first man to step onto the planet Mars, he left NASA in July 1970 to enter private business. He logged more than 225 hours in space.

Cunningham, R. Walter (civilian), was born on 16 March 1932 in Creston, Iowa. At the age of 19 he dropped out of college to enter the US Navy, completed pilot training, received his wings in 1953, and for three years served as a Marine Aviator at air bases in the United States and in Japan. He was detached from duty in August 1956, but remained a Marine Reservist until his retirement with the rank of Colonel in 1975. In 1960 he received a BA, and in 1961 he received an MA in physics from the University of California at Los Angeles. He worked on his doctorate while employed as a scientists by the RAND Corporation. At RAND he worked on submarine launch ballistic missile defence, and studies of the magnetosphere, which led to an experiment flown on the first Orbiting Geophysical Satellite. Selected to NASA in October 1963, he completed training and was assigned to Apollo, and served as CapCom Gemini 4 and Gemini 8. In September 1966 he was assigned to Apollo 2, although he had already been working on the flight for several months. By December the flight was cancelled, and he was reassigned as back-up Apollo 1. Following the first Apollo he was named to the crew of Apollo 7 in May 1968, and flew on the first manned Apollo in October 1968. Following his spaceflight he was assigned to the Chief of the AAP office of CB, expecting the command of the first mission. He was replaced as Chief of Skylab by Pete Conrad in August 1970, and after he was assigned not to the second crew but to the back-up crew he left NASA on 1 August 1971. Since he left NASA he has worked in private business, in engineering, and in business investment funding. His autobiography, *All-American Boys*, was published in 1977. He logged more than 260 hours in space, in one flight.

Eisele, Donn F. (1930–1987) (Captain USAF), was born on 30 June 1930 in Columbus, Ohio. He entered the US Naval Academy and graduated with a BSc degree in 1952, and in 1960 received an MS degree in astronautics from the Air Force Institute of Technology. He chose the USAF after graduating from Annapolis, received his wings in 1954, and served as an interceptor pilot, stationed at Ellsworth AFB, South Dakota, until 1958. He then entered the USAF Institute of Technology and, after graduating, remained there as a rocket propulsion and weapons engineer. He is a graduate of the ARPS school and served there as an instructor before being assigned as a test pilot at the Air Force Special Weapons Center, Kirkland AFDB, Mexico. Selected to NASA in Group 3, he was assigned to Apollo early in training, and worked on the testing of the prototype lunar EVA suit. In late 1965 he was assigned, with Grissom and Chaffee, to the crew of the first manned Apollo, but just prior to Christmas he injured his shoulder in a trampoline accident and was replaced on the crew by Ed White. When Eisele recovered, he replaced White on Schirra's Apollo 2 crew until the mission was cancelled in December 1966, after which he was reassigned as back-up Pilot Apollo 1. As a result of his shoulder injury he escaped the fate of the Apollo 1 astronauts in the fire of 27 January 1967, and was assigned to the crew of the first manned Apollo when flights resumed. Apollo 7 was his only spaceflight, and after serving as back-up CMP Apollo 10 he was not offered a second mission and so left the Astronaut Office to become Technical Assistant for Manned Spaceflight at NASA Langley Research Center, Hampton Field, Virginia until resigning from NASA and retiring from the USAF in July 1972. He became a

Director of the Peace Corps in Thailand, and then sales manager for Marion Power Shovel Company, as well as being an executive with the Oppenheimer & Company investment firm in Fort Lauderdale. Finally, he set up his own consultancy, Space Age America. On 1 December 1987, while in Tokyo to open a space camp for children, he died of a heart attack. He logged more than 260 hours in space.

Freeman, Theodore C. (1930–1964) (Captain USAF), was born on 18 February 1930 in Haverford, Pennsylvania. In 1953 he graduated with a BSc from the US Naval Academy at Annapolis. In 1960 he received an MS degree in aeronautical engineering from the University of Michigan. Freeman completed USAF pilot training after graduation from the Naval Academy, and then served as a pilot in the Pacific and at George AFB, California. After earning his MSc degree he was assigned to Edwards AFB, California, as an aerospace engineer, and later graduated from ARPS. At the time of his selection to NASA, as a member of Group 3, he was an instructor at ARPS. He was the first American astronaut to lose his life in a training accident when, on 31 October 1964, his T-38 struck a flock of geese, and he died in the crash. As the first of his group to fly, he was a leading contender for an early assignment to a Gemini mission.

Glenn, John H., Jr (Lieutenant Colonel USMC), was born on 18 July 1921 in Cambridge, Ohio. He entered the Naval Aviation Cadet Program and received pilot training, and spent the final year of World War II flying F4U fighters in the Pacific, completing 59 combat missions. After the war he remained in the USMC and became a flight instructor, and during the Korean conflict flew 63 combat missions with Marine Fighter Squadron 311 and a further 27 missions as an exchange pilot with the USAF. During the last nine days of the war he shot down three enemy fighters. In 1954 he attended the US Navy Test Pilot School at Patuxent River, Maryland, and remained there as an instructor for two years. He was then assigned to the Naval Bureau of Aeronautics in Wahington, where he conceived, planned and flew as pilot for Project Bullet, and on 16 July 1957 set a transcontinental record by flying an F8U from Los Angeles to New York in 3 hours 23 minutes. Glenn was an early volunteer for centrifuge tests, and was one of the Group 1 astronauts selected by NASA in April 1959. He served as back-up to both Shepard and Grissom, and onboard Friendship 7 on 20 February 1962 he became the first American to orbit the Earth. He was afterwards assigned to Apollo but was ruled out of a second flight by President Kennedy, and decided to leave NASA and the USMC to run for the Senate. An injury prevented his participation in the 1964 selection, and so he entered private business, continued to push for office, finally won a seat in 1974. He remained in the Senate until 1998. Despite a failed attempt to be selected as US President in 1984, he remained a popular figure, and campaigned for a return to space. This was finally realised when, at the age of 77, he flew on the Space Shuttle (STS-95) in 1998. He retired from US government duty – after more than 50 years' service – in 1999. Glenn logged more than 218 hours on two missions separated by 36 years.

Gordon, Richard F., Jr (Lieutenant Commander USN), *back-up Pilot Gemini 8, Pilot Gemini 11*, was born on 5 October 1929 in Seattle, Washington. After gainin a BSc he joined the US Navy in 1951. He received his wings in 1953, and attended the all-weather fighter school and jet transitional training school before reporting to the All-Weather Squadron at Jacksonville NAS, Florida. He graduated from the US Navy Test Pilot school, and served as a flight test pilot there until 1960. After serving as a flight instructor in the F4H at NAS, Miramar, California, he helped to introduce the aircraft to the Atlantic and Pacific fleets, and also served as flight safety officer with Fighter Squadron 96 at Miramar, and assistant operations officer and ground training officer. In May 1961 he won the Bendix Trophy Race from Los Angeles to New York, establishing a new speed record of 869.74 mph and a transcontinental speed record of 2 hours 47 minutes. In October 1962 he was selected to NASA in Group 3. After Gemini he was assigned to Conrad's Apollo, serving as back-up to the second and then the third Apollo that eventually became Apollo 9, before serving as CMP Apollo 12 in November 1969 and flying in lunar orbit. He was back-up Commander Apollo 15, and was to have been Commander Apollo 18, but the flight was cut in the budget restrictions, and he lost the chance of being the thirteenth man on the Moon. After working on early Shuttle development issues, he left NASA in 1973 to enter private business. He logged more than 315 hours on two spaceflights, including 2 hours 45 minutes on EVA.

Grissom, Virgil I. ('Gus') (1926–1967) (Captain USAF), *Command Pilot Gemini 3, back-up Command Pilot Gemini 6*, was born on 3 April 1926 in Mitchell, Indiana. He was an aviation cadet in the US Army Air Corps from 1994 until his discharge in November 1945. In 1956 he obtained a degree in mechanical engineering, and returned to aviation cadet flight training. He received his wings in March 1951, and flew 100 combat missions during the Korean conflict, after which he became a jet pilot instructor at Bryan AFB, Texas. He then studied aeronautical engineering at the Air Force Institute of Technology at Wright-Patterson AFB, Ohio. Between October 1956 and May 1957 he attended the USAF Test Pilot School, Edwards AFB, California, and following graduation was assigned to its fighter branch as a test pilot. He was selected to NASA in the first astronaut group announced on 9 April 1959, and in July 1961 flew as Pilot of Mercury Redstone 4 (Liberty Bell 7) – the second sub-orbital flight – and almost drowned when the capsule took on water during recovery operations. He became the Astronaut Office branch chief for Gemini, being one of the first of the team to work on the programme. As Command Pilot of Gemini 3 he became the first American astronaut to fly in space twice. After his Gemini assignments he was assigned as Commander for the first manned Apollo mission – Apollo 1 (AS-204). He was in training for a February 1967 mission when, with fellow astronauts Ed White and Roger Chaffee, he died in a flash fire inside the capsule on Pad 34 during a simulated countdown on 27 January 1967. He logged more than 5 hours in space.

Lovell, James A., Jr (Lieutenant Commander USN), *back-up Command Pilot Gemini 4, Pilot Gemini 7, back-up Command Pilot Gemini 10, reassigned back-up Command*

Pilot Gemini 9, Command Pilot Gemini 12, was born on 25 March 1928 in Cleveland. He graduated from the US Naval Academy in 1952, and after training as a pilot he completed a number of assignments as a naval aviator, including a three-year tour as a test pilot at the Naval Air Test Center, Patuxent, Maryland. He served as Programme Manager with F4JH Weapons Systems Evaluation, and as a flight instructor and safety officer with Fighter Squadron 101 at NAS Oceana, Virginia. Selected to the astronaut programme in September 1962, he had been short-listed to the first NASA selection, but was not selected due to a minor liver condition at the time. After Gemini he was assigned to the back-up crew on Apollo 3, and then to Apollo 8. After Michael Collins was grounded from that mission, he replaced him on the flight crew, and during Christmas 1968 he became one of the first three men to orbit the Moon. He was also back-up Commander to Neil Armstrong on Apollo 11, and should have commanded Apollo 14 to the Moon, but replaced Shepard's crew on Apollo 13 when they needed more training. As Commander of Apollo 13 he should have been the fifth man on the Moon, but an in-flight explosion led to the abortion of the mission, and Lovell lost the chance to complete the third lunar landing. After administrative roles he left NASA in 1973 to enter private business. His autobiography, *Lost Moon*, was published in 1994, and in 1995 it was used as the basis of the highly successful film *Apollo 13*. Lovell logged more than 715 hours in space, and was the first person to complete four spaceflights.

McDivitt, James A. (Captain USAF), *Command Pilot Gemini 4*, was born on 10 June 1929 in Chicago, Illinois. He gained an MSc in aeronautical engineering in 1959. He had joined the USAF in 1951, and during the Korean conflict he flew 145 combat missions in F-80 and F-86 aircraft. He is also a graduate of the USAF Test Pilot School and ARPS, and was eligible for assignment to the X-15 rocket plane programme, but instead chose NASA. He was selected to the astronaut programme with the second class in September, and following Gemini he was assigned to Apollo, originally as back-up Commander Apollo 1, and then reassigned Commander Apollo 3, and from 1966 to 1969 worked on the development of the first manned flight test of the Apollo LM. After the Apollo 1 pad fire he was assigned as Commander second manned Apollo, only to be reassigned to the third flight due to the delay in LM delivery. Apollo 9 flew in March 1969, and McDivitt became the first person to fly the LM in Earth orbit. After Apollo 9 he was assigned an administrative role as Apollo spacecraft manager from 1969 to 1972, until leaving NASA and the USAF to enter private industry. He logged 338 hours in space.

Schirra, Walter M. ('Wally'), Jr (Lieutenant Commander USN), *back-up Command Pilot Gemini 3, Command Pilot Gemini 6*, was born on 12 March 1923 in Hackensack, New Jersey. In 1945 he graduated from the US Naval Academy, and following flight training in Pensecola he served with Navy Fighter Squadron 71 before becoming an exchange pilot with the 154th Air Force Fighter–Bomber Squadron, with which he flew 90 combat missions during the Korean conflict. Schirra flew the F-84E, and is credited with one downed MiG and a second probable kill. Following the Korean War he was assigned to the development of the

Sidewinder missile and as a project pilot for the F7U3 Cutlass, in addition to serving as an instructor pilot for the Cutlass and the J3 Fury. He subsequently became the Operations Officer for the 124th Fighter Squadron onboard the *Lexington* in the Pacific, and later completed test pilot training at the Naval Air Test Center, Patuxent, Maryland. He was one of seven original Mercury astronauts selected by NASA on 9 April 1959, and served as back-up pilot on Mercury Atlas 7 (Carpenter – Aurora 7) before flying a 'textbook' six-orbit mission on Mercury Atlas 8 (Sigma 7) in October 1962. Following Gemini assignments in early 1966, he was named Commander Apollo 2, but was reassigned as back-up Commander Apollo 1 when his mission was cancelled later that year. After the pad fire his crew was named as the first to fly the modified Apollo, which became Apollo 7 in October 1968. In July 1969 he retired from NASA to enter private business. He was the only NASA astronaut to fly in each of the pioneering US manned spacecraft programmes – Mercury, Gemini and Apollo. He logged more than 295 hours on three missions.

Schweickart, Russell L. ('Rusty') (civilian), was born on 25 October 1935 in Neptune, New Jersey. In 1956 he gained a BS in aeronautical engineering from MIT, and for the next four years served as a pilot in the USAF until returning to MIT to study for his MS in aeronautics and astronautics, which he obtained in 1963. In 1961 he had been recalled to active duty for a year, and subsequently served with the Air National Guard. At the time of his selection to NASA as one of the Group 3 astronauts on 17 October 1963, he was a scientist at the Experimental Astronomy Laboratory at MIT, researching into the physics of the upper atmosphere, and star tracking. He was assigned to technical assignments, and worked on scientific experiments for Gemini and Apollo missions prior to his assignment as LMP on the three-man back-up crew for Apollo 1 in 1966. After several reassignments, the same crew eventually flew as the prime crew on Apollo 9, during which Schweickart tested the Apollo lunar EVA suit on a one-hour EVA. He was not assigned to an Apollo lunar mission, and was transferred to Skylab as back-up Commander for the first manned mission, after which he transferred to NASA HQ as Director of User Affairs in the Office of Applications. In November 1975 he returned to JSC to work on Shuttle payload policies, and the following year he took a leave of absence to serve on the Californian Governor's committee staff as assistant for science and technology. He resigned from NASA in July 1979, and has since served on the Californian Commission for Energy, the US Energy Commission, and the US Antarctic Program Safety Review Panel. A founder member of the Association of Space Explorers, he has also served as President of NSR Communications and as Executive Vice President of CTA. He is a frequent lecturer on space and the environment. Schweickart logged 241 hours in space, including an hour of EVA.

Scott, David R. (Captain USAF), *Pilot Gemini 8*, was born on 6 June 1932 in San Antonio, Texas. He graduated fifth in a class of 633 graduates from the US Military Academy at West Point, and later earned an MA degree in aeronautics and astronautics. He also graduated from both the USAF Test Pilot School and ARPS. He joined NASA as a member of Group 3 in October 1963, and after Gemini 8 he was assigned to the McDivitt Apollo crew, initially as back-up Apollo 1, and then as prime CMP Apollo 2, training for an Earth orbit test flight of the combined CSM/LM for the first time. Delays with the LM led to the exchange of the second and third crews, and Scott was reassigned as CMP Apollo 9, which flew in March 1969. During this mission, Scott completed a 1-hour stand-up EVA. After assignment as back-up Commander Apollo 12 he Commanded Apollo 15 and became the seventh man to walk on the Moon, logging 66 hours 54 minutes on the surface and 19 hours 7 minutes on four EVAs (including one stand-up EVA). He was assigned as back-up Commander Apollo 17, but was replaced and reprimanded for carrying unauthorised first-day covers during Apollo 15, and lost the chance of completing a fourth flight on an early Shuttle mission. In 1973 Scott left the Astronaut Office to become Director of DFRC in California, supporting the Shuttle ALT programme in 1977 before leaving NASA to enter private industry. He logged more than 546 hours in space, including more than 20 hours on EVA.

See, Elliot M. (1927–1966) (civilian), *back-up Pilot Gemini 5, original Command Pilot Gemini 9,* was born on 23 July 1927 in Dallas, Texas. He gained a BS degree from the US Merchant Marine Academy in 1949, and an MS degree in engineering from the University of California at Los Angeles in 1962. From 1949 to 1953 he worked for General Electric, and after duty as a pilot with the US Navy from 1953 to 1956 he returned to General Electric where, until 1962, he was a flight test engineer, group leader, and experimental test pilot. He served as a project pilot for the J79-8 engineering development programme, conducted power plant flight tests on the J-47, J-73, J-79 and CJ805 aft-fan engines, and he was involved in flying the F-86, XF4D, F-104, F-11, F-1, RB-66, F4H and T-38 aircraft. He was a member of Group 2, selected by NASA in September 1962. He died in the crash of his T-38 in February 1966.

Shepard, Alan B., Jr (1931–1998) (Lieutenant Commander USN), *original Command Pilot Gemini 3, Chief of Astronaut Office,* was born on 18 November 1923 in East Derry, New Hampshire. He graduated from the US Naval Academy in 1944, and earned a BSc. During the final year of World War II he served onboard *Cogswell* in the Pacific, and following pilot training he served as a naval aviator with Fighter Squadron 42 and 193 onboard carriers in the Mediterranean and the Pacific. He graduated from the US Navy Test Pilot School in 1950, where he afterwards completed two tours as an instructor. He was selected in the first group of NASA astronauts in 1959, and on 5 May 1961 was the first American to fly in space, onboard Freedom 7. He was back-up on the last Mercury flight, and was a lead contender for a subsequently cancelled three-day Mercury mission. He was the original Command Pilot for Gemini 3, but was grounded because of an inner ear ailment, which forced him to take an administrative role. He became Chief of the Astronaut Office, and (with Slayton) was instrumental in the selection of all American flight-crews from 1964 to 1974. He was restored to flight status in 1969, and secured the command of Apollo 13 over Gordon Cooper, but limited experience in the flight-crew resulted in an exchange with the Apollo 14 crew commanded by Jim

Lovell. As Commander of Apollo 14, Shepard became the fifth man to walk on the Moon (and the only Mercury astronaut to do so), logging 33 hours 30 minutes on the lunar surface, including 9 hours 22 minutes on two EVAs. He afterwards resumed his role as Chief Astronaut, and in 1974 he retired with the rank of Rear Admiral, to engage in private business. Shepard's business and investments made him a millionaire. He died on 21 July 1998, and his wife Louise died a few weeks later. He logged more than 216 hours in space, including 9 hours 22 minutes on EVA.

Slayton, Donald K. ('Deke') (1924–1993) (Captain USAF), *Director of Flight Crew Operations*, was born on 1 March 1924 in Sparta, Wisconsin. He entered the Army Air Corps at the age of 18, and became a pilot at the age of 19. During World War II he was a B-25 bomber pilot, and flew 56 combat missions over southern Europe and six over Japan. He earned a BS in aeronautic engineering in 1949, and for the next two years worked as an aeronautical engineer with the Boeing Company in Seattle, Washington, before being recalled to active duty during the Korean conflict. He served in Germany before attending Edwards AFB Test Pilot School in 1955, and was a test pilot there when selected by NASA for the first astronaut group in 1959. Slayton was to have flown the second Mercury orbital flight, but due to a long-standing heart condition he was grounded three weeks before the scheduled launch. He resigned his USAF commission, and remained at NASA as Director Flight Crew Operations, controlling astronaut selection, flight assignments (subject to HQ approval) and aircraft operations until 1981. He was restored to flight status in 1972, and in 1975 he flew in space as a member of the Apollo crew on the US/USSR Apollo–Soyuz Test Project. He subsequently served as Manager for the Shuttle ALT and OFT programmes, and in 1982 left NASA to become President of Space Service Inc. He died of brain cancer on 13 June 1993. He logged 217 hours in space.

Stafford, Thomas P. (Captain USAF), *back-up Pilot Gemini 3, Pilot Gemini 6, back-up Command Pilot Gemini 9, reassigned Command Pilot Gemini 9*, was born on 17 September 1930 in Weatherford, Oklahoma. He graduated from the US Naval Academy in 1952, and transferred his commission to the USAF. He flew fighter interceptors in the US and in Germany, and graduated from the USAF test pilot school, Edwards AFB. He was Chief Performance Branch of the ARPS, responsible for the supervision and administration of the flying curriculum for students, and he was the author of two handbooks. He was selected in the second group of astronauts in September 1962. After his Gemini assignments, Stafford was assigned to Apollo – first to Borman's crew, and then as back-up Commander to what became Apollo 7. As Commander Apollo 10, in May 1969 he completed the dress rehearsal lunar flight, taking the LM to within nine miles of the Moon. He served as Active Chief of the CB from 1969–1971, while Shepard trained for and then flew Apollo 14. Stafford did not seek a Skylab mission or a return to the Moon, and was considering leaving NASA and the USAF to run for the Senate; but he stayed to take the Command seat on the US/USSR Apollo–Soyuz Test Project in July 1975. He resigned from NASA in 1975, to return to the Air Force, until 1979, when he entered private business and continued as an aerospace advisor and as a member of numerous space study groups. He was perhaps the most experienced rendezvous and docking astronaut in the early groups. He logged more than 507 hours in space on four missions.

White, Edward H., II (1930–1967) (Captain USAF), *Pilot Gemini 4, back-up Command pilot Gemini 7*, was born on 14 November 1930 in San Antonio, Texas. He graduated from US Military Academy in 1952, and narrowly missed selection to the 1952 US Olympic team as a 400-m hurdlist. In 1959 he gained an MSc degree in aeronautical engineering, and after West Point he received flight training in Florida and Texas and then completed a 3½-year assignment in Germany, flying F-86 and F-100 aircraft. He graduated from the USAF Test Pilot School, Edwards AFB, California, and served as an experimental test pilot in the Aeronautical Systems Division at Wright-Patterson AFB, Ohio, completing flight tests for research and weapons systems development. He applied for selection as a member of NASA's Group 1, but was not short-listed, and was one of the Group 2 astronauts. After his Gemini assignments he was assigned to senior Pilot position on Apollo 2, and when the senior Pilot of Apollo 1 (Eisele) injured his shoulder, White replaced him. He was in training for Apollo 1 when he was killed in the pad fire at the Cape on 27 January 1967. He had logged more than 97 hours in space, including a 20-minute EVA, and was considered by some to be a leading contender for an early command of an Apollo landing crew – possibly the first.

Williams, Clifton C. (1932–1967) ('CC') (Captain USMC), *back-up pilot Gemini 10*, was born on 26 September 1932 in Mobile, Alabama. He gained a BSc in mechanical engineering before joining the USMC and attending the US Navy Test Pilot School at Patuxent, Maryland, and completed three years as a test pilot in the Carrier Suitability Branch of the Flight Test Division at Pax River, including land-based and ship-borne tests of the F8E, TF8E, F8E (Attack), A4E and automatic carrier systems. Williams was selected to NASA in October 1963, and was assigned to Conrad's Apollo crew to serve as back-up on the second manned Apollo. He was killed in the crash of his T-38 in October 1967, and was replaced on Apollo by Bean, who had been his Commander on the Gemini 10 back-up crew. Had Williams lived he would probably have been back-up LMP Apollo 9, and, on Apollo 12, may well have become the fourth man to walk on the Moon.

Young, John W. (Lieutenant USN), *Pilot Gemini 3, back-up Pilot Gemini 6, Command Pilot Gemini 10*, was born on 24 September 1930 in San Francisco, California. He gained a BSc in aeronautical engineering, and entered the US Navy in June 1952. He served a tour on the destroyer *Laws* before serving as a test pilot and programme manager of the F4H 'Phantom' weapons systems project, and then maintenance officer for the All-Weather Fighter Squadron 143 at NAS Miramar, California. In 1962 he set time-to-climb records for 1,800 feet and 15,500 feet in the US Navy F4B fighter in Project High Jump. In September 1962 he was selected as a member of NASA's Group 2, and after his Gemini assignments he was assigned as Command

Module Pilot on Stafford's Apollo crew, which eventually served as back-up crew on Apollo 7. In May 1969 he flew as CMP on Apollo 10, the second lunar orbital flight, and became the first person to fly solo in lunar orbit, while Stafford and Cernan test-flew the LM to within nine miles of the surface. His next assignment was as back-up Commander on Apollo 13, and he then rotated to Command Apollo 16 to the Moon in April 1972. He became the ninth man to walk on the Moon, logging 71 hours 2 minutes on the surface and 20 hours 14 minutes during three EVAs. He replaced Scott as back-up Commander on Apollo 17 – the last Apollo mission – before being assigned Chief of the Shuttle Branch of the Astronaut Office in 1973. He took over from Al Shepard as Chief of the Astronaut office in 1974, and held this position until 1987. During his tenure of thirteen years he oversaw the activities of the Apollo–Soyuz Test Project, the Shuttle Approach and Landing Test series, 25 Shuttle launches, and the recovery from the *Challenger* accident. His fifth mission was as Commander of STS-1 – the maiden flight of the Shuttle, on 12 April 1981 – and a record sixth flight as Commander of Shuttle 9, carrying Spacelab 1, in November/December 1983. He was assigned to Command STS-61J to deploy the Hubble Space Telescope, but after the *Challenger* accident he was reassigned to administrative, engineering and safety support positions at JSC. Although Young has not been assigned to a crew since 1987, he has retained his astronaut status and maintained his flying proficiency and simulator time; and forty years after being selected to NASA he was still (2001) eligible to command a Shuttle flight. He remains the longest-serving active astronaut, and was first person to fly six missions and to log seven rocket launches – one of them from the Moon. In six missions (two Gemini, two Apollo and two Shuttle) John Young logged more than 835 hours, including more than 20 hours of EVA.

The MOL astronauts

Abrahamson, James A. (Major USAF), *Group 3*, was born 19 May 1933 in Williston, North Dakota. He obtained a BS in aeronautical engineering in 1955, and an MS degree in aerospace engineering in 1961. He entered the USAF in 1955, and was selected for the MOL in 1967. When the MOL was cancelled he continued his USAF career, and was NASA Associate Administrator for Space Flight from 1981 to 1984. He retired from the USAF, with the rank of Lieutenant General, in 1989, and continued as an executive and consultant in the aviation industry.

Adams, Michael J. (1930–1967) (Major USAF), *Group 1*, was born 5 May 1930 in Sacramento California. He earned a BS degree in aeronautical engineering in 1958. He entered the USAF in 1950, and flew 59 combat missions during the Korean conflict. He later earned a BS degree, and graduated from test pilot school and from ARPS in 1962. He chose to remain with the USAF instead of applying to NASA, and after selection to the MOL in 1965, he left on 20 July 1966 to fly the X-15. He flew the aircraft seven times, and on the last flight – on 15 November 1967 – flew it to 50 miles, and attained the USAF Astronaut Pilot rating. Unfortunately, on that same flight he lost control of the aircraft and was killed in the crash.

Bobko, Karol J. (Captain USAF), *Group 2*, was born 23 December 1936. He received his BS degree from the USAF Academy in 1959. He was a member of the first graduating class, and obtained an MS degree in aerospace engineering in 1970. He completed pilot training in 1960, and after the MOL was cancelled he joined NASA, to work on support assignments for Skylab (crew-member, 56-day SMEAT ground test in 1972), ASTP, and Shuttle ALT. He flew on three Shuttle missions – once (STS-5) in 1982, and twice (STS 51-D and STS 61-J) in 1985. In 1989 he left NASA to enter the aerospace industry. He logged more than 386 hours in space.

Crews, Albert H., Jr (Major USAF), *Group 1*, was born on 23 March 1929 in El Dorado, Arkansas. He earned a BS degree in chemical engineering in 1950 and an MS degree in aeronautical engineering in 1960. In 1950 he joined the USAF and became a pilot, and graduated from test pilot school in 1960. In 1962 he was one of six pilots named to fly the X-20 Dyna Soar space-plane, which was cancelled in 1963. He was a member of the first class of MOL pilots, and when the project was cancelled in 1969 he was over the age limit to become a NASA astronaut. and he instead joined the aircraft operations directorate in Houston He became one of the most experienced NASA aerospace research and aircraft pilots of the ensuing three decades, and often supported Shuttle flights, aircraft ferry operations and CB support duties.

Crippen, Robert L. ('Crip') (Lieutenant USN), *Group 2*, was born 11 September 1937 in Beaumont, Texas. He received a BS degree in aeronautical engineering in 1960. He then entered the US Navy, and in 1969 he transferred to NASA after cancellation of the MOL. He was Commander of the SMEAT ground-test simulation of 1972, and was support crew for all three Skylab and ASTP missions. He flew on the first Shuttle flight (STS-1) in April 1981, and commanded three more missions – one of them (STS 7) in 1983, and two (STS 41-C and STS 41-G) in 1984. He was assigned to command the first launch (STS 62-A) out of Vandenberg (from where he would have flown a MOL flight), before it was cancelled as a result of the *Challenger* accident in 1986. He then assumed administrative roles in NASA, and become Director of KSC from 1991 to 1995, after which he left NASA to enter the private aerospace industry in executive roles. He logged more than 565 hours in four spaceflights.

Finley, John L. (Lieutenant USN), *Group 1*, was born on 22 December 1935 in Winchester, Massachusetts. In 1957 he graduated with a BSc from the US Naval Academy, and then served on *Ticonderoga* for four years. He was selected for the MOL in 1965, and in April 1968 the US Navy reassigned him at his own request. He continued his career with the US Navy until retiring with the rank of Captain in May 1980, and subsequently engaged in private business.

Fullerton, C. Gordon (Captain USAF), *Group 2*, was born on 11 October 1936 in Rochester, New York. He earned a BS in 1957 and an MS in mechanical engineering in 1958. After working as a mechanical design engineer for Hughes Aircraft Corporation, and while studying for his Master's degree, he joined the USAF. He became a B-47 bomber pilot with the Strategic Air Command (SAC), and then a bomber test pilot, and was later selected for the MOL. In 1969 he transferred to NASA, where he worked as a support crew-member for Apollo 14 and Apollo 17, and as CapCom for Apollo 14 through Apollo 17. He was a crewman on the ALT programme and, with Fred Haise, completed the first free flight of *Enterprise* in 1977. He flew two Shuttle missions – the first (STS-3) in 1982, and the second (STS 51-F /Spacelab 2) in1985. In October 1986 he left the Astronaut Office to become an aerospace research pilot at NASA Dryden, where he flew a variety of aircraft. He logged more than 382 hours in space.

Hartsfield, Henry W. ('Hank'), Jr (Captain USAF), *Group 2*, was born on 21 November 1933 in Birmingham, Alabama. He gained an BSc in physics in 1954, and an MS in engineering science in 1971. He began serving in the USAF in 1955 and transferred to NASA in 1969, serving on the support crew for Apollo 16 and all three Skylab missions. In 1977 he resigned from the USAF with the rank of Colonel, and continued as a civilian astronaut, serving as back-up Pilot on STS-2 and STS-3. He flew as Pilot on STS-4 – the fourth and final test flight – before assuming the command of STS 41-D in 1984 and STS 61-A (Spacelab D1) in 1985. He has since continued with administrative roles in operations, safety and training for the International Space Station. He logged more than 482 hours in space.

Herres, Robert T. (Lieutenant Colonel USAF), *Group 3*, was born on 1 December 1932 in Denver, Colorado. He earned a BSc in 1954, an MS in electrical engineering in 1960, and a second MS in public administration in 1965. As well as serving as a pilot, Herres worked as an intelligence officer in West Germany. When the MOL was cancelled he declined to transfer to NASA, and returned to a USAF career that included a number of staff appointments and a combat tour in Vietnam. He retired with the rank of General on 1 February 1990, and entered into private business.

Lawrence, Robert H., Jr (1935–1967) (Major USAF), *Group 3*, was born on 2 October 1935 in Chicago, Illinois. He gained a BS degree in Chemistry in 1956, and a PhD in nuclear chemistry in 1965. He was a member of the last MOL selection in June 1967, and was killed in the crash of an F-104 jet at Edwards AFB, California, on 8 December 1967. Lawrence was the first African American to be selected for spaceflight training. He would have been eligible to join NASA in 1969, and could have flown the Shuttle as a Pilot and possibly as a Commander.

Lawyer, Richard E. (Captain USAF), *Group 1*, was born on 8 November 1932 in Los Angeles, California. He gained a BS in aeronautical engineering in 1955. He served with the Tactical Air Command, and completed a brief tour in Vietnam. For the MOL, his speciality was pressure suits and EVA. After the MOL was cancelled he returned to active duty with the USAF until his retirement with the rank of Colonel in 1983. He has since continued his career in the aviation industry.

Macleay, Lachlan (Captain USAF), *Group 1*, was born on 13 June 1931 in St Louis, Missouri. He gained a BS degree in 1954, and an MBA degree in 1971. On graduating from the US Naval Academy in 1954, he chose to serve in the USAF, and afterwards graduated test pilot school serving as a Special Projects (U2) test pilot. After serving as an advisor to the Republic of Korea Air Force, he joined the MOL. Following the cancellation of the MOL he continued with his USAF career until retiring with the rank of Colonel in June 1978. He has since worked on missile programmes at Hughes Aircraft.

Neubeck, F. Gregory (Captain USAF), *Group 1*, was born on 11 April 1932 in Washington DC. After gaining a BS degree from the US Naval Academy in 1955, he chose to serve in the USAF, and worked on the development of weapons systems and as a flight instructor. His classmates at ARPS were future NASA astronauts Mike Collins and Joe Engle. After the MOL was cancelled he resumed his USAF career, includin a combat tour in south-east Asia, until his retirement in 1982. Since then he has worked in the aerospace industry, has become an author and a holder of a patent, and has entered politics.

Overmyer, Robert F. (Captain USMC), *Group 2*, was born on 14 July 1936 in Lorain, Ohio. He earned a BS in physics in 1958, and an MS in aeronautics in 1964. After the MOL was cancelled he transferred to NASA and worked on technical issues on Skylab, after which he served on the support crew and as CapCom for Apollo 17 and the ASTP, including a term at TsUP (Moscow Mission Control) during the flight in July 1975. He was support crew and CapCom chase for Shuttle ALT in 1977, and deputy vehicle manager for preparing *Columbia* for the first spaceflight, afterwards completing two spaceflights in 1982 (STS 5) and 1985 (STS 51-B/Spacelab 3). In 1986 he left NASA to form his own consultancy firm Mach 25, worked with McDonnell, and worked as a writer. In 1995 he left McDonnell to join Cirris (a builder of small aircraft), and on 22 March 1996 was killed while flight testing one of their aircraft. He logged 290 hours in space.

Peterson, Don H. (Major USAF), *Group 3*, was born on 22 October 1933 in Winona, Mississippi. He graduated from the US Military Academy with a BSc degree in 1955, and earned an MS degree in nuclear engineering in 1962. In 1967 he was selected as one of the third group of MOL astronauts, and in 1969 he transferred to NASA to serve on the support crew for Apollo 16, and later to work on Shuttle development issues. He retired from USAF in January 1980, although he remained as a civilian astronaut with NASA, and flew one mission on the Shuttle (STS-6) in April 1983. In 1985 he left NASA to become an aerospace consultant. He logged more than 120 hours in space.

Taylor, James M. (1930–1970) (Major USAF), was born on 27 November 1930 in Stamps, Arkansas. He was awarded an Associate of Arts degree in 1950, and in 1951 he enlisted in the USAF. After flight training, he gained his wings in 1953. He gained a BS degree in electrical engineering in 1959. He was a flight test engineer, was one of the first selected for the MOL, and was a leading candidate for command of the first MOL manned mission. When the programme was cancelled in 1969 he returned to the USAF as an instructor at ARPS at Edwards AFB, and was killed in a T-38 crash on 4 September 1970.

Truly, Richard H. (Lieutenant USN), *Group 1*, was born on 12 November 1937 in Fayette, Mississippi. He gained a BS degree in aeronautical engineering in 1959. He was selected for the MOL on his 28th birthday, and transferred to NASA in 1969. He was support crew and CapCom for all three Skylab and ASTP missions, and was Pilot on the Shuttle ALT programme when, with Joe Engle, he flew *Enterprise* off the back of a Jumbo Jet. He was back-up Pilot on STS-1, and on 12 November 1981 – his 44th birthday – he was launched into space as Pilot on STS-2. In 1983 he flew as Commander of STS-8, and afterwards left NASA to head the US Navy Space Command in Virginia. In February 1986 he returned to NASA as Associate Administrator for Manned Spaceflight, and in January 1989 he became Administrator of NASA. In February 1992 he retired from NASA, and also retired from the US Navy with the rank of Vice Admiral. He then became a professor at Georgia Tech, and has recently been appointed head of the National Renewable Energy Laboratory in Colorado. He logged more than 199 hours in space.

Career experience of astronauts (Groups 1–3)

	Year Selected	NASA Group	Spaceflight/Gemini				Career missions	Total flights h:m	Number of EVAs	Total EVA h:m
			Gemini Flights	Mission h:m	Total EVAs	EVA h:m				
Aldrin	1963	3	1	94:34	3	05:37	2	289:54	4	08:09
Anders	1963	3	0				1	147:01		
Armstrong	1962	2	1	10:41			2	206:00	1	02:31
Bassett	1963	3	0							
Bean	1963	3	0				2	1,671:45	3	10:30
Borman	1962	2	1	330:35			2	477:36		
Carpenter	1959	1	0				1	4:56		
Cernan	1963	3	1	72:20	1	2:10	3	566:16	4	24:13
Chaffee	1963	3	0							
Collins	1963	3	1	70:46	1	01:29	2	266:06	1	01:29
Conrad	1962	2	2	262:12			4	1,179:39	4	11:33
Cooper	1959	1	1	190:55			2	225:15		
Cunningham	1963	3	0				1	260:09		
Eisele	1963	3	0				1	260:09		
Freeman	1963	3	0							
Glenn	1959	1	0				2	218:35		
Gordon	1963	3	1	71:17	2	02:41	2	315:53	2	02:41
Grissom	1959	1	1	04:52			2	05:09		
Lovell	1962	2	2	425:09			4	715:05		
McDivitt	1962	2	1	97:56			2	338:57		
Schirra	1959	1	1	25:51			3	295:13		
Schweickart	1963	3	0				1	241:01	1	01:07
Scott	1963	3	1	10:41			3	546:54	5	20:14
See	1962	2	0							
Shepard	1959	1	0				2	216:17	2	09:22
Slayton	1959	1	0				1	217:28		

Name	Year								
Stafford	1962	2	2	98:11	507:44	00:21	4	00:21	
White	1962	2	1	97:56	97:56	1		1	
Williams	1963	3	0						
Young	1962	2	2	75:38	835:42		6		20:14	3

Aldrin — Second man to walk on theMoon (Apollo 11)

Anders — First manned lunar flight (Apollo 8)

Armstrong — First man to walk on the Moon (Apollo 11)

Bassett — Died 28 February 1966 (jet crash)

Bean — Fourth man to walk on the Moon (Apollo 12)

Borman — First manned lunar flight (Apollo 8)

Carpenter — Mercury astronaut (MA-7)

Cernan — Eleventh man to walk on the Moon (Apollo 17)

Chaffee — Died 27 January 1967 (Apollo 1)

Collins — CMP Apollo 11

Conrad — Third man to walk on the Moon (Apollo 12); died 8 July 1999 (injuries sustained in motorcycle crash)

Cooper — Mercury astronaut (MA-9)

Cunningham — LMP Apollo 7 (no LM flown)

Eisele — CMP Apollo 7; died 1 December 1987 (heart attack)

Freeman — Died 31 October 1964 (jet crash)

Glenn — First American in orbit; flew the Shuttle at the age of 77

Gordon — CMP Apollo 12

Grissom — Died 27 January 1967 (Apollo 1)

Lovell — CMP Apollo 8; CDR Apollo 13

McDivitt — CDR Apollo 9

Schirra — Mercury astronaut (MA-8); CDR Apollo 7

Schweickart — LMP Apollo 9

Scott — Seventh man to walk on the Moon (Apollo 15)

See — Died 28 February 1966 (jet crash)

Shepard — First American in space; fifth man to walk on the Moon; died 21 July 1998 (leukaemia)

Slayton — DMP Apollo (18), Apollo–Soyuz Test Project; died 13 June 1993 (brain cancer)

Stafford — CDR Apollo 10; CDR Apollo (18) Apollo–Soyuz Test Project

White — First American on EVA; died 27 January 1967 (Apollo 1)

Williams — Died 5 October 1967 (jet crash)

Young — Tenth man to walk on the Moon (Apollo 16); CDR of first and ninth Shuttle

USAF Manned Orbiting Laboratory (Gemini B)

	Year Selected	MOL Group	Career Missions	Total Flights h:m	Notes
Abrahamson	1967	3	0		NASA Associate Administrator for Spaceflight, 1981–1984
Adams	1965	1	0		Died 15 November 1967; X-15 crash; posthumously awarded USAF Pilot Astronaut wings
Bobko	1966	2	3	386:04	STS 6; 51-D; 51-J
Crews	1965	1	0		Former X-20 selectee; worked for NASA Aircraft Directorate from 1969
Crippen	1966	2	4	565:48	STS-1; 7; 41-C; 41-G; Director of KSC, 1991–1995
Finley	1965	1	0		Resigned from MOL in April 1968
Fullerton	1966	2	2	382:51	STS-3; 51-F
Hartsfield	1966	2	3	482:51	STS-4; 41-D; 61-A
Herres	1967	3	0		Retired with the rank of General USAF, 1 February 1990
Lawrence	1967	3	0		Died 8 December 1967; crash of F-104
Lawyer	1965	1	0		Retired with the rank of Colonel USAF, 1983
Macleay	1965	1	0		Retired with the rank of Colonel USAF, 1978
Neubeck	1965	1	0		Retired with the rank of Colonel USAF, 1982
Overmyer	1966	2	2	290:23	STS-5; 51-B; died July 1996 (aircraft crash)
Peterson	1967	3	1	120:24	STS-6; the only former MOL astronaut to perform an EVA – 1 x 4h 17m
Taylor	1966	1	0		Candidate for first MOL command; died September 1970 (jet crash)
Truly	1965	1	2	199:22	STS-2; 8

There were no MOL manned spaceflights, and this list reflects only the spaceflight experience of the NASA Group 7 transfers *after* cancellation of the MOL

Bibliography

In addition to the author's own archive, and assistance from individuals quoted in the Acknowledgements, from 1988 significant use was made of the NASA Gemini Collection (Boxes 201–446) originally held at Rice University, Houston, and now (2001) located at NARA in Fort Worth, Texas. The post-flight Mission Report document from each Mission File, as well as several other major documents and reports from that collection (identified as GC-NARA), together with other significant sources, are listed below. A comprehensive list of source material is available upon request.

Taped interviews including information relating to Gemini
Gene Cernan, Houston, Texas, August 1988; Karl Henize, Houston, Texas, April 1989; Walt Cunningham, London, England, July 1989; James McBarron, JSC, Houston, August 1989; Ron Wood, KSC, Florida, November 1990; Joe Kosmo, JSC, Houston, July 1991; and John Young, JSC, Houston, Texas, June 1992.

1959   *Man in Space: The USAF Program for Developing the Spacecraft Crew*, Ed. Lt Col Kenneth F. Gantz, USAF, Duell, Sloan and Pearce

1963   *Project Mercury: A Chronology*, James Grimwood, NASA SP-4001, 1963
        Air Force Participation in Gemini, NASA/DoD Joint Ad Hoc Study Group, Final Report, 6 May 1963, Copy 16 of 100, declassified 1 December 1978. (GC-NARA)

1964   *Preliminary Technical Development Plan for the Manned Orbiting Laboratory*, as of 30 June 1964, declassified 1975
        Astronautics and Aeronautics, 1963 , NASA SP-4004
        Gemini 1 Mission Report, MSC-R-G-64-1, May, NASA (GC-NARA)

1965   *Final Report of Paraglider Research and Development Program*,
        North American Aviation Inc., SID65-196, (copy no. 11), 19 February (GC-NARA)
        Gemini 2 Mission Report, MSC-G-R-65-1 February, NASA (GC-NARA)
        Gemini 3 Mission Report, MSC-G-R-65-2 April, NASA (GC-NARA)
        Gemini 4 Mission Report, MSC-G-R-65-3 June, NASA (GC-NARA)
        Gemini 5 Mission Report, MSC-G-R-65-4 October, NASA (GC-NARA)
        Astronautics and Aeronautics, 1964, NASA SP-4005

1966   *Gemini 7 Mission Report*, MSC-G-R-66-1 January, NASA (GC-NARA)
       Gemini 6 Mission Report, MSC-G-R-66-2 January, NASA (GC-NARA)
       Gemini 8 Mission Report, MSC-G-R-66-4 April, NASA (GC-NARA)
       Gemini 9 Mission Report, MSC-G-R-66-(July), NASA (GC-NARA)
       Gemini 10 Mission Report, MSC-G-R-66-(August), NASA (GC-NARA)
       Gemini 11 Mission Report, MSC-G-R-66-8 October, NASA (GC-NARA)
       This New Ocean, A History of Project Mercury, Loyd Swenson,
       James Grimwood, Charles Alexander, NASA SP-4201
       Space Suit Development Status, Richard Johnston, James Correale and
       Mathew Radnofsky, NASA TN D-3291, February (GC-NARA)
       Astronautics and Aeronautics, 1965, NASA SP-4006
1967   *Gemini 12 Mission Report* MSC-G-R-67-1 January, NASA (GC-NARA)
       Gemini Summary Conference, 1–2 February, NASA SP-138, MSC
       Houston, Texas
       Summary of Gemini Extravehicular Activity, Ed. Reginald M. Machell,
       MSC, Houston, NASA SP-149
       Gemini Land Landing System Development Program (two volumes) NASA
       TN D-3869, March 1967 (GC-NARA)
       Final Report Gemini Space Suit Program, David Clark Inc., June
       (GC-NARA)
       Astronautics and Aeronautics, 1966, NASA SP-4007
1969   *Project Gemini: A Chronology*, Jim Grimwood, Barton Hacker,
       Peter Vorzimmer. NASA SP-4002
1970   *Vanguard – a History* Constance McLaughlin Green and Milton Lomask,
       NASA SP-4202
1971   *A History of Canaveral District, 1950–1971*, South Atlantic Division,
       US Army Corps of Engineers, from the KSC History Archive
1972   *From Sand to Moondust: a narrative of Cape Canaveral, then and now*, USAF
       and Pan American World Airways Inc., from the KSC history Archive
1973   *Return to Earth*, Buzz Aldrin with Wayne Warga, Random House
1974   *Carrying the Fire: an Astronaut's Journeys*, Michael Collins, Farrar
       Straus Giroux
       A New Environment, Petersen's Book of Man In Space, Volume 2
1975   *Foundations of Space Biology and Medicine*, (three volumes), joint US/USSR
       Publication, NASA edition
1976   *Origins of NASA Names*, Helen Wells, Susan Whiteley and
       Carrie Karegeannes, NASA SP-4402
1977   *All American Boys*, Walter Cunningham, Macmillan
       On The Shoulders Of Titans, A History of Project Gemini, Barton Hacker
       and Jim Grimwood, NASA SP-4203
1980   *The Manned Orbiting Laboratory*, Curtis Peebles, *Spaceflight*, 22 (4), April
       1980; 22 (6), June 1980; 24 (6), June 1982
1981   *The History of Manned Spaceflight*, David Baker, New Cavendish Books
1982   *Managing NASA in the Apollo Era*, Arnold S. Levine, NASA SP-4102
1983   *Manned Spaceflight Log*, Tim Furniss, Jane's Publishing

1985   *Yeager*, Chuck Yeager and Leo Janos, Century Books
         The Human Factor, John Pitts, NASA SP-4213

1986   *Space Patches, from Mercury to the Space Shuttle*, Judith Kaplan and
         Robert Muniz, Sterling Publishing
         America's Astronauts and their Indestructible Spirit, Dr Fred Kelly,
         Aero (Tab) Books

1987   *Guardians – Strategic Reconnaissance Satellite*, Curtis Peebles, Ian Allan

1988   *Countdown*, Frank Borman with Robert J. Sterling, Silver Arrow Books
         Lift-off: The Story of America's Adventure in Space, Michael Collins,
         Grove Press
         NASA Historical Data Book Volume 1 (NASA Resources 1958–1968);
         Volume 2 (Programs and Projects 1958–1968) NASA SP-4012)
         Schirra's Space, Walter Schirra with Richard Billings Quinlan Press

1993   *'Suddenly Tomorrow Came...' A History of the Johnson Space Center*,
         Henry Dethloff, NASA SP-4307

1994   *Deke! US Manned Space from Mercury to Space Shuttle*, Donald K. Slayton
         with Michael Cassutt, Forge Books

1998   *Paraglider: Land Landing for Gemini*, Ed Hengeveld, X-Planes
         Monograph-3, HPM Publications

1999   *The Last Man on the Moon*, Eugene Cernan with Don Davis,
         St Martin's Press
         Who's Who in Space: The International Space Station Edition,
         Michael Cassutt, Macmillan

2000   *Failure is not an Option: Mission Control from Mercury to Apollo 13 and
         Beyond*. Gene Kranz, Simon and Schuster
         A Leap of Faith, Gordon Cooper with Bruce Henderson, Harper Collins
         Gemini 6, The NASA Mission Reports, Ed. Robert Godwin, Apogee Books
         Disasters and Accidents in Manned Spaceflight, David J. Shayler,
         Springer–Praxis

2001   *The Rocket Men: Vostok and Voskhod, The First Soviet Manned
         Spaceflights*, Rex Hall and David J. Shayler, Springer–Praxis
         Skylab: America's Space Station, David J. Shayler, Springer–Praxis

In addition, the unpublished paper *Science Training of the Apollo Astronauts*, by William C. Phinney, also provided information on the NASA astronaut science training schedule of 1962–1965.

Index